Human Toxoplasmosis

HUMAN TOXOPLASMOSIS

Edited by

DARREL O. HO-YEN

and

ALEX W. L. JOSS

Scottish Toxoplasma Reference Laboratory
Inverness

OXFORD NEW YORK TOKYO
OXFORD UNIVERSITY PRESS
1992

Oxford University Press, Walton Street, Oxford OX2 6DP

Oxford New York Toronto
Delhi Bombay Calcutta Madras Karachi
Petaling Jaya Singapore Hong Kong Tokyo
Nairobi Dar es Salaam Cape Town
Melbourne Auckland
and associated companies in
Berlin Ibadan

Oxford is a trade mark of Oxford University Press

Published in the United States
by Oxford University Press, New York

A catalogue record for this book is available from the British Library

Library of Congress Cataloging in Publication Data
Human toxoplasmosis / edited by Darrel O. Ho-Yen, Alex W. L. Joss.
p.   cm. — (Oxford medical publications)
Includes index.
1. Toxoplasmosis.   I. Ho-Yen, Darrel O.   II. Joss, Alex W. L.
III. Series.
[DNLM: 1. Toxoplasmosis.   WC 725 H918]
RC186.T75H84   1992   616.9'36—dc20   92–1057
ISBN 0 19 8547501

Typeset by Graphicraft Typesetters Ltd, Hong Kong
Printed in Great Britain by Biddles Ltd,
Guildford & King's Lynn

Preface

The writing of this book has involved all of the staff of the Scottish Toxoplasma Reference Laboratory. Although individual chapters were predominantly written by one author, there has been great editorial involvement and comments by all authors of related chapters. We hope that these frequent consultations have resulted in a book which could be read as a whole and which has a minimum of repetition.

As a Reference Laboratory, we have always been consulted by medical practitioners (both local and in other hospitals in Scotland and Northern Ireland), other professional staff (nurses, social workers, etc.) and the lay public. In recent years, enquiries from the latter two groups have dramatically increased. Information for all three groups is usually available separately, but we have found that there is considerable overlap in the questions asked of us. Although this book is principally aimed at medical practitioners and other professional staff, we hope that other groups may also find it readable. As many individuals may be interested in only the chapter on toxoplasmosis in pregnancy, all aspects (clinical features, diagnosis, and treatment) are considered in this chapter. In addition we have included an appendix on the most common questions that we are asked about toxoplasmosis in pregnancy.

Although it would have been nice to include chapters from the many experts in other Reference Laboratories, we felt that long-distance collaboration in writing has too many disadvantages. Nevertheless, many of our attitudes to toxoplasmosis are a result of the numerous meetings we have had with members of the other British Toxoplasma Reference Laboratories. We are also grateful for much help, in particular to Mrs I. L. Moir, Ms L. J. Skinner, Dr M. H. Elia, Dr J. A. M. Macrae, Dr R. Rankin, Dr E. Walker, Dr D. G. Goff, Dr W. H. Haining, and Mrs A. Ellingford. Our Librarian Mrs R. Higgins, and her staff have shown remarkable patience. The Medical Illustration Department, especially Mr J. G. D. McGhie, Mr A. MacLeister, and Mrs K. Crawford, have been particularly helpful. Mr Alan McGinley has produced almost all the illustrations in the book and has shown tremendous skill, insight, and originality. We are tremendously indebted to our secretary, Ms Vivian MacFarquhar; she has shown both fortitude and forgiveness for our misdemeanours in the many drafts of this book.

We owe much to Dr Harry Williams and his wife, Dr Kathleen Williams. They were responsible for the creation of the Scottish Toxoplasma Reference Laboratory and for establishing a tradition of good research. Since his retirement, Dr Williams has been a constant source of advice and encouragement for which we are grateful. Lastly, we would like to thank our families, colleagues, and friends who have shown great understanding and sympathy while we have been involved in the preparation of this book.

Inverness   D. O. H.-Y.
August 1991   A. W. L. J.

Contents

Contributors

Scottish Toxoplasma Reference Laboratory
Microbiology Department
Raigmore Hospital
Inverness IV2 3UJ
Scotland

David Ashburn
Jean M. W. Chatterton
Marilyn M. Davidson
Roger Evans
Darrel O. Ho-Yen
Alex W. L. Joss

1

History and general epidemiology

DAVID ASHBURN

The first appreciation that an organism is a new entity is an important mile-stone. This can be difficult when the morphology of the organism is similar to that of others described earlier. Thus before *Toxoplasma gondii* was classified, the problems of similarity to sarcocystis and encephalitozoon[1] had to be overcome. All of these organisms are oval, crescentic, or elongated tear-drop shaped, but there are significant differences in size, both of the vegetative and cystic forms of the parasite. Although on first examination these parasites may be confused with each other, on detailed measurement encephalitozoon is found to be only half the size of toxoplasma,[2] whereas sarcocystis is similar in one dimension but can extend to three times the length in the other.[3] Similarly toxoplasma pseudocysts are twice the size of encephalitozoon cysts.[1,2] However, the differences in tissue cysts of toxo-plasma and sarcocysts are much greater. It is not uncommon to find sarco-cysts which are easily visible to the unaided eye.[1,3] Although toxoplasma tissue cysts are much smaller than sarcocysts, they are up to three times the size of the encephalitozoon cysts[2,4] (Table 1.1).

The problems associated with poor description also serve to compound the identification and classification process. In 1913, toxoplasma-like organisms were found in a 14-year-old Sinhalese boy suffering fever and splenomegaly.[5] However, the author failed to produce adequate illustrations of the finding and to furnish details about parasite aggregation. In addition, lesions in the affected organs were not described. Also the possible infection in an anaemic child with splenomegaly and hepatomegaly[6] must be questioned. Although the illustrations and descriptions are informative, the author believed the structures described to be analogous to those found in the case described above. Because of the uncertain nature of the parasite observed in the Sinhalese boy, this description and any which refer to it are suspicious.

In 1927, toxoplasma-like organisms were found in a neonate who died with convulsions two days after birth.[7] The parasite was classified as an encephalitozoon. This judgement was based on the exclusive intracellular location of the organism and the fact that the organism found was smaller than toxoplasma. However, toxoplasma also exists in extracellular forms[1,8] and the comparison was being made with an organism observed in sections from the neonate and films prepared from rabbit material. It is probable that under similar conditions there would have been no difference in observed size. A third influence on this decision was probably the desire to conform with previous observations. Like some other authors,[9] there seemed to be a

Table 1.1 Comparison of toxoplasma, sarcocystis, and encephalitozoon[1-4]

Characteristic	Toxoplasma	Sarcocystis	Encephalitozoon
Coccidian	Yes	Yes	No
Vegetative parasite	Crescentic or elongated tear-drop shaped, $4-7 \times 2-5$ μm	Bow shape $5-12 \times 1-4$ μm	Oval or pyriform 2.5×0.5 μm
Intracellular	Yes	Yes	Yes
Extracellular	Yes	No	Yes
Type of division	Endodyogeny, binary fission, schizogony	Endodyogeny, binary fission	Multiple fission
Types of aggregate	Pseudocysts $50-60$ μm, tissue cyst up to 100 μm	Sarcocyst up to 5 mm	Cysts 20×30 μm
Incomplete host	Birds and warm-blooded animals including humans	Reptiles, birds, and animals including humans	Rodents, rabbits, and humans
Complete host	Cats	Carnivores/ominivores (including humans)	

degree of reluctance to acknowledge the existence of a new organism. As a syncytial phase was believed not to have been previously observed in toxoplasma,[7] toxoplasmosis was discounted from the diagnosis.

Despite initial difficulties in identification[7,9,10] and some arbitrary descriptions[5,6] it is now appreciated that toxoplasma is common in both animals and humans.[11] A great deal has been learnt about diagnosis and treatment (Chapters 4 and 5) and also about the implication of toxoplasmosis in different patient groups (Chapters 6 and 7). This chapter is structured to depict the major discoveries of toxoplasma in animals and humans. Sections on modes of transmission and the discussion of historically significant cases in humans are followed by sections on the development of diagnostic tests and the treatment of infection. Finally a section on general epidemiology describes the prevalence of toxoplasma in communities world-wide and factors which influence this prevalence.

TOXOPLASMOSIS IN ANIMALS

The first observation of toxoplasma was in 1900 by Laveran,[12] (Fig. 1.1). The parasite was found in sections of the spleen and bone marrow of Java sparrows (*Padda oryzivora*) which were infected with *Haemamoeba danilewskyi*. Although what were seen were recognized to be parasites, they were believed to be reproductive forms of *H.danilewskyi*. However, it is probable that what was seen was toxoplasma. The definitive description of toxoplasma was given in 1908 by Nicolle and Manceaux[8] (Fig. 1.1). They

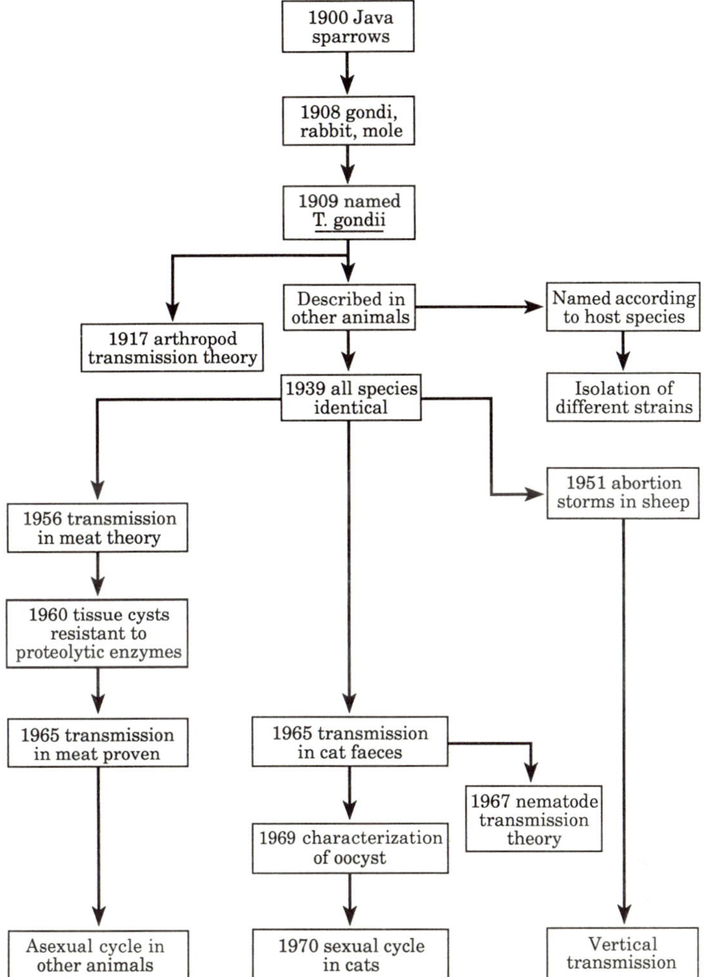

Fig. 1.1   Animal toxoplasmosis: landmarks and transmission.

reported organisms in the blood, spleen, and liver of North African rodents (*Ctenodactylus gondii*), which had been captured in southern Tunisia. The previous year the same authors had described another organism, *Piroplasma quadrigeminum*, in the same species of animal.[8] Piroplasma is smaller and has a second chromatin body (kinetoplast). In addition the new organism possessed a distinct nucleus and was found in several organs whereas piroplasma has a vesicular nucleus and only inhabits erythrocytes.[8] The parasite more closely resembled leishmania and so was named *Leishmania gondii, gondii* being taken from the species name of the host in which they found the parasite. However, the new parasite is excluded from the genus

Leishmania due to its lack of a kinetoplast.[13] As the parasite also did not resemble either trypanosoma or the haemogregarines, in 1909 Nicolle and Manceaux proposed a new genus to accommodate their finding.[13] This name was *Toxoplasma* (toxon = bow or arc, plasma = form, from the Greek), which describes the crescentic shape of the parasite. The species name was retained and the new parasite became *Toxoplasma gondii*.

History has shown that 1908 was a very productive year for observations of toxoplasma-like organisms. The initial observations were all in small animals. In Brazil, Splendore found organisms in a rabbit which had died with paralysis[14] (Fig. 1.1). This finding was examined by Nicolle and Manceaux[13] who considered them to be morphologically identical to the parasite which they had described. A third observation in 1908 occurred in Japan, where Mine found parasites in a mole which were retrospectively recognized as toxoplasma.[15] In the succeeding years there were more observations in animals both small and large, wild and domesticated.[16] These observations served to demonstrate the world-wide distribution of toxoplasma and the range of animals affected. The names given to the various findings were usually in accordance with the host species, e.g. *T.cuniculi* (rabbit), *T.talpae* (mole), *T.canis* (dog), etc. It was not until 1939 that Sabin[17] provided evidence that all of these strains had identical characteristics.

Classification of toxoplasma

Although toxoplasma has been recognized as being protozoan[8,12] it has been allocated to several diverse genera. Laveran[12] classified it as haemamoeba, whereas Nicolle and Manceaux[13] deliberated on piroplasma, leishmania, trypanosoma, and the haemogregarines before proposing a new genus toxoplasma. At that time the known life cycle was incomplete which made it difficult to classify toxoplasma in one particular group. However, by far the most frequent position allocated was *incertae sedis* (uncertain seat, Latin).[18] The advent of the electron microscope and its use in 1954 revealed details of the ultrastructure of toxoplasma and so facilitated a more definite classification. The information gained was used to show similarities with plasmodium, eimeria, isospora, and sarcocystis.[4,18] The greatest interest was generated by the similarity of tachyzoites of toxoplasma and the zoites of coccidian isospora and eimeria.[18] The breakthrough came in the early 1970s after the discovery of the sexual cycle in cats.[19,20] The product of this cycle is an oocyst. The oocyst, by the process of sporulation, forms eight sporozoites (Chapter 2). The oocyst was noticed to be similar to that of *Isospora bigemina*,[19] and is typical of other coccidia, besnoitia, sarcocystis, and frenkelia.[3] Identification of a sexual stage in its life cycle confirmed toxoplasma as a sporozoan, thus resulting in the positioning of the genus as shown in Table 1.2.[18,21]

Table 1.2   Classification of *Toxoplasma gondii*[18,21]

Subphylum: Apicomplexa
  Class: Sporozoasida
    Order: Eucoccidiorida
      Suborder: Eimeriorina
        Family: Eimeriidae
          Genus: *Toxoplasma*
            Species: *T.gondii*

TRANSMISSION OF TOXOPLASMA

The question of transmission was first raised by workers who had made the early discoveries of toxoplasma.[13] From a group of 45 gondis captured in the Matmata region of Tunisia only one animal was infected.[13] Using the infected spleen material from this animal, 15 more animals were inoculated, five of which were gondis, three rhesus monkeys, three rats, and four guinea-pigs. All of the animals were inoculated intraperitoneally with the exception of one rhesus monkey which was inoculated subcutaneously. None of the animals showed any sign of infection except the five gondis which died within 11 days. It is possible that the infective dose for the other animals used in the experiment was lower than for the gondis, or the other animals used were already immune. Gondis are strict herbivores and could encounter toxoplasma by oocyst contamination of soil. However, the natural habitat of the gondi is arid and would prove hostile to toxoplasma oocysts (Chapter 2) unless protected within another host.[4] Thus the gondis may have been susceptible[21] whereas other animals which were captured from the wild may have been immune.

Similar difficulties in determining the effects of infection were obtained by Darling.[22] Six guinea-pigs were fed muscle tissue believed to be infected by sarcocystis although this may have been toxoplasma. When killed six months later two of these animals were infected with parasites similar to those described in the human case of toxoplasmosis which had been diagnosed as sarcosporidiosis.[9] However, at that time it was not known that guinea-pigs are naturally subject to infection by toxoplasma and sarcocystis.[23] Therefore it is probable that the infection had been acquired naturally prior to the experiment.

Despite the initial difficulties in transmitting toxoplasma to other animals, subsequent experiments have successfully infected a variety of animals after intraperitoneal inoculation.[24] Toxoplasma was not transmitted to all animals used, and although the origins of all of the animals are not stated, some were taken from the wild. Therefore as with previous experiments, the number with past toxoplasma infection is unknown.

Arthropod vector

During a prolonged study between 1907 and 1917, the tissues of 471 gondis were examined microscopically for toxoplasma.[25] All of the animals used were captured in southern Tunisia. Four hundred of these were examined within one month of capture and two were found to be infected. The 71 remaining animals were not examined until after more than a month in captivity. Of these animals 33 were infected, thus suggesting that the infection was not a natural one but that the animals had contracted the infection while in captivity. However, microscopic examination of the animals' tissues limits the sensitivity of diagnosis. More accurate results might have been obtained if toxoplasma specific serological tests had been available. During the ten-year period of the study, a seasonal variation of infection was noted. [25] The number of deaths from toxoplasmosis rose from one or two during the summer months to six or eight during the colder winter months. There were two possible explanations: the lower temperatures reduced the animals' resistance against what might be a latent infection; or, transmission was via an intermediate vector such as an insec . Toxoplasma tachyzoites had been observed in the blood.[8] Parasitaemia led to the suggestion that entry to the host may have been via blood-sucking insects transferring parasites during blood meals.[4] Later experiments demonstrated that toxoplasma was ingested by many species of blood-sucking insects[26] but transmission by a biting insect was never found. It must therefore be concluded that transmission by blood-sucking insects is not significant.

Infection from meat

Toxoplasma is known to encyst in tissues of both animals and humans, and each cyst contains many viable parasites (Chapter 2). It is interesting to note that in 1923 frequent eating of rabbit meat during pregnancy was considered a contributory factor to what was probably congenital toxoplasmosis.[27] Other authors also suggested meat as a possible source of infection.[28] However, it was not until 1956 after tissue cysts were demonstrated in pork that a possible link between toxoplasmosis and ingestion of tissue cysts was suggested[29] (Fig. 1.1). Tissue cysts have since been demonstrated in other meat animals but to a lesser extent (Chapter 2). Recognizing the implication of these observations on transmission, experiments were carried out to examine the effects on both tachyzoites and tissue cysts of storage, extremes of temperature, and artificial gastric juice.[30] The tissue cyst was found to be more resistant to the various challenges, thus supporting the theory that it is a major factor in transmission of toxoplasma. Proof of the theory was provided in 1965[31] when it was demonstrated that seroconversion in children admitted to a tuberculosis hospital in France increased from 50 per cent to 100 per cent per year when raw lamb replaced beef or horse meat in their diet (Fig. 1.1).

However, tissue cysts are susceptible to heat and are destroyed by adequate cooking.[30] Carnivorism by animals and eating of partially cooked infected meats by humans accounts for some infection,[32] but does not explain infection of vegetarians or herbivorous animals. Therefore it was realized that in order for toxoplasma to persist in so many different animals, there must be another more resilient stage in its life cycle.[32]

Infection from cats

Evidence of a more resilient stage of toxoplasma was presented in 1965, the same year that infection from meat was demonstrated, when toxoplasma was transmitted in the faeces of cats (Fig. 1.1).[33] Infectivity was retained for over a year even when the faeces were stored in ordinary tap water.[33] Microscopic examination of the faeces showed bacteria, fungi, oocysts of *Isospora* species, and ova of the cat nematode *Toxocara cati*. Because infectivity was resistant to external factors which would normally kill toxoplasma, Hutchinson proposed the nematode transmission theory (Fig. 1.1).[32] It was suggested that the toxocara ova conferred some protection on toxoplasma by incorporating it in the shell of the ova.[32,34] The mechanisms postulated for this were ingestion of a cystic form of toxoplasma by the adult female toxocara or by active penetration of the ova. Thus toxoplasma could safely pass to the external environment. When the ova were ingested and subsequently hatched, toxoplasma would be liberated into the intestine of the host which would then become infected.[32]

Over the next few years a flurry of experiments were carried out which appeared to confirm the theory.[34–6] Although some authors reported transmission in the absence of toxocara,[34,35,37,38] they were reluctant to discount the importance of nematode eggs without conclusive proof that the cats used in the experiment were free of toxocara infection. The theory was also supported by observations that removal of toxocara ova by filtration destroyed the infectivity of the inoculum.[32] It was not then appreciated that toxoplasma oocysts have a tendency to aggregate on other particles too large to pass through the filter.[39] In addition enzymes produced by the mature toxocara larvae are necessary for hatching of the ova,[40] yet toxoplasma was transmitted in experiments with unembryonated or immature eggs.[38,39] This discredited the nematode transmission theory and suggested that there was an independent resistant form of toxoplasma. In many experiments oocysts were observed,[33,35,41] most frequently resembling those of *Isospora bigemina*.[19] The presence and numbers of oocysts correlated very well with the infectivity of the faeces.[19,41] In addition exposure to physical and chemical factors had corresponding effects on infectivity and oocysts alike.[42]

Following the discovery of the toxoplasma oocyst,[19,41,42] in 1970 several authors identified the cat as the complete host of toxoplasma.[19,20,43] The oocyst is produced as a result of the sexual cycle which takes place in the gut

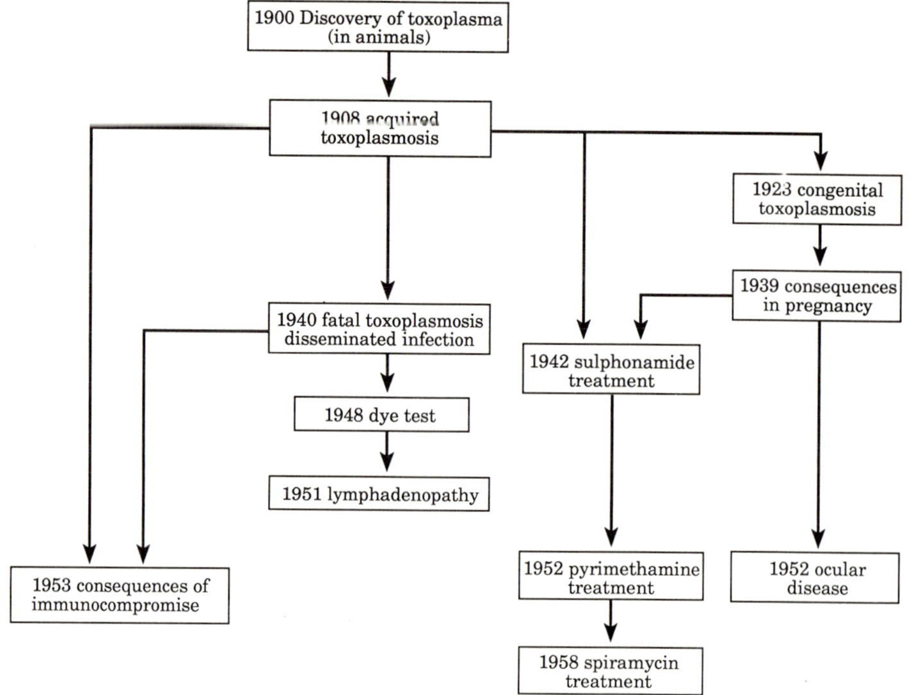

Fig. 1.2   Historical landmarks in human toxoplasmosis.

epithelium of the cat (Chapter 2). An infected cat may shed many millions of oocysts which contaminate vegetables, animal feed, and grazing land and which when ingested proceed to infect other animals. Discovery of the sexual cycle proved to be the key not only to the classification of toxoplasma (Table 1.2), but also to the explanation of how infection of vegetarians and herbivores occurs. Once again toxoplasma was recognized to be entangled in the food chain but now at a much earlier stage than previously known.

TOXOPLASMOSIS IN HUMANS

As well as the observation of toxoplasma in animals in 1908, the first human case of toxoplasmosis was described[9] (Fig. 1.2). A 20 year old Barbadian Negro was admitted to hospital in Panama suspected of suffering from typhoid fever. Darling took muscle biopsies with the intention of demonstrating trichinosis, a condition caused by the nematode *Trichinella spiralis*. On examination, two successive muscle biopsies showed what were probably toxoplasma tissue cysts, but neither parasites nor cysts were found in a third taken after the patient had recovered. Thus he felt that the diagnosis was neither typhoid fever nor trichinosis, but possibly sarcosporidiosis. This conclusion was influenced by the definitive finding in 1894 of sarcosporidiosis in man.[44] Although Darling made the diagnosis of sarcosporidiosis, he noted several differences between his findings and those previously described for sarcocystis (Table 1.1).

Over the following thirty years there was further confusion world-wide with toxoplasma being mistaken for encephalitozoon[7,10] (Table 1.1). In each of these cases, toxoplasma-like organisms were found in several organs including the brain. Classification could only be made on morphological grounds and the lesions caused by the organism. However in 1939 an extensive review[16] compared the morphology of encephalitozoon with that of toxoplasma. They concluded that the causative agent in the cases reviewed was probably toxoplasma. These cases were all infants presenting with neurological disorders which, because of the clinical picture and age of onset, were probably congenital toxoplasmosis.

Congenital toxoplasmosis

The first recognized case of congenital infection was probably in 1923[27] (Fig. 1.2). Although the child was not medically examined until he became blind at 3 months, the mother noted left microphthalmia only 3 days after birth. In addition she described the lower eyelids as covering the eyes and a nurse noticed an unusually large separation of the sutures on the head. After admission to hospital a detailed physical examination revealed severe hydrocephalus which progressed over the following months. Ocular examination showed bilateral abnormalities in the choroid, and a white area in the macular regions of the retinae. The child suffered irregular generalized convulsions with nystagmoid movement of the eyes. On examination several months later there were also spastic contractions of upper and lower limbs. Death occurred when the child was approximately 1 year old. At the post-mortem it was found that the aqueduct of sylvius was completely destroyed. There were also several other lesions in the brain. The right eye was of normal size whereas the left eye was smaller. External appearances of both eyes were normal but there were extensive lesions on both retinae. The retinal lesion of the right eye contained several parasitic cysts which Jankû referred to as 'sporocysts'. The parasite was cautiously categorized as protozoan in the class sporozoa. This classification was made on the basis of the structure of the 'sporocyst' and the 'sporozoites' within it. Jankû was also of the opinion that these parasites were the causative agents of the lesions observed and that infection had probably been acquired by the mother during pregnancy (Chapter 6). This case was reviewed in 1939[16] and it is probable that the parasites were not sporozoan but were toxoplasma. Without more specific tests this cannot be stated for certain but if so, this was the first reported case of congenital toxoplasmosis.

Since this first case of congenital toxoplasmosis, several other cases have been described.[7,10,16,45] Clinical signs which may present in a typical case were described in 1939:[16] hydrocephalus, cerebral calcification, retinochoroiditis, and generalized involvement of the central nervous system, but one or more or these symptoms may be absent (Chapter 6).

Acquired toxoplasmosis

Congenital toxoplasmosis can have severe consequences (Chapter 6), as can acquired toxoplasmosis in adults, especially if immunocompromised (Chapter 7). This was first demonstrated by Pinkerton and Weinman[1] in 1940 when they reported a case observed three years earlier (Fig. 1.2). A young Peruvian man working in a hospital in Lima presented with complaints of weakness, pallor, and fever. On admission the patient was suffering from a bartonella infection which virtually cleared in two days. Despite this apparent improvement the patient died one week after hospitalization. At post-mortem it emerged that the lymph nodes and the spleen were considerably enlarged. The heart was dilated but otherwise normal in size. The spleen and heart both had several yellowish areas of necrosis which were found to contain pseudocysts and some free parasites. Although parasitaemia was not evident either before or after death, parasites were also demonstrated in the lungs, liver, brain, and skin. These were contained within similar lesions to those described for the heart and spleen. Leishmania and *Trypanosoma cruzi* were ruled out as causative agents due to lack of a kinetoplast. This was the same conclusion as had been reached 31 years before in one of the early descriptions of toxoplasma in animals.[13] Encephalitozoon was also rejected because of its smaller size. Sarcocystis was considered but the extent of tissue damage in this case was much greater than had been reported to be caused by sarcocystis.[44] Finally after comparing the organism found with that in laboratory animals and also comparing histological lesions with those known to be caused by toxoplasma, the diagnosis of toxoplasmosis was made. Despite finding toxoplasma in most of the organs the concurrent infection with bartonella made it difficult to ascertain the original cause of the illness. Indeed, it was considered that a latent toxoplasma infection may have been reactivated as a result of bartonella infection.[1] However a latent infection may also be reactivated if the patient is immunocompromised in some other way (Chapter 7). Therefore discovery of the bartonella infection may have been coincidental. The following year two further unequivocal cases having a similar course were reported. These also proved to be fatal.[46] Despite toxoplasma having the potential to kill its host, it should be stressed that the vast majority of acquired infections in healthy individuals are not fatal. It is possible that there may have been immunocompromise in these reported cases.

Lymphadenopathy

Although lymphadenopathy was observed at post-mortem in the first reported case of fatal acquired toxoplasmosis[1] it was not recognized as a characteristic sign of toxoplasmosis until 1951 when it was described during pregnancy.[47] This case was a pregnant woman suffering from a pyrexial illness which subsided four days later. However the patient still had a few tender areas with slight swelling in the neck. Over the following months lymph nodes in the neck, axillae, and groin became grossly enlarged. Lymphadenopathy is characteristic of other illnesses some of which may be serious. As the aetiology of the enlarged glands was not clear, a lymph node from the side of the neck was removed but histological studies did not assist in diagnosis. Two months after the initial illness the lymph nodes were less swollen but still enlarged.[47] Although toxoplasmosis had not been suspected, a routine toxoplasmin skin test performed at a maternity clinic gave a strong positive reaction. This provoked a re-examination of the previously excised lymph node. Some structures morphologically similar to toxoplasma were observed but it was not possible to confirm that they were toxoplasma. At term of an otherwise normal pregnancy the child was born without any signs of congenital toxoplasmosis. The same authors also referred to two other cases of toxoplasmosis with lymphadenopathy.[47] One of these cases was in a laboratory worker and neither individual was pregnant. A second report in 1951 also described lymphadenopathy in non-pregnant individuals.[48] Lymphadenopathy is now known to be a common symptom in acquired toxoplasma infection in non-pregnant and pregnant individuals (Chapter 3).

Ocular disease

Infiltration of macular lesions by toxoplasma was first noticed in a congenitally infected baby in 1923.[27] Two subsequent cases of congenital toxoplasmosis did not report the presence of ocular lesions.[7,45] Further reports in 1942[49] and 1943[50] both described retinochoroiditis in patients who showed a positive neutralizing antibody test. Despite being suggestive of toxoplasma as a causative agent in ocular disease, the unreliability of the neutralizing antibody test rendered the results inconclusive.[51] Using the dermal sensitivity test, evidence linking toxoplasmosis with retinochoroiditis and uveitis in adults was presented in 1949.[52] However, retinochoroiditis is found in diseases other than toxoplasmosis. Syphilis, tuberculosis, and cytomegalovirus may produce similar ocular lesions.[53] Indeed in 1952 in a study performed on eyes removed from 53 adults suffering retinochoroidal lesions, nine of these had previously been diagnosed as tuberculosis.[54] Morphological examination of the eyes demonstrated toxoplasma not in the choroid as had previously been suspected but in the retinal lesions. Thus the correct description of the uveitis was not chorioretinitis, but retinochoroiditis.[54] Two years later serological studies were performed on a large group of individuals with retinochoroiditis.[55] Of these patients 64 per cent had positive skin tests and 82 per cent had positive dye tests. The dye test titres in patients with chronic lesions were not particularly high but when related to the presence of retinal scars even low titres were considered diagnostic.[55] This suggested that toxoplasmosis may be a common cause of retinochoroiditis and that serological tests might support or exclude this diagnosis. Whether most cases of ocular toxoplasmosis are a result of congenital or acquired infection is still undecided (Chapter 3).

Immunocompromised states and malignancy

Since encystment of toxoplasma may occur in areas such as the brain where the immune response is less effective,[56] reactivation of a latent infection results in early involvement of the brain. Toxoplasmic encephalitis was described in 1953,[57] in a patient suffering from Hodgkin's disease and sarcoidosis since 1949. Following a course of X-ray irradiation therapy over a period of 28 months, he was given chemotherapy and 4 months later died of toxoplasmic encephalitis.[57] It was not clear if the toxoplasma infection was latent or recently acquired. A series of animal experiments suggested that immunosuppressive treatment was more likely to provoke reactivation of a latent infection.[57] Further studies of toxoplasma related to malignancy were made in the early 1960s.[58] Immunocompromised states are associated with malignancy, organ transplant, and infection. Because of the human immunodeficiency virus, which depresses the immune response, the number of such patients is likely to increase in the future (Chapter 7).

DIAGNOSIS AND TREATMENT OF TOXOPLASMOSIS

Early attempts to diagnose toxoplasmosis in both animals and humans relied on histological demonstration of parasites in the tissues, often after the death of the host.[1,7,10,16,45] However, in 1939[17] it was demonstrated that animals inoculated with human toxoplasma developed immunity against animal toxoplasma. Similarly those inoculated with animal toxoplasma became immune to human toxoplasma. Thus it was concluded that toxoplasma found in animals and humans were immunologically identical. Demonstration of antibodies to toxoplasma formed the basis for the development of many diagnostic tests,[59-68] and possible drug treatment.

Development of diagnostic tests

The first diagnostic test for toxoplasma, the neutralizing antibody test was described in 1937[59] (Table 1.3). Intracutaneous inoculation of non-immune animals with toxoplasma tissue cysts produces typical necrotic skin lesions, fever, or death, but if the inoculum is mixed with serum containing anti-

Table 1.3 Landmarks in diagnosis of toxoplasmosis

Year	Test
1937	Neutralization antibody test
1942	Complement fixation test
1948	Dermal hypersensitivity test
	Methylene blue dye test
1957	Indirect haemagglutination test
1962	IgG immunofluorescent antibody test
1968	IgM immunofluorescent antibody test
1983	Western blot analysis
1989	Polymerase chain reaction

toxoplasma antibodies before inoculation, no illness is observed. However, if the serum does not contain antibodies then the usual response is observed.[17] Thus the test could be used to determine previous exposure to toxoplasma. Unfortunately this test suffers from several drawbacks which include the use of live parasites and test animals with a delay of up to seven days before a result is readable.[60] Some of these problems were overcome, when in 1942 a simple and rapid serological procedure for measuring complement-fixing toxoplasma antibodies was described[60] (Table 1.3). The complement fixation test used toxoplasma antigen prepared from infected rabbit brain and complement from either fresh or lyophilized guinea-pig serum. Unfortunately the rabbit brain antigen can produce non-specific reactions.[60,61] Therefore, in 1948, a toxoplasma antigen cultivated from the chorioallantoic membrane of embryonated eggs was used.[61] The chorioallantoic membrane of chick embryos 6–12 days old will support the growth of toxoplasma. Death usually occurs by the eighth day after infection when the membrane is removed and prepared for use. Using antigen derived from this source the complement fixation test correlates well with the neutralizing antibody test.[61] Complement fixation antibodies persist for a shorter period of time in serum than neutralizing antibody.[60] Therefore the complement fixation test is of more use in the diagnosis of active or recent toxoplasmosis.[60]

The dermal hypersensitivity test which was described in 1948[62] (Table 1.3) uses antigen derived from chick embryos or peritoneal exudates of mice. Infection with toxoplasma stimulates a cell-mediated response (Chapter 7). Therefore intradermal inoculation of a toxoplasma antigen stimulates a local inflammatory response in an immune patient. Maximum response is attained within 48–72 hours of inoculation and the results correlate well with serological tests.[63] The test is simple to perform and has been used as a screening test,[51,55] but a hypersensitivity reaction may not be detectable until several years after a toxoplasma infection.[63] Therefore its use in diagnosis of recently acquired infection is severely limited (Chapter 4).

The most significant development in the diagnosis of toxoplasmosis was the Sabin–Feldman dye test, also described in 1948[64] (Table 1.3). As with many discoveries the development of the dye test was more by accident than design.[65] During studies of the neutralizing antibodies of toxoplasma it was noticed that toxoplasma mixed with serum containing neutralizing antibodies lost the refractility seen with normal serum. Also when mixtures were allowed to dry on a slide, subsequent addition of Wrights stain failed to stain those toxoplasma which had been treated with immune serum.[64] A similar observation was made when alkaline methylene blue was added to the mixtures; only parasites in non-immune serum stained properly. When immune serum was heated to 56 °C for 30 minutes, its effect on parasites was lost. This observation was not due to destruction of the antibody but of a labile accessory factor necessary for the reaction. Addition of more accessory factor in the form of fresh non-immune serum re-enables killing and restores the effect on staining of parasites. It was later demonstrated that the accessory factor was complement.[66] Despite the necessity for live tachyzoites and fresh serum for accessory factor, the dye test is now into its fifth decade of use and remains the standard by which other serological tests for toxoplasma are measured.[67]

During the span of the dye test, other tests have been developed. Most have some drawbacks but are generally accepted for routine use. Indirect haemagglutination was described in 1957[68] but the time course for detection of antibody is longer than that of the dye test and therefore is not suitable as a diagnostic test in pregnant women or newborns.[67] During the 1960s specific immunofluoresence antibody tests for IgG and IgM were developed (Table 1.3) so enabling distinction between acute and chronic infection. These were followed in the 1970s by enzyme-linked immunosorbent assays (ELISA), and in the 1980s by the IgM immunosorbent agglutination assay (Chapter 4). More recently these techniques have been supplemented by new techniques such as Western blotting and polymerase chain reaction (Chapter 8).

Development of treatment

Early attempts to demonstrate the effects of known anti-protozoal drugs against toxoplasma *in vitro* looked promising but they had no therapeutic effect in infected mice and rabbits.[69] In contrast certain sulphonamides were ineffective *in vitro*, but were shown to have anti-toxoplasma effects in mice and rabbits.[69] Later, in 1952 the anti-malarial drug pyrimethamine was also shown to protect experimentally infected mice.[70] Both sulphonamides and pyrimethamine are potentially toxic but, because they act synergistically,[71] they can be used in combination at lower concentrations, which retain therapeutic efficacy but reduce the risk of toxicity. However, neither drug can eliminate tissue cysts, so reactivation of a latent infection cannot be prevented (Chapter 5). Because of the possible teratogenic effects of pyrimethamine, this drug must be used with care during pregnancy and only when fetal infection is proven or suspected (Chapter 6). Spiramycin, a suitable alternative for preventing fetal infection and which has less serious side effects (Chapter 6), was described in 1958.[72] Since these early descriptions few other drugs have been described with anti-toxoplasma activity at a tolerable dose (Chapter 5). Therefore the mainstay of treatment still relies on discoveries made nearly 50 years ago (Chapter 5). The urgent need to find alternatives to treat the increasing number of immunocompromised patients with toxoplasmosis is now stimulating research on new drugs (Chapter 5).

GENERAL EPIDEMIOLOGY

During the 90 years since toxoplasma was first described it has been realized that it is prevalent in most areas of the world. Furthermore its versatility is such that it infects most warm blooded birds and animals, including humans, as incomplete hosts.[73] Only one group of animals, the cats, can be infected as complete hosts for toxoplasma (Chapter 2). This lack of host specificity has resulted in widespread toxoplasma infection in humans. Approximately one-third of the entire world population possesses antibodies to toxoplasma although there is considerable variation between and even within countries.[74] Such widespread dispersal can be attributed to the many ways in which toxoplasma may infect a host. Major routes of infection are: ingestion of tissue cysts or oocysts (Chapter 2), transplantation of infected organs (Chapters 2 and 7), and congenital infection (Chapter 6). Infection may also occur as a result of inoculation of tachyzoites (Appendix 2). Much of the epidemiology of toxoplasma centres around the ability of the infective stages to survive in the environment (Chapter 2). Areas in which the climate is not conducive to survival of an infective stage tend to show a lower prevalence of infection.[74] Geographic factors are not the only influence on prevalence of infection. Host factors which result in increased exposure to toxoplasma are also a major consideration.

Infection from oocysts

Toxoplasmosis in humans has been reported throughout the world.[74] Much of the distribution of toxoplasma, especially in herbivores and vegetarians is dependent on the prevalence of cats and the resulting spread of oocysts.[74] This is demonstrated by the near absence of toxoplasma on Pacific atolls which are devoid of cats.[75] Under favourable conditions oocysts will sporulate within 24 hours of deposition in cat faeces (Chapter 2) after which they may remain infective for more than a year.[73] The most amicable environment for survival of the oocyst is a warm and moist one (Chapter 2). Therefore countries with a tropical climate and many cats have a high prevalence

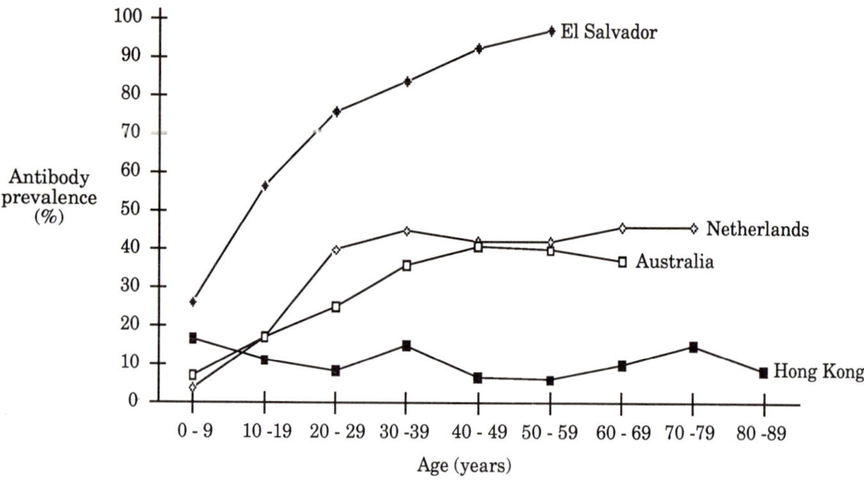

Fig. 1.3 Variations in the prevalence of toxoplasma antibody with age and geography.[76,79,81,83]

of infection. El Salvador is typical of such a country.[76] Soil texture is such that it retains moisture, therefore conditions are ideal for harbouring oocysts produced by the many cats. Hygiene is poor and children are infected, from contaminated soil, at a very early age (Fig. 1.3). Antibody prevalence in El Salvador is 26 per cent in children reaching 57 per cent in teenagers and more than 90 per cent by the age of 50 years[76] (Fig. 1.3). Similarly high figures have been found in the adult populations of Guatemala (94%), Costa Rica (88%) and Tahiti (85%)[77] all of which have tropical climates. In contrast, in Iceland which has a cold climate, antibody prevalence is low (11%)[21]. However in Alaska a large number of cats are kept as domestic pets and antibody prevalence is 28 per cent, but there may also be infection from tissue cysts.[78]

Desiccation of oocysts also renders them non-infective.[74] Therefore in environments with low rainfall or in which soil does not retain moisture there is less risk of infection from oocysts. In southern Australia the rate of antibody acquisition is much less than that of El Salvador[76,79] (Fig. 1.3). Because of the low level of rainfall in southern Australia oocysts are unlikely to survive. Infection is probably a result of tissue cyst infection rather than oocyst infection. In the dry Andean region of Peru antibody prevalence is only 9 per cent compared to 42 per cent in the coastal region.[80] However this difference may be due to the differences in altitude between the two areas.[73] Altitude has been shown to be inversely related to antibody prevalence.[65] In Columbia, 53 per cent of the population living between 1000 and 2500 metres elevation had toxoplasma antibodies. This was reduced to 43 per cent in those living between 2500 and 4500 metres.[65] This may be due to the climate in mountainous regions being unfavourable for oocyst development or there being few cats in the area.[73]

In highly populated areas contact with oocyst-contaminated soil is less likely. Thus in areas of high population density the possibility of infection from oocysts is more remote. This is demonstrated on Hong Kong[81] where there are few cats and therefore few oocysts. Antibody prevalence in all age groups of the Chinese community of Hong Kong is 6–15 per cent[81] (Fig. 1.3). In a group of travelling people in Scotland antibody prevalence was 28 per cent, randomly distributed in all age groups.[82] Because of the nature of their lifestyle these people rarely own cats. Additionally the areas in which they camp are usually removed from residential areas where there may be domestic cats. Because of the low contact with cats infection from oocysts is unlikely. The most probable source of infection for these people is tissue cysts in undercooked meat.[82]

Infection in the Netherlands is from both oocysts and tissue cysts.[83] There are a large number of cats and conditions are ideal for sporulation of oocysts. Thus there is a high rate of infection in children as a result of ingestion of oocysts (Fig. 1.3). However in the age group 15–35 years seroconversion is probably a result of infection by tissue cysts.[83]

Infection from tissue cysts

Infection from tissue cysts results from ingestion of infected meat (Chapter 2). However tissue cysts are destroyed at temperatures attained by thorough cooking. The rates of infection of ethnic groups in a community have been shown to differ.[84] However it is probable that such differences are due to customs rather than the inherent resistance to toxoplasma of a particular group.[73] Some religions may dictate which animals, if any, may be eaten so reducing or eliminating the threat of infection from meat. The differences in prevalence may also be due to cooking methods employed or preference for undercooked meat. The different ethnic groups which co-exist in Hawaii acquire antibody at different rates.[84] The percentage of Japanese school children that are antibody positive (14%), was less than one-third (45%) that of the Filipinos. These differences in antibody prevalence can be attributed to differences in affluence and eating habits.[84] The Japanese in Hawaii are affluent and enjoy a high income and standard of living. In addition the Japanese custom of cooking meat in small pieces has been retained so it is eaten well cooked thus minimizing the risk of infection from meat. Contrary to this, the Filipinos are economically underpriviledged with a lower standard of living. Filipinos have a preference for eating raw meat[84] and are therefore at greater risk from infection by tissue cysts than the Japanese. These results contrast with the prevalence of antibody in the ethnic groups in New York, USA.[85] Compared to black people on low income, the affluent Caucasian adult population of the city have a high rate of antibody aquisition.[85] This was attributed to the culturally acquired habit of eating raw meat. A similarly high rate of antibody aquisition is demonstrated in Paris.[31] Antibody prevalence exceeds 50 per cent before the age of 10 years. This again is due to the Parisian taste for undercooked meat.

Age and sex

The sex of an individual appears to have little overall influence on the incidence of infection.[80,86] However one study of toxoplasmic lymphadenopathy in different age groups did suggest that there was some variation between the sexes.[86] Males up to the age of 15 years have a higher incidence of infection than females. This may be because boys have more outdoor activities than girls which result in higher risk of infection from oocysts. Conversely more women than men over 25 years are infected and this was attributed to women having more contact with cats, and raw meat during cooking, than men. Thus the increased prevalence is probably due to higher levels of exposure to toxoplasma.[86]

Infection in different age groups shows a much more noticeable trend. Babies are unlikely to be infected as a result of oocysts in soil and their food is carefully prepared, thus the rate of infection is low. Once children become mobile they are more likely to come into contact with toxoplasma and so the infection rate increases. However in general, seroconversion occurs most frequently in the age group 15–35 years.[74] Although infection by oocysts through outdoor activities may be important, a major factor accounting for the seroconversion is, as with all age groups, ingestion of tissue cysts by consumption of raw or undercooked meat.[74] The risk of infection from meat may be increased in this age group either by careless cooking practice[83] or a desire to experiment with cuisine. Antibody acquisition in different age groups is dependent on the overall prevalence of toxoplasma in the environment. Thus in areas where there is a high prevalence of toxoplasma, children of a particular age group will show a higher rate of infection than those of the same age group living in areas of lower toxoplasma incidence.[11,87]

Occupational hazards

People whose occupations increase the potential contact with toxoplasma are more at risk of infection. Those most at risk are people such as abattoir workers, butchers, and meat inspectors.[88,89] In Brazil, the antibody prevalence in meat inspectors and abattoir workers was found to be 92 and 65 per cent respectively[89] compared with 14–32 per cent in military recruits.[65] The duties of the inspectors involve close examinations of suspicious lesions on meat and they may be infected due to this close contact with the meat.[89] Similar figures (68%) were found in Japanese slaughter house workers[88] and contrasts sharply with 30 per cent prevalence in farmers used as controls. However only 34 per cent of abattoir workers in Britain were found to have antibody to toxoplasma compared with 27 per cent of males over 20 years in the general population.[90] This may be due to different prevalences of infection in animals from different countries[21] or better hygiene. In addition to the hazards of contamination of hands with tissue cysts from raw meat and subsequent ingestion of the tissue cysts, those working with raw meat are also subject to risk of infection by bradyzoites.[88,89] Tissue cysts may become ruptured during cutting and preparation of meat so releasing bradyzoites. Therefore infection may occur as a result of bradyzoites penetrating minor cuts or abrasions. Likewise a knife used for cutting contaminated meat may itself become contaminated. A subsequent cut from such a knife will effectively amount to inoculation of parasites.

As toxoplasmosis is more common in animals than humans[91] animal contact has been implicated in transmission of toxoplasma to humans.[34,65] Veterinarians have quite intimate contact with animals and this is reflected in their antibody prevalences. In the UK 34 per cent of veterinary surgeons had antibodies to toxoplasma (compared with 27% in the control group).[90] In the USA this level varied from 8.3 to 43.7 per cent[74] (14% of a group of military recruits).[65] The variation may be due in part to climatic factors in different areas or different animal contact between city and rural practice. Cats are less of a risk because it is contact with the faeces, not the cat, which is the risk factor.[21] Occupational risk associated with animals may also be expected in farm workers.[92] This study showed a negative association with animals. Only 9.4 per cent of those handling stock demonstrated antibody whereas 39.6 per cent of those working on the land showed evidence of previous toxoplasma infection.[92] Therefore it is obvious that those working on the land acquired infection by means of oocyst contamination of soil and agricultural implements.

SUMMARY

Recognizing toxoplasma as a new entity was hindered by its morphological similarity to other parasites. Many early descriptions were reported as an encephalitozoon or sarcocystis. In addition some observations which were classified as toxoplasma are suspicious because of the inadequate information provided.

The first description of toxoplasma was in 1900 in Java sparrows. In 1908 it was simultaneously described in gondis, a rabbit, a mole, and a human. In the following year the organism was named *Toxoplasma gondii*. It was subsequently realized that toxoplasma infected most birds and warm-blooded animals, including humans, and that it was present in most countries world-wide. Appreciation of such widespread distribution directed attention towards the mechanisms of toxoplasma infection between hosts. Attempts to demonstrate involvement of blood-sucking insects were unsuccessful. Then in 1956, following the demonstration of tissue cysts in pork, the theory of infection from meat was postulated. Tissue cysts were later shown to be resistant to proteolytic enzymes in gastric secretions, and seroconversion was shown to increase when children were fed raw meat. However, tissue cysts are destroyed by adequate cooking and such transmission could not explain infection of herbivorous animals and vegetarians.

In 1965, the same year that infection from meat was proven, toxoplasma was shown to be transmitted from the faeces of cats. Two years later the nematode transmission theory was proposed. This theory which associated transmission of toxoplasma with the nematode *Toxocara cati* was later discredited. The oocyst is better able to withstand high temperatures and desiccation which destroy other infective forms. Thus the oocyst is responsible for infection of non-carnivorous animals. Then in 1970 the sexual cycle in cats was described which confirmed the taxonomic position of toxoplasma in the coccidia.

Although the first case of congenital toxoplasmosis had been described much earlier, it was not until 1939 that toxoplasma was recognized as the causative agent. The recognition of toxoplasma as a cause of disease in humans paralleled the development of diagnostic tests and treatment for toxoplasma. The dye test, developed in 1948, and other reliable tests were used to facilitate association of lymphadenopathy and ocular disease with toxoplasma. Reactivation of latent infections in immunocompromised patients was also observed. Despite development of more sophisticated tests the dye test still remains the standard by which tests are judged. Similarly, few drugs have been found to be effective against toxoplasma and the synergistic combination of sulphonamides and pyrimethamine is still used today.

Serological tests were also used to gather epidemiological data. Approximately one-third of the entire world population has toxoplasma antibody but there is considerable variation in antibody prevalence between and even within countries. Climate plays a major role in survival of oocysts. Thus countries with a tropical climate have a very high prevalence of antibody but those with cold or dry climates and areas of high altitude tend to have a much lower prevalence. Infection may also result from ingestion of tissue cysts in meat. Tissue cysts are easily destroyed by cooking, therefore differences in antibody acquisition in ethnic groups within communities are usually explained by the extent to which meat is cooked. The sex of an individual has some influences on antibody prevalence. Very young children have a low antibody prevalence but this increases when they start to explore the environment and are infected from oocysts. Seroconversion is most frequent in the age group 15–35 years. This may be due to contact with oocysts in outdoor activities, or from experimentation with different cuisine.

People who are subject to increased exposure to toxoplasma in their occupation are more at risk of infection, especially those who work on the land or with animals. Thus, in the last 90 years, the confusion in the initial description of toxoplasma infection has been resolved, but much needs to be done.

REFERENCES

1 Pinkerton, H. and Weinman, D. (1940). Toxoplasma infection in man. *Arch. Pathol.*, **30**, 374–92.

2 Levaditi, C., Nicolau, S., and Schoen, R. (1924). L'Étiologie de l'encéphalite épizootique du lapin, dans ses rapports avec l'étude expérimentale de l'encéphalite léthargique *Encephalitozoon cuniculi* (nov. spec.). *Annu. Inst. Pasteur*, **38**, 651–712.

3 Frenkel, J. K. (1973). Toxoplasmosis: parasite life cycle, pathology, and immunology. In *The Coccidia: Eimeria, Isospora, Toxoplasma, and related genera*. (ed. D. M. Hammond and P. L. Long), pp. 343–410. University Park Press, Baltimore.

4 Jacobs, L. (1974). *Toxoplasma gondii*: parasitology and transmission. *Bull. N.Y. Acad. Med.*, **50**, 128–45.

5 Castellani, A. (1913). Protozoa-like bodies in a case of protracted fever with splenomegaly. *J. Ceylon Br. Brit. Med. Assoc.*, **10**, 20–1.

6 Fedorovitch, A. I. (1916). Hémoparasites trouvés dans un cas de fièvre chronique. *Annu. Inst. Pasteur*, **30**, 249–50.

7 Torres, C. M. (1927). Sur un nouvelle maladie de l'homme, caractérisée par la présence d'un parasite intracellulaire, très proche du toxoplasma et de l'encephalitozoon, dans le tissu musculaire cardiaque, les muscles du squelette, le tissu cellulaire sous-cutané et le tissue nerveux. *C.R. Soc. Biol.*, **97**, 1778–81.

8 Nicolle, C. and Manceaux, L. (1908). Sur une infection à corps de Leishman (ou organismes voisins) du gondi. *C.R. Acad. Sci.*, **147**, 763–6.

9 Darling, S. T. (1908). Sarcosporidiosis: with report of a case in man. *Proc. Canal Zone Med. Assoc.*, **1**, 141–52.

10 Wolf, A. and Cowen, D. (1937). Granulomatous encephalomyelitis due to an encephalitozoon (encephalitozoic encephalomyelitis). *Bull. Neurol. Inst. N.Y.*, **6**, 306–71.

11 Frenkel, J. K. (1973). Toxoplasma in and around us. *Bioscience*, **23**, 343–52.

12 Laveran, M. (1900). Au sujet de l'hématozoaire endoglobulaire de *Padda oryzivora*. *C. R. Soc. Biol.*, **52**, 19–20.

13 Nicolle, C. and Manceaux, M. (1909). Sur un protozoaire nouveau du gondi. *C. R. Acad. Sci.*, **148**, 369–72.

14 Splendore, A. (1908). Uno nuovo protozoa parasite dei conigli: incontrato nelle lesioni anatomiche d'una malattia che ricorda in molti punti il Kala-azar dell'uomo. *Rev. Soc. Sci. Sao Paulo*, **3**, 109–12.

15 Mine, N. (1911). Observation on protozoan parasites in Japan. *Gun-I-Dan Zasshi*, **27**, 1–68. Quoted from[88].

16 Wolf, A., Cowen, D., and Paige, B. H. (1939). Toxoplasmic encephalomyelitis III. A new case of granulomatous encephalomyelitis due to a protozoon. Am. J. Pathol., **15**, 657–95.

17 Sabin, A. B. (1939). Biological and immunological identity of toxoplasma of animal and human origin. *Proc. Soc. Exp. Biol.*, **41**, 75–80.

18 Levine, N. D. (1977). Taxonomy of toxoplasma. *J. Protozool.*, **24**, 36–41.

19 Frenkel, J. K., Dubey, J. P., and Miller, N. L. (1970). *Toxoplasma gondii* in cats: fecal stages identified as coccidian oocysts. *Science*, **167**, 893–6.

20 Hutchison, W. M., Dunachie, J. F., Siim, J. C., and Work, K. (1970). Coccidian-like nature of *Toxoplasma gondii*. *Br. Med. J.*, **1**, 142–4.

21 Dubey, J. P. and Beattie, C. P. (1988). *Toxoplasmosis of animals and man*. CRC Press, Florida.

22 Darling, S.T. (1910). Experimental sarcosporidiosis in the guinea pig and its relation to a case of sarcosporidiosis in man. *J. Exp. Med.*, **12**, 19–28.

23 Kean, B. H., Major, M. C., and Grocott, B. S. (1945). Sarcosporidiosis or toxoplasmosis in man and guinea-pig. *Am. J. Pathol.*, **21**, 467–83.

24 Laveran, A. (1915). Nouvelle contribution à l'etude du *Toxoplasma gondii*. *Bull. Soc. Pathol. Exot.*, **8**, 58–63.

25 Chatton, E. and Blanc, G. (1917). Notes et reflexions sur le toxoplasme et la toxoplasmose du gondi (*Toxoplasma gondii* Ch. Nicolle et Manceaux 1909). *Arch. Inst. Pasteur Tunis*, **10**, 1–41.

26 Woke, P. A., Jacobs, L., Jones, F. E., and Melton, M. L. (1953). Experimental results on possible arthropod transmission of toxoplasmosis. *J. Parasitol.*, **39**, 523–32.

27 Jankû, J. (1923). Pathogenesa a pathologická anatomie tak nazvaného vrozeného kolombu žluté skvrany voku normálně velikem a microphthalmickem s nálezem parasitu v sitnici (Pathogenesis and pathologic anatomy of coloboma of the macula lutea in an eye of normal dimensions, and in a microphthalmic eye, with parasites in the retina). *Casopis Lekarew Ceskyck*, **62**, 1021–143. Quoted from.[10]

28 Sabin, A. B. (1941). Toxoplasmic encephalitis in children. *JAMA*, **116**, 801–7.

29 Weinman, D. and Chandler, A. H. (1956). Toxoplasmosis in man and swine — an investigation of the possible relationship. *JAMA*, **161**, 229–32.

30 Jacobs, L., Remington, J. S., and Melton, M. L. (1960). The resistance of the encysted form of *Toxoplasma gondii*. *J. Parasitol.*, **46**, 11–21.

31 Desmonts, G., Couvreur, J., Alison, F., Baudelot, J., Gerbeaux, J., and Lelong, M. (1965). Étude épidémiologique sur la toxoplasmose: de l'influence de la cuisson des viandes de boucherie sur la fréquence de l'infection humaine. *Rev. Franc Étud Clin. Biol.*, **10**, 952–8.

32 Hutchison, W. M. (1967). The nematode transmission of *Toxoplasma gondii*. *Trans. R. Soc. Trop. Med. Hyg.*, **61**, 80–9.

33 Hutchison, W. M. (1965). Experimental transmission of *Toxoplasma gondii*. *Nature*, **206**, 961–2.

34 Hutchison, W. M., Dunachie, J. F., and Work, K. (1968). The faecal transmission of *Toxoplasma gondii*. *Acta Pathol. Microbiol. Scand.*, **74**, 462–4.

35 Jacobs, L. (1967). Toxoplasma and toxoplasmosis. In *Advances in parasitology*, vol. 5. (ed. Ben Dawes), pp. 1–45. Academic Press, London.

36 Dubey, J. P. (1968). Feline toxoplasmosis and its nematode transmission. *Vet. Bull.*, **38**, 495–9.

37 Dubey, J. P. (1968). Isolation of *Toxoplasma gondii* from the feces of a helminth free cat. *J. Protozool.*, **15**, 773–5.

38 Hutchison, W. M. and Work, K. (1969). Observations on the faecal transmission of *Toxoplasma gondii*. *Acta Pathol. Microbiol. Scand.*, **77**, 275–82.

39 Frenkel, J. K., Dubey, J. P., and Miller, N. L. (1969). *Toxoplasma gondii*: fecal forms separated from eggs of the nematode *Toxocara cati*. *Science*, **164**, 432–3.

40 Rogers, W. P. (1960). The physiology of infective processes of nematode parasites; the stimulus from the animal host. *Proc. R. Soc. Lond. (Biol.)*, **152**, 367–86.

41 Work, K. and Hutchison, W. M. (1969). The new cyst of *Toxoplasma gondii*. *Acta Pathol. Microbiol. Scand.*, **77**, 414–24.

42 Dubey, J. P., Miller, N. L., and Frenkel, J. K. (1970). Characterization of the new fecal form of *Toxoplasma gondii*. *J. Parasitol.*, **56**, 447–56.

43 Hutchison, W. M., Dunachie, J. F., Work, K., and Siim, J. C. (1971). The life cycle of the coccidian parasite *Toxoplasma gondii* in the domestic cat. *Trans. R. Soc. Trop. Med. Hyg.*, **65**, 380–99.

44 Baraben, L. and Saint-Remy, M. G. (1894). Sur un cas de tubes psorospermiques observés chez l'homme. *C. R. Soc. Biol.*, **10**, 201–2.

45 Richter, R. (1936). Meningo-encephalomyelitis neonatorum anatomic report of a case. *Arch. Neurol. Psychiatr.*, **36**, 1085–100.

46 Pinkerton, H. and Henderson, R. G. (1941). Adult toxoplasmosis: a previously unrecognised disease entity simulating the typhus-spotted fever group. *JAMA*, **116**, 807–14.

47 Gard, S. and Magnusson, J. H. (1951). A glandular form of toxoplasmosis in connection with pregnancy. *Acta Med. Scand.*, **141**, 59–64.

48 Siim, J. C. (1951). Acquired toxoplasmosis: report of seven cases with strongly positive serologic reactions. *JAMA*, **147**, 1641–5.

49 Sabin, A. B. (1942). Toxoplasma neutralizing antibody in human beings and morbid conditions associated with it. *Proc. Soc. Exp. Biol. Med.*, **51**, 6–10.

50 Vail, D., Strong, J. C., and Stephenson, W. V. (1943). Chorioretinitis associated with positive serologic tests for toxoplasma in older children and adults. *Am. J. Ophthalmol.*, **26**, 133–41.

51 Frenkel, J. K. and Jacobs, L. (1959). Ocular toxoplasmosis. Pathogenesis, diagnosis, and treatment. *A. M. A. Arch. Ophthalmol.*, **59**, 260–79.

52 Frenkel, J. K. (1949). Uveitis and toxoplasmin sensitivity. *Am. J. Ophthalmol.*, **32**, 127–35.

53 O'Connor, G. R. (1978). Current concepts in ophthalmology: uveitis and the immunologically compromised host. *N. Engl. J. Med.*, **299**, 130–2.

54 Wilder, H. C. (1952). Toxoplasma chorioretinitis in adults. *A. M. A. Arch. Ophthalmol.*, **48**, 127–36.

55 Frenkel, J. K. (1954). Chorioretinitis associated with positive tests for toxoplasmosis. *Acta XVIII Cong. Ophthalmol.*, **3**, 1965–76.

56 Veins, P. and Morisset, R. (1975). Toxoplasmosis and the compromised host. *Int. J. Clin. Pharmacol.*, **11**, 361–5.

57 Frenkel, J. K., Nelson, B. M., and Arias-Stella, J. (1975). Immunosuppression and toxoplasmic encephalitis: clinical and experimental aspects. *Hum. Pathol.*, **6**, 97–110.

58 Vietzke, W. M., Gelderman, A. H., Grimley, P. M., and Valsamis, M. P. (1968). Toxoplasmosis complicating malignancy. *Cancer*, **21**, 816–27.

59 Sabin, A. B. and Olitsky, P. K. (1937). Toxoplasma and obligate intracellular parasitism. *Science*, **85**, 336–8.

60 Warren, J. and Sabin, A. B. (1942). The complement fixation reaction in toxoplasmic infection. *Proc. Soc. Exp. Biol. Med.*, **51**, 11–4.

61 Warren, J. and Russ, S. B. (1948). Cultivation of toxoplasma in embryonated egg. An antigen derived from chorioallantoic membrane. *Proc. Soc. Exp. Biol.*, **67**, 85–9.

62 Frenkel, J. K. (1948). Dermal hypersensitivity to toxoplasma antigens (toxoplasmins). *Proc. Soc. Exp. Biol. Med.*, **68**, 634–9.

63 Remington, J. S. and Desmonts, G. (1990). Toxoplasmosis. In *Infectious diseases of the fetus and newborn infant*, 3rd edn. (ed. J. S. Remington and J. O. Klein), pp. 89–195. W. B. Saunders, Philadelphia.

64 Sabin, A. B. and Feldman, H. A. (1948). Dyes as microchemical indicators of a new immunity phenomenon affecting a protozoon parasite (toxoplasma). *Science*, **108**, 660–3.

65 Feldman, H. A. (1974). Toxoplasmosis: an overview. *Bull. N. Y. Acad. Med.*, **50**, 110–27.

66 Feldman, H. A. (1980). Maxwell Finland lectures: to establish a fact. *J. Infect. Dis.*, **141**, 525–9.

67 Wilson, M., Ware, D. A., and Juranek, D. D. (1990). Serological aspects of toxoplasmosis. *J. Am. Vet. Med. Assoc.*, **196**, 277–81.

68 Jacobs, L. and Lunde, M. N. (1957). A haemagglutination test for toxoplasmosis. *J. Parasitol.*, **43**, 308–14.

69 Sabin, A. B. and Warren, J. (1942). Therapeutic effectiveness of certain sulfonamides on infection by an intracellular protozoon (toxoplasma). *Proc. Soc. Exp. Biol. Med.*, **51**, 19–23.

70 Eyles, D. E. and Coleman, N. (1952). Tests of 2,4–diaminopyrimidines on toxoplasmosis. *Public Health Rep.*, **67**, 249–52.

71 Eyles, D. E. and Coleman, M. (1953). Synergistic effect of sulfadiazine and daraprim against experimental toxoplasmosis in the mouse. *Antibiot. Chemother.*, **3**, 483–90.

72 Garin, J. P. and Eyles, D. E. (1958). Traitment de la toxoplasmose expérimentale de la souris par la spiramycine. *Presse Méd.*, **66**, 957–8.

73 Beattie, C. P. (1982). The ecology of toxoplasmosis. *Ecol. Dis.*, **1**, 13–20.

74 Jackson, M. H. and Hutchison, W. M. (1989). The prevalence and source of toxoplasma infection in the environment. *Adv. Parasitol.*, **28**, 55–105.

75 Wallace, G. D. (1969). Serologic and epidemiologic observations on toxoplasmosis on three pacific atolls. *Am. J. Epidemiol.*, **90**, 103–11.

76 Remington, J. S., Efron, B., Cavanaugh, E., Simon, H. J., and Trejos, A. (1970). Studies on toxoplasmosis in El Salvador. Prevalence and incidence of toxoplasmosis as measured by the Sabin–Feldman dye test. *Trans. R. Soc. Trop. Med. Hyg.*, **64**, 252–67.

77 Wright, W. H. (1957). A summary of the newer knowledge of toxoplasmosis. *Am. J. Clin. Pathol.*, **28**, 1–17.

78 Peterson, D. R., Cooney, M. K., and Beasley, R. P. (1974). Prevalence of antibody to toxoplasma among Alaskan natives: relation to exposure to the Felidae. *J. Infect. Dis.*, **130**, 557–63.

79 Johnson, A. M., Roberts, H., and McDonald, P. J. (1980). Age–sex distribution of toxoplasma antibody in the south Australian population. *J. Hyg.*, **84**, 315–20.

80 Cantella, R., Colichon, A., Lopez, L., Wu, C., Goldfarb, A. Cuadra, E. *et al.* (1974). Toxoplasmosis in Peru: geographic prevalence of *Toxoplasma gondii* antibodies in Peru studied by indirect fluorescent antibody technique. *Trop. Geogr. Med.*, **26**, 204–9.

81 Ko, R. C., Wong, F. W., Todd, D., and Lam, K. C. (1980). Prevalence of *Toxoplasma gondii* antibodies in the Chinese population of Hong Kong. *Trans. R. Soc. Trop. Med. Hyg.*, **74**, 351–4.

82 Jackson, M. H., Hutchison, W. M., and Siim, J. C. (1987). A seroepidemiological survey of toxoplasmosis in Scotland and England. *Ann. Trop. Med. Parasitol.*, **81**, 359–65.

83 Van der Veen, J. and Polak, M. F. (1980). Prevalence of toxoplasma antibodies according to age with comments on the risk of prenatal infection. *J. Hyg.*, **85**, 165–74.

84 Wallace, G. D. (1976). The prevalence of toxoplasmosis on pacific islands, and the influence of ethnic group. *Am. J. Trop. Med. Hyg.*, **25**, 48–53.

85 Kimball, A. C., Kean, B. H., and Fuchs, F. (1974). Toxoplasmosis: risk variations in New York city obstetric patients. *Am. J. Obstet. Gynecol.*, **119**, 208–14.

86 Beverley, J. K. A., Fleck, D. G., Kwantes, W., and Ludlam, G. B. (1976). Age–sex distribution of various diseases with particular reference to toxoplasmic lymphadenopathy. *J. Hyg.*, **76**, 215–28.

87 Stagno, S. and Thiermann, E. (1973). Acquisition of toxoplasma infection by children in a developing country. *Bull. W. H. O.*, **49**, 627–31.

88 Komiya, Y., Kobayashi, A., and Koyama, T. (1961). Human toxoplasmosis, particularly on the possible source of its infection in Japan: a review. *Jap. J. Sci. Biol.*, **14**, 157–72.

89 Riemann, H. P., Brant, P. C., Behymer, D. E., and Franti, C. E. (1975). *Toxoplasma gondii* and *Coxiella burneti* antibodies among Brazilian slaughterhouse employees. *Am. J. Epidemiol.*, **102**, 386–93.

90 Beverley, J. K. A., Beattie, C. P., and Roseman, C. (1954). Human toxoplasma infection. *J. Hyg.*, **52**, 37–46.

91 Frenkel, J. K. (1990). Toxoplasmosis in human beings. *J. Am. Vet. Med. Assoc.*, **196**, 240–8.

92 French, J. G., Messinger, H. B., and MacCarthy, J. (1970). A study of *Toxoplasma gondii* infection in farm and non-farm groups in the same geographic location. *Am. J. Epidemiol.*, **91**, 185–91.

2

Life cycle and animal infection
ROGER EVANS

Toxoplasma gondii is an organism that can parasitize most, perhaps all, animals and birds. As it has a large host range toxoplasma can be found throughout the world. Although toxoplasma is a common infection in animals, including humans (Chapter 1), it is relatively unheard of outside the veterinary and medical professions. This anonymity may be mostly due to the infection not producing a characteristic illness. Similarly, in animals, toxoplasma infection does not produce a distinctive illness apart from abortion. However, the consequences of animal infection are not negligible: infected meat animals can be a serious potential public health hazard; and toxoplasmosis accounts for 10–20 per cent of sheep abortions in the UK[1] and the USA[2] resulting in severe economic loss to the livestock owners. This chapter will deal with the parasite's life cycle, the sources of infection, and the results of infection in various animals. Finally the importance of strain variation of toxoplasma will be discussed and how it may affect infection in animals and humans.

LIFE CYCLE

Nomenclature

As with most parasites the life cycle of toxoplasma can be arbitrarily divided into two parts: the sexual cycle and the asexual cycle. During each of these cycles the parasite passes through quite distinct developmental stages. The sexual cycle is found only in the complete host, the cat. The asexual cycle can take place in most animals, including cats and birds (Fig. 2.1). The identification of the various stages of the toxoplasma life cycle has not been easy. The problem has principally been due to the relatively late discovery of the sexual cycle[3] of the organism and the realization that this occurs only in domestic cats and other Felidae.[4] Differences in nomenclature have not helped and even now some stages are referred to by different names in the medical and veterinary literature. Thus the medical literature frequently refers to 'trophozoite' whereas the veterinarians tend to use 'tachyzoite'. Although the precise reasons for these differences are difficult to ascertain, it is probably due to the fuller acceptance of Frenkel's classification[5] by the veterinarians. Since more controversy has surrounded the tachyzoite and bradyzoite, they will be dealt with in greater detail than the other stages.

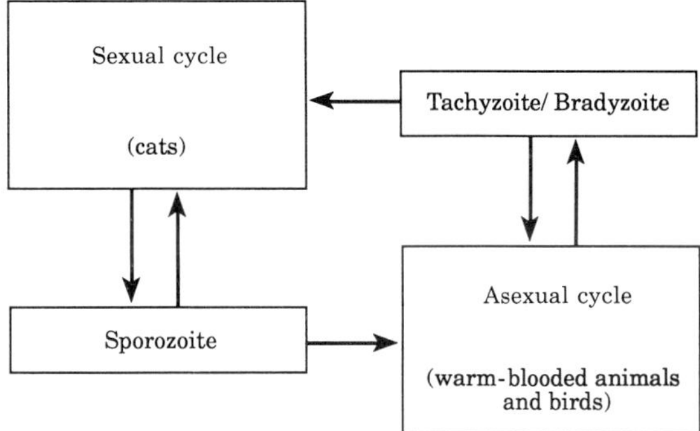

Fig. 2.1   Simplified toxoplasma life cycle.

Tachyzoite

This is the rapidly multiplying stage of the parasite (tachos = speed, Greek) forming part of the asexual cycle (Fig. 2.2). It measures 2–4 × 4–7 μm and is often crescentic in shape being slightly pointed anteriorly and rounded posteriorly. It divides within host cells by repeated endodyogeny. Endodyogeny[6] (endon = within/inside, dyo = two, genesis = birth, Greek) is the formation of two daughter cells within a mother cell, division being completed with the destruction of the mother cell by its progeny. Traditionally this stage has been known as the 'trophozoite' but after the discovery of the sexual cycle two new terms were introduced. Hoare[7] accepted the name 'endozoite', referring to the stage's type of division, whilst a year later, in 1973, Frenkel[5] put forward the term 'tachyzoite' because of the stage's high rate of division. Tachyzoite is possibly the more appropriate term since other stages of toxoplasma also divide by endodyogeny. Similarly many terms have been given for the aggregates of tachyzoites within host cells: pseudocysts,[3] groups,[5] or terminal colonies.[3] Although Frenkel[5] criticized the use of the term pseudocyst since it had been used for both 'cystic' stages of the toxoplasma asexual cycle, Jacobs,[8] in a review, encouraged its use because the term was familiar and its previous misuse could now be prevented by redefining it. The pseudocyst (pseudo = false, Greek) contains the aggregates of tachyzoites and its wall has been shown to be made from both the host and the parasite,[9] unlike a true parasite cyst whose wall is of parasitic origin only.[10] Therefore retention of the term pseudocyst would seem appropriate.

Bradyzoite

The bradyzoite is the slowly dividing form (brady = slow, Greek) in the asexual cycle of the parasite (Fig. 2.2). Although slightly smaller than the tachyzoite it is structurally very similar. However there are distinct differences biologically and antigenically (Chapter 8). Bradyzoites have a much shorter prepatent period (3–10 days) than the tachyzoites (19–48 days).[11]

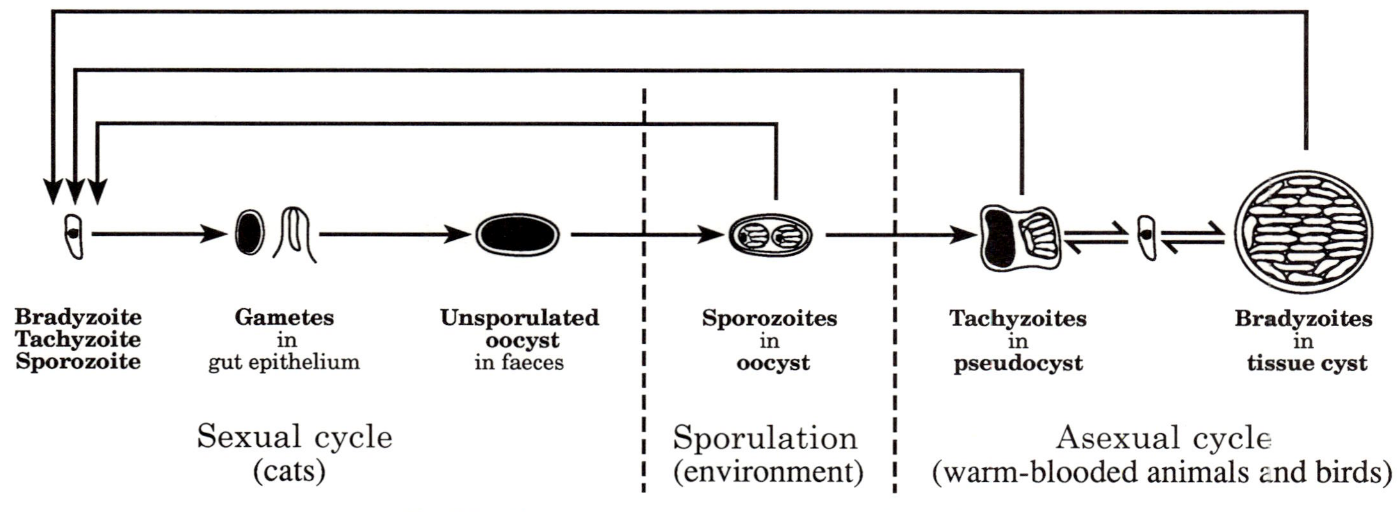

Fig. 2.2 The major stages of the toxoplasma life cycle.

The prepatent period of toxoplasma is the time interval from the ingestion of an infective stage, by the complete host, to the appearance of oocysts in the faeces. They also appear to resist peptic digestion better than the tachyzoite.[12] Previously called 'zoite' this term was recognized as being too general and was therefore succeeded by Hoare's[7] 'cystozoite' and Frenkel's[5] 'bradyzoite'. Although both types are as accurate in their derivation, the term bradyzoite is better as it compliments tachyzoite. The bradyzoites are enclosed within a parasite-derived wall which forms the tissue cyst.

Others

The sporozoite (2×8 μm) is similar to the tachyzoite both morphologically and in that it has a long prepatent period. It is an infective stage of the parasite resulting from the sporulation of a fertilized egg within an oocyst (Fig. 2.2). The micro- and macrogametocytes are precursors of the male and female stages the microgamete (3–5 μm long) and the macrogamete ($7–8 \times 4–7$ μm) respectively. The development of these stages is called gametogony.

Sexual cycle

Hutchison's discovery[13] of the transmission of toxoplasma infection in the faeces of the cat in 1965 was a major breakthrough in the history of toxoplasmosis (Chapter 1). At last, questions concerning the parasite's status, parasitology and transmission could begin to be answered. However it took a further five or six years before the sexual cycle in the intestinal epithelium of the cat was fully described.[3,14] Entry into the cat is usually by the ingestion of one of the infective stages (Fig. 2.2). As all these stages can lead to oocyst production, the sexual cycle must have a common pathway. The prepatent period of the tachyzoite and sporozoite differ from that of the bradyzoite in being much longer.[11] It was suggested that the tachyzoite and sporozoite underwent asexual development within the tissues of the host before invading the intestinal epithelium as a bradyzoite.[15] Thereafter the common pathway of the sexual cycle is clearly recognized as a two phase development of schizogony then gametogony.[3,14]

Schizogony

Parasites which rely on the faecal–oral route as a mode of transmission can increase their chances of success by producing a large number of infective stages. To achieve this, toxoplasma multiplies by schizogony. Schizogony is the repeated division of the nucleus of a mother cell before the cytoplasm divides producing many progeny from one cell. In cats fed tissue cysts five distinct stages or 'types' of toxoplasma (A–E) have been identified and defined according to various characteristics (Table 2.1). As well as developing sequentially types B–E can multiply into several generations before passing to the next stage.[14] However when these experiments were repeated no stage

Table 2.1 Characteristics of stages in toxoplasma sexual cycle[3,14]

Characteristics of sexual stages

Stage	Time of infection	Individual		Group		Intestinal site
		Size (μm)	Shape	Number of individuals	Shape	
A	12–18 hours	1–2×2	Round	1–3	Round	Jejunum
B	12–54 hours	2–3×1–2	Oval	2–30	Irregular	Jejunum, ileum
C	24–54 hours	2–4×1	Elongated	16–40	Round	Jejunum, ileum
D	32 hours–15 days	3–5×1	Elongated	2–35	Round	Jejunum, ileum, colon
E	3–15 days	3–5×1	Elongated	4–24	Round	Jejunum, ileum, colon
Macrogametocyte	67 hours–15 days	7–8×4–7	Oval			Small intestine
Microgametocyte	67 hours–15 days	3–5 (length)	Elongated	6–32	Oval	Small intestine

or 'type' of the parasite could be found until 4 days after infection.[16] Thereafter the identification of the stages appeared to be quite similar. To account for the lack of parasites 0–4 days after infection a hypothesis was proposed[16] which involved an extra-intestinal, pregametogonic stage (EIPS). The EIPS was suggested as a precursor of the schizogonic phase. However in this study[16] weaned cats were used which are less susceptible to tissue cyst infection[17] and it might have been possible to miss the relatively low numbers of the early stages in the intestine. This may account for the inability to find early types A, B, and C. Nevertheless, it has been well established that the parasite multiplies by schizogony progressively throughout the small and large intestine in preparation for the next phase of development.

Gametogony

Gametogony is the development of the sexual stages, the micro- and macrogamete. Although these stages are known to be present in the intestine between 3–15 days after tissue cyst infection (Table 2.1), this may vary since oocysts, the products of fertilization, have been passed in cat faeces only 60 hours after infection.[14] It is unlikely that types A, B, or C are precursors for gametocytes since they are not present when gametocytes appear in the intestinal epithelium whereas both types D and E are present. The macrogametocyte grows quite large without nuclear division, its development being relatively straightforward. Contrary to this the microgametocyte matures to contain a number of microgametes (6–32).[3,14] The microgametes are elongate in shape and have two whip-like organelles called flagella which are used for locomotory purposes. The relatively low percentage of microgametocytes (2–4%)[14] of the total number of gametocytes is presumably compensated by the development of many microgametes. Fertilization has not been observed but it is assumed that after the release of the microgametes they move toward and fuse with the macrogamete. The result of fertilization is a zygote or fertilized egg which becomes encapsulated within the tough oocyst wall. It remains in the epithelium for a while before being shed and passed into the faeces.

Sporulation

The unsporulated oocyst is roughly spherical in shape (10 × 12 μm). Enclosed within the tough oocyst wall an infective stage, the sporozoite, develops (Fig. 2.2). The zygote passes through an intermediate stage by dividing to form two binucleate sporoblasts which, in turn, become ellipsoidal sporocysts.[18,19] Inside each sporocyst, by schizogonic division, four sporozoites and a residual body are produced.[20] The net result of sporulation is eight sporozoites, four in each sporocyst, which are encapsulated within the oocyst wall. Development is not synchronous[20] but is rapid, sporozoites being seen as early as 24 hours after being shed from the cat.[20,21]

Asexual cycle

Unlike its sexual cycle, the asexual cycle of toxoplasma is not restricted to only one family of animals.[4,5,22] Indeed a large variety of warm-blooded animals and birds are found throughout the world that act as incomplete hosts. Therefore it is not surprising to find that the initial discovery of toxoplasma as a parasite was by the identification of one of its asexual stages (Chapter 1). As more information became available there was a realization that the asexual cycle involved a two-stage developmental process. This is characterized by the 'acute' phase of the tachyzoite followed by the 'latent' resting phase of the bradyzoite.

Tachyzoite

Once they have penetrated a host tachyzoites invade cells very quickly, in a quarter of the time (15–40 s) that a phagocytic cell would take to engulf and internalize a parasite.[23] All the infective stages require specialized organelles adapted for cell penetration. The conoid, positioned anteriorly, can move in all directions and at the beginning of the invasive process is positioned adjacent to the host cell wall. The sac-like rhoptries open into the conoid and are thought to excrete penetration-enhancing factor to facilitate cell penetration.[24] By a combination of active chemical and mechanical processes on the part of the parasite and some host cell participation, the parasite can enter almost any cell. When the tachyzoite enters a host cell a pseudocyst forms of host and parasite origin[9] (Fig. 2.2). This prevents destruct'on of the tachyzoite by host cell lysosomes, which for some unknown reason(s), cannot fuse with the pseudocyst wall (Chapter 8). Therefore the tachyzoite is able to multiply within a protected environment. Multiplication is by endodyogeny[6] and is rapid. Within 4–6 hours the pseudocyst is filled with tachyzoites and when the host cell disintegrates there can be invasion of further cells. Since the host cell can rupture even with a small number of parasites within the pseudocyst, disintegration is not entirely dependent upon the rapid multiplication of the parasite.[25] Tachyzoites engulfed by phagocytic cells become widely distributed throughout the body which may quickly disseminate infection.

Bradyzoite

The precursor of the bradyzoite is the tachyzoite. During the 1950s it had been suggested[26] that the immune system played a role in the transformation of tachyzoites to bradyzoites. However tissue cysts still formed in immuno-compromised hosts. Also as early as 3 days after infecting mice with tachyzoites, before a full immune response develops, tissue cysts have already begun to form. Bradyzoite multiplication is slow and, like tachyzoites, is by endodyogeny. Tissue cysts range in size from a few microns to 100 μm in diameter and can contain many thousands of bradyzoites[27] (Fig. 2.2). They have been found particularly in skeletal and heart muscle, the brain, and other tissues of the central nervous system although bradyzoites will invade any organ of the host. Generally spherical, tissue cysts may also adhere to the shape of the invaded cell, being elongate in the spindle-shaped cells of skeletal muscle. The tissue cyst frequently remains within the host cell and so is not directly exposed to the host's immune system. Perhaps this is why there is often a lack of a cellular immune response around them. Although the tissue cysts can have lifelong viability within the host, the cyst wall can disintegrate and the subsequent release of bradyzoites results in reactivation of latent toxoplasma infection (Fig. 2.2). More tissue cysts may form from these bradyzoites. As reactivation of a latent infection is often associated with immunocompromised people (Chapter 7) it is plausible that the immune system plays an important role in the maintenance of the latent tissue cyst infection.

SOURCES OF INFECTION

Part of the success of parasites such as toxoplasma is the ease of transmission of infection between hosts. When this fails, the spread of the parasite and therefore infection is prevented. The high prevalence of antibody to toxoplasma in populations of man and animals would suggest that it is a successful parasite. Infection by any of the three infective stages, the oocyst, tachyzoite, or tissue cyst, is dependent upon distinct factors that relate to either the host or the parasite. Knowledge of the routes of transmission has taken some time to acquire. Initial studies on insects and other arthropods as a means of transmission proved fruitless (Chapter 1). Although the discovery[28] in 1923 of an infected baby pointed to congenital infection, it was not until the 1960s that the ingestion of tissue cysts in undercooked meats was recognized as one of the main routes of human toxoplasmosis.[29,30] The full picture became evident when the cat was confirmed as the complete host in the early 1970s. This section considers the relative importance of each of the three infective stages, their sources, the routes of infection and factors that affect their survival in transmission.

Oocyst

Enclosed within oocysts the sporozoites are transmitted by the faecal–oral route. Mechanically breaking the oocyst wall exposes the sporocysts, which have been shown to rupture in an excysting medium of pH 7.3.[31] This mechanism is likely to reflect the *in vivo* situation where once the oocyst wall is broken down in the stomach, the sporocysts on entering the environment of the duodenum rupture to release the sporozoites. The sporozoites are then free to enter the cells of the gut of the host. During primary infection cats can shed many millions of oocysts.[25] These oocysts can remain infective in soil for months. Under the experimentally favourable conditions of a temperate, moist environment oocysts were still viable after a year.[32] More remarkably they survived the extremes of 18 months (and two winters) in Kansas soil.[33] Oocysts are not restricted to soil only, as contaminated water has also been implicated as a source of toxoplasma infection[34] (Fig. 2.3). The resistance of the oocyst to many extreme conditions is a marked feature of this stage of the parasite. Acids, alkalis, and common laboratory detergents fail to destroy it. However exposure to 10 per cent ammonium hydroxide for 5 minutes[21] or exposure to heat greater than 55 °C for 30 minutes killed all oocysts.[21] Oocyst-contaminated material must be handled with great care. A group of laboratory personnel seroconverted two to three years after starting work with oocysts despite having worked on average 10 years with tachyzoites and tissue cysts without acquiring infection.[4] However as the evidence was circumstantial other sources of infection cannot be excluded. Nevertheless it could mean that oocysts are the most infective of the three stages. Even so, they may not be the most frequent means of natural infection when host-dependent factors also need to be taken into account.

Tachyzoite

Of the three infective stages the tachyzoite is the most sensitive to extremes of temperature[21] and thus does not easily survive outside its host. Nevertheless tachyzoites in the peritoneal exudate of rats and mice, stored at 4 °C, can remain viable for up to 10 days. Tachyzoites have been found to survive in the body fluids of their host: blood, milk, saliva, urine, tears,[35] and semen.[36] Of these, reports have only implicated blood[37,38] and milk[39] in toxoplasma infection (Fig. 2.3). Epidemiological evidence also suggests that sexual transmission plays no part in the spread of infection in humans.[40,41] Since tachyzoites generally enter nucleated cells[23] the white cell fraction of blood is more likely to be the major source of infection in blood transfusion. There is a record of a white cell transfusion resulting in infection with toxoplasma.[37] However platelet transfusions have also been implicated in the transmission of infection.[38] Giving a large volume of blood to a patient, as in a major surgical procedure, increases the potential infective dose.

Unpasteurized goat's milk has been identified as a source of infection.[39] In this study there was evidence of current and past infection in two brothers, but both parents were seronegative. On further investigation it was found that both sons drank glasses of goat's milk while their parents only had small amounts in tea and coffee, the temperature of which would be sufficient to kill any tachyzoites present. The goat herd was found to be seropositive and a milk sample produced a positive isolation. Raw goat's milk was also implicated as the source of infection in a seven month old child.[42] Since tachyzoites are easily destroyed by gastric juices[12] the stomach wall is less

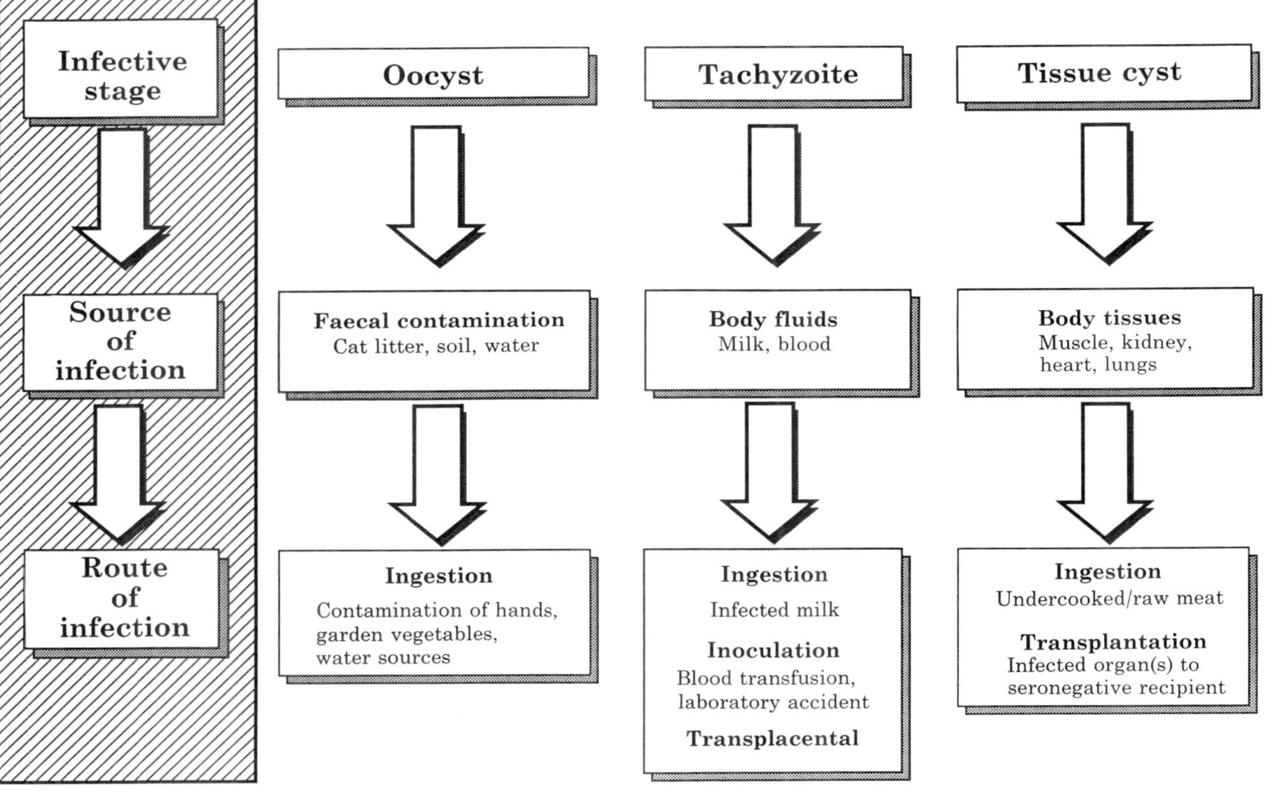

Fig. 2.3 The main stages, sources, and routes of toxoplasma infection.

likely to be the site of host cell invasion. Alternatively the tachyzoite need not enter the stomach if it can penetrate the host's buccal mucosa. Obviously the means of entry into the host is important and direct inoculation of tachyzoites in a laboratory needle-stick injury frequently results in infection (Appendix 2). Of wider public interest and concern is congenital infection. If a woman acquires toxoplasmosis during pregnancy parasites may pass across the placenta to her fetus. The earlier in pregnancy this happens the greater the possible damage to the fetus (Chapter 6).

Tissue cyst

Bradyzoites are transmitted collectively within the tissue cyst. The efficacy of tissue cyst transmission is indicated by the relatively low infective dose needed and by its more successful rate of infection compared with the other two stages.[11] Tissue cysts can remain viable for up to 3 hours incubation in digestive fluid.[12] This brief period can allow the bradyzoite to enter the host via the stomach or intestinal epithelial lining. Because of this characteristic, carnivorism was suspected as a major route of transmission. Further experiments confirmed the hypothesis so that the eating of raw or undercooked meat is now a proven route of transmission (Chapter 1). The tissue cyst's ability to remain viable in tissues for the lifetime of the host maximizes the time it is available for transmission. Tissue cysts stored at 4 °C have remained viable for 68 days,[12] but freezing usually, but not always, destroys their infectivity.[12,43] Heating meat (pork) to 67 °C for 3 minutes renders tissue cysts non-infective.[44] Therefore if meat is well cooked the risk of infection is minimal. Similarly, prolonged exposure to salt, as in the salting of meat, destroys tissue cysts.[12] Various methods are used commercially for the control of infection in meat (Chapter 9). A new route of toxoplasma infection has evolved with the practice of organ transplantation where seronegative recipients are at risk from infected donor organs (Chapter 7). The tissue cyst's sources and routes of infection are summarized in Fig. 2.3.

CATS

As the complete host of toxoplasma the cat has created much academic interest. Cats can play a major role in the spread of toxoplasma infection especially as there are an estimated one million stray cats in Britain.[45] However the cat is also a common pet and questions concerning cat ownership and toxoplasma infection must be considered. Therefore the prevalence and results of infection in cats will be discussed, concluding with possible methods of controlling infection.

Table 2.2   Prevalence of toxoplasma antibody and faecal oocysts in cats[46,47,49,54–56]

	Oocyst		Antibody	
	No. examined	% positive	No. examined	% positive
Hawaii	1604	0.8	1568	14
Kansas	510	0	667	27
Ohio	1000	0.7	ND	ND
Maryland	185	0.5	650	14
Beirut	313	9.9	324	78
Glasgow	ND	ND	158	19

Prevalence of infection

Acute toxoplasma infection can be demonstrated by finding oocysts in the faeces of a cat. They can be detected microscopically or by mouse passage. The results of a number of such studies are summarized in Table 2.2. Excluding the Lebanese study the prevalence is low (<1%). The elevated result of the Lebanese study was considered to be mainly due to the suitable conditions for transmission of infection at the time the study was undertaken.[46] Cats had increased exposure to infected meat, prey, and contaminated soil. A result of interest from the Hawaiian study[47] was the later reidentification of one of the 'positive' toxoplasma strains isolated from a cat as a very similar protozoan organism called *Hammondia hammondi*.[48] The reduced figure of 0.68 per cent is very similar to that obtained in another study in Ohio (Table 2.2).[49] The low prevalence is due to the short period of time (7–20 days)[14] during which oocysts are shed from a primary infected cat. However over this time millions of oocysts can be shed, each of which is very hardy and can remain viable for many months in the soil. Therefore although the prevalence is low, the potential infection risk is substantially greater than would be expected. When a primary infection results in oocyst shedding cats usually acquire immunity and rarely reshed oocysts.[50] This can imply that once a cat has been infected it will not be a further infection risk. However if a latently-infected cat is also infected with *Isospora felis*, a protozoan parasite common to cats, the latent toxoplasma infection is reactivated with further shedding.[51] Therefore a previously infected cat is not necessarily a 'safe' cat.

Newborn kittens of infected mothers can have maternal antibody to toxoplasma which has been passed across the placenta during pregnancy. This maternal antibody may not be protective,[52,53] and as it does not reflect infection of the kitten, it can produce falsely high seroprevalence results. Thus, the true prevalence of antibody to toxoplasma is greater in older cats.[47,54,55] Kittens may become infected by oocysts (Fig. 2.1), however when they begin to eat prey brought to them by their mothers or begin hunting themselves they can be exposed to tissue cyst infection. Tissue cyst infection may be the main cause for the increase in the seroprevalence rate amongst the older cats. In those studies that differentiated between stray and domiciled cats there was a higher percentage of positive stray cats than domiciled cats.[46,54–56] Stray cats are more likely to be infected by ingestion of infected prey than domiciled cats. The worldwide prevalence of antibody to toxoplasma in cat populations is given in Table 2.2.

Cats usually prey on small mammals and birds, particularly mice.[57] The prevalence of infection in small mammals and birds is similar to that of cats.[58,59] These animals are infected by ingesting oocysts that are deposited on the soil by the cat. Cats only partially bury their faeces with soil. Instead of being restricted to the site of deposition, oocysts can be spread over a large area. Rain splash can transport helminth ova several metres after a particularly heavy downpour.[60] Transport hosts, like earthworms, cockroaches, and flies can harbour oocysts[59] and pass on infection when eaten by birds and other insectivores. Therefore their movements in and over the soil does not limit the oocyst to the site where it was deposited.

Results of infection

Although there is a relatively high prevalence of antibody to toxoplasma in cat populations (Table 2.2) there have been comparatively few recorded cases of clinical toxoplasmosis.[52] The result of infection is age dependent.[14] Newborn kittens can die from toxoplasmosis whereas older cats usually become immune with little or no clinical symptoms. Of those cases that have been reported many of the diagnoses were confirmed only after death.[53,61] The organ most commonly affected is the lung but the liver, spleen, mesenteric lymph nodes, and pancreas can also be involved.[53] In a study of 15 infected cats,[62] five presented with fever, five with weight loss, five with respiratory disease, and nine with ophthalmic disease. Other reports show fever, anorexia, lethargy, dyspnoea, and pneumonia as common findings.[53,61,63] Other signs were icterus, vomiting, diarrhoea, muscle pain, seizures, and paralysis. The high frequency of ocular toxoplasmosis in the first study[62] is likely to be due to sampling bias. The actual figure is probably much lower.[52] Ophthalmic disease presents primarily as retinochoroiditis,[53,64] a condition also found in humans (Chapter 3).

Clinical signs alone are not diagnostic of toxoplasmosis. X-rays of the cat's lungs are useful,[53] since pneumonia is often associated with toxoplasma infection. Haematological results vary: neutrophilic leucocytosis, lymphocytosis, neutropenia, and eosinophilia have been found.[62] Tests for total bilirubin, amylase, alanine aminotransferase, and serum protein levels may be of use. Serum enzymes, such as creatinine kinase and lactate dehydrogenase can be raised according to the extent of tissue damage. Faecal examination may not be helpful as oocysts are shed for a short period.[65] Oocysts are very small (12 μm) and difficult to identify particularly to the inexperienced eye. Serological tests are available: indirect haemagglutination test,[66] indirect fluorescent antibody test,[66] dye test, and others (Chapter 4) are all of some value but have their limitations. After infection detectable levels of antibody can remain for months.[52] Therefore detection of antibody is not necessarily indicative of current infection. An IgM enzyme-linked immunosorbent assay technique (Chapter 4) has been developed which may be of use in diagnosing acute infection.[67]

The drugs available for the treatment of toxoplasmosis are quite limited. A combination of sulphadiazine and pyrimethamine is usually used but pyrimethamine is toxic in cats,[61] as in humans (Chapter 5). Unless clinical signs are evident or diagnosis has been confirmed this treatment is not indicated. Clindamycin hydrochloride has been used in the treatment of eye disease.[62] In this study all but one of the cats with acute retinochoroiditis responded to treatment. Corticosteroids given to the unresponsive cat achieved 100 per cent resolution. Cats presenting with anterior uveitis were also treated with corticosteriods. However 3/9 (33%) of cats with anterior uveitis did not completely recover. Two others, which initially recovered had a recurrence of anterior uveitis at a later date. Recent work has elucidated a strain of toxoplasma that could be used as a potential vaccine. The benefits of vaccination are obvious (Chapter 9).

Control of infection

Cats are a potential source of infection because they shed oocysts. Oocysts play a major role in the transmission of toxoplasma infection in humans. A survey of vegetarians and non-vegetarians in Bombay, India[68] showed that the prevalence of infection was not different in the two groups. Vegetarians are exposed to infection by oocysts and, rarely, tachyzoites. Similarly, human toxoplasma infection was rare on an island with no cats, yet present in islands that had cats.[69] In contrast to this is the high prevalence of antibodies in the adult Parisian population (>80%)[29] where the eating of rare or lightly cooked meat is renowned. However these results do not imply that the tissue cyst was the only causative infective stage because cats are also common in Paris. As the predominant source of oocysts, cats must be a major infection risk. However it can be argued that it is water or soil contaminated by cat's faeces and not the cat itself that is the problem. The implication is that a person who does not own a cat may be at risk from his/her neighbour's cat. Investigations attempting to answer this or similar questions have produced equivocal results. A study of some families in Sweden could find no correlation between owning cats and toxoplasma infection,[70] whilst cat breeders in England showed a high antibody prevalence rate.[71] The actual handling of a cat is unlikely to result in infection as it keeps itself quite clean. However, the individual relationship between an owner and his/her cat may be close.

Likewise, the handling of litter trays, a rich source of oocysts, is likely to increase the risk of infection. Cat owners should be aware that cats are a potential source of infection to themselves and the general public. Owners have a responsibility to minimize the risk of infection not only to themselves but to their neighbours.

Fortunately there are simple but quite effective means of minimizing the infection risk from cats. Litter trays should be cleaned daily thus disposing of the oocysts before they become infective.[5] Whenever this job is being done gloves should be worn and hands washed afterwards. Because of the possible consequences of infection in pregnancy, pregnant women should avoid handling litter trays (Chapter 6). Although there is no ideal method for the disposal of cat litter, sealing it in a strong polythene bag for collection by refuse collectors is probably the most convenient and practical solution. Stray cats and house cats that roam may defaecate anywhere. Thus hands should be washed after gardening and root vegetables should be thoroughly washed. Similarly, precautions can be taken to minimize infection of cats from tissue cysts. Tinned pet food or well-cooked meat is not a risk.

DOMESTIC ANIMALS

A food survey completed in 1988 showed that just over 1 kg of meat was eaten per capita per week in Britain.[72] Pig meat (pork, bacon, ham) and beef each made up one-third and poultry about one-fifth of the total meat eaten. In relation to other European countries Britain was one of the lowest meat consumers, France being the highest.[72] It is tempting to conclude that the high seroprevalence of toxoplasma in the adult population of Paris[29] is related to their high consumption of meat. Unpasteurized milk, too, is a potential risk, particularly goat's milk (Fig. 2.3). Therefore consideration will be given here to the prevalence and results of toxoplasma infection in a number of domestic animals: sheep, pigs, cattle, goats, poultry, and horses. Lastly, present and future methods of controlling infection will be discussed.

Sheep

Prevalence of infection

Sheep are commonly infected with toxoplasma. Results of seroprevalence studies are given in Table 2.3. These results were obtained using various tests. Prevalence appears to increase with age,[52,73,74] a trend which does not support congenital transmission as the main route of infection. Twice as many ewes have detectable antibodies to toxoplasma as lambs,[52] although a survey of sheep from New Jersey,[74] USA showed a reversal of this trend (42% of the sheep tested, and 48% of the lambs tested were positive for

Table 2.3 Seroprevalence in domestic animals, average per cent (range)[52,55]

	Sheep	Pigs	Cattle
USA	43 (4–100)	33 (1–69)	49 (1–100)
UK	39 (15–92)	8 (4–12)	6 (3–8)
France	36 (12–72)	23 (10–38)	22 (4–71)

antibody to toxoplasma). There was no obvious reason for this change. Sheep from all other areas examined at the same time had a higher prevalence in ewes than lambs. Sheep from five flocks in Yorkshire, UK,[75] were identified as having toxoplasma infection. The prevalence of antibody in the hill flock was lower than the grassland or worral flocks. Similarly, 5 per cent of Navajo sheep were positive for antibody compared with 56 per cent of Kentucky sheep.[75] The Navajo sheep are reared in dry, mountainous areas whereas Kentucky sheep are reared on rich grassland. These trends are indicative of oocyst infection; the presence of cats and the survival of oocysts in those areas with lower prevalence rates being less likely. Toxoplasma has been isolated from tissues of naturally infected sheep.[52,75] As many as 23/34 (68%) of sheep have had toxoplasma isolated from skeletal muscle and brain tissue,[52] parasites being found more commonly in the musculature than the brain. The relatively high prevalence of toxoplasmosis in sheep may point to an important role in the transmission of human infection. However, in Britain only one-tenth of an average person's meat consumption is mutton and lamb and its popularity appears to have decreased in the last 10 years.[72]

Results of infection

Although toxoplasma infection is relatively common in adult sheep (Table 2.3) it rarely presents as a clinical disease. However a dramatic result of toxoplasma infection is seen when non-immune ewes become infected during pregnancy. Abortion and fetal or neonatal death can be the result.[1,52,75] The extent of the disease in the fetus or newborn is mainly dependent upon the status of the fetal immune system.[1] Therefore infection of the mother early in pregnancy quickly leads to death of the fetus. Later in the pregnancy, the fetus can be infected but is more likely to survive due to its own immune response. The newborn lamb will then be immune. Lambs that survive the first week after parturition are likely to reach adulthood and produce toxoplasma-free newborn.[52] The pregnant ewe itself is often asymptomatic but may present with lymphadenopathy and slight fever.[1] The annual incidence of abortion in breeding ewe populations is estimated to be between 1–2 per cent in Britain,[1] a considerable economic loss. However it is abortion 'storms' that may have the gravest consequences to the individual farmer. This is when a high proportion of non-immune ewes become infected during pregnancy and the resulting loss of lambs is quite severe.

Pigs

The seroprevalence of pigs for toxoplasma is shown in Table 2.3. Different serological tests, types of pig, and management of the animals may account for the variations in prevalence. Only three small studies accounted for the seroprevalence range of 4–12 per cent in Britain (Table 2.3). Thus this result may not be representative of the seroprevalence in the pig population of Britain. Toxoplasma has been isolated from various tissues of naturally infected pigs: skeletal muscle, heart, and brain.[76] It was not isolated from the liver or the kidneys. Organs of pigs killed at 267 and 357 days after inoculation were infective to mice.[76] In particular, various meat cuts such as bacon, ham, and spare ribs were found to be infected. Most infections in pigs are subclinical.[52] Where disease does occur it is more likely to be in younger pigs.[77] Eight to ten day old pigs infected with tissue cysts of a low virulence strain of toxoplasma developed an immune response and survived although there was slight damage to tissues. When a similar infective dose was given to one day old pigs some of them died.[77] Tissue damage was quite severe. It is believed that improvement of the management of swine has partly accounted for the decrease in toxoplasma infection over the years in different countries.

Cattle

Some doubt has been cast over the accuracy of serological testing in cattle. The dye test is too unreliable for the diagnosis of toxoplasmosis in cattle as false positive results are caused by bovine serum protein.[78] This means that the true prevalence rate is likely to be lower than the actual figures reported (Table 2.3). Although the prevalence of tissue cysts in cattle is low[52] this result can be misleading. The quantities of tissue used in the isolation technique would be small relative to the whole animal and thus it may give falsely low results. As beef is one of the most common meats eaten[72] the potential infection risk may be quite substantial.

Clinical cases of bovine toxoplasmosis are rare.[52] Conflicting evidence exists concerning the diagnosis of cases of toxoplasmosis in cattle[78,79] and it is suspected that disease has been caused not by toxoplasma but by similar protozoan parasites. Abortions in cattle have been recognized as being caused by *Sarcocystis cruzi* and *Neospora caninum*.[52,78] Both these organisms are morphologically similar to toxoplasma. When congenital transmission has been attempted experimentally in cattle it has not been very successful.[79] Tachyzoites in cow's milk have rarely been found. One of 2058 samples in experimentally infected animals was positive by isolation.[79] However as cow's milk is usually pasteurized, toxoplasma is not a great risk to humans.

Goats

Studies from California[80] and northwest USA[81] showed a prevalence of toxoplasma infection in goats of 23 and 22 per cent respectively. Both sample sizes were of 1000 or more goats. The most commonly infected tissues of the goat are skeletal muscle, diaphragm, intestine, heart, kidneys, and liver.[82] Although goats are commonly infected with toxoplasma, the risk of infection in Britain from eating infected meat is not as important as consumption is low. Toxoplasma infection in adult goats rarely develops into disease;[83] the fetus is at greatest risk. Naturally occurring cases of abortion or congenitally infected kids caused by toxoplasma have been reported.[84–87] Unlike sheep, infection has been found in the fetuses from previously infected mothers.[88] Thus the infection in does can have serious economic consequences as future pregnancies may be affected. Although toxoplasma was not detected in the milk of 29 goats,[80] only small samples (1 ml) of milk were used during testing. Isolations have been made from milk of two experimentally infected goats.[82] Furthermore, tachyzoites in goat's milk have been implicated as the infective agent in cases of human toxoplasmosis[39,42] (Fig. 2.3).

Others

As little information is known about toxoplasma infection in poultry and horses, both these animals will be considered together in this subsection. Results of seroprevalence studies in commercially raised poultry are of little use as the serological tests used are not sensitive enough to detect antibody to toxoplasma.[52] An IgG enzyme-linked immunosorbent assay method (Chapter 4) has been proposed as a possible screening method.[89] Over the last two or three decades there has been a steady increase in the consumption of poultry by the public. Although it is not yet possible to accurately assess the prevalence of infection in poultry its increasing popularity as a food will increase its importance. In one study, five of a flock of 40 chickens were positively identified for toxoplasma by histological methods.[90] Poultry present with anorexia, emaciation, paleness, shrinking of the comb, occasional diarrhoeic episodes, and behaviour suggestive of blindness.[90] Toxoplasma could not be isolated from embryonated eggs of these chickens. However when a large dose of oocysts (50 000) was fed to chickens, decreased egg production was observed with an increased mortality of embryonated eggs.[89] Thus, the consequences of infection may be substantial for the birds, and poultry may become of increasing importance as a source of infection for humans.

Ten per cent of 500 horses slaughtered at a meat-packing plant in Ohio, USA were seropositive to toxoplasma.[91] Parasites were isolated from the tissues of horses but unfortunately the tissues were pooled therefore the prevalence rates could not be determined. Horses present with a slight fever

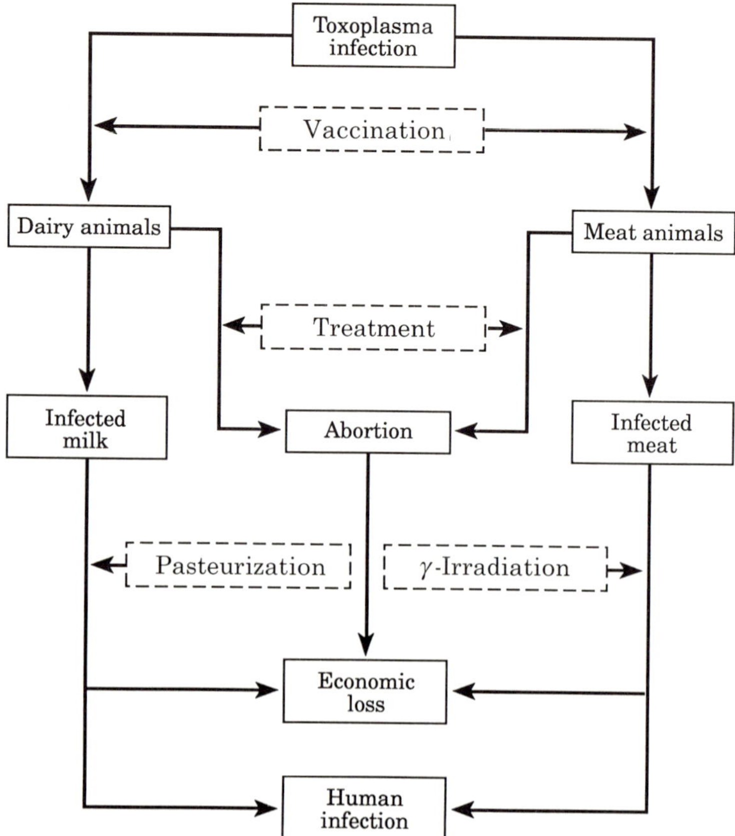

Fig. 2.4 Possible measures for controlling the effects of toxoplasma infection in domestic animals.

about a week after inoculation.[92] These horses were the youngest tested. Tissue cysts can persist in horse tissues for at least a year.[92] In countries where horse meat is consumed, rare steaks may therefore be a significant risk.

Control of infection

There are two major purposes for controlling infection in domestic animals: to minimize human infection and reduce economic loss (Fig. 2.4). The two main sources of human infection from domestic animals are infected meat and milk. Fortunately cow's milk is usually made safe by pasteurization. However raw milk is a risk and should be heated before consumption. Irradiation of carcasses by gamma rays has been used against *Trichinella spiralis* infection of meat and it is possible that this method could be adopted for toxoplasma (Chapter 9). A simpler, more practical solution that can be used in every household is cooking meat well. Infected pork is rendered non-infectious after heating to 67 °C for more than 3 minutes.[44]

As most domestic animals rarely produce clinical signs with toxoplasmosis, the first indication of infection may be an aborted fetus. Unfortunately by this stage, treatment by drugs may be of limited use as pathological changes are likely to have already taken place in the other animals.[93] However the treatment of animals during pregnancy in an attempt to prevent infection may be helpful. Pregnant ewes treated with monensin have had fewer abortions than untreated ones.[52] A limited immunity to toxoplasma has been shown in goats inoculated with *Hammondia hammondi* (Chapter 9). Ideally vaccination is the best method of control as it is aimed directly against infection (Fig. 2.4). Unfortunately there are many drawbacks associated with the development of suitable vaccines (Chapter 9). A live, attenuated strain of toxoplasma has given promising results in sheep in New Zealand,[1] but it is still not currently available for use in many countries. Finally it is known that the cat is a major source of infection for domestic animals,[1,52,73,94] thus methods of controlling infection in cats would be of benefit to controlling the effects of toxoplasmosis in domestic animals.

PETS AND EXOTICS

After the cat was recognized as the complete host of toxoplasma it was realized that other pets, particularly the dog, had little or no role in the spread of infection to man. However these animals are susceptible to toxoplasmosis and so the prevalence and results of infection in dogs and other pets will be discussed in this section. Secondly as toxoplasma has a wide host range, animals from any part of the world can be infected. Thus prevalence, and possible sources and routes of infection in exotic animals will be discussed.

Dogs and other pets

From a total of 64 reports 0–94 per cent of dogs had antibody to toxoplasma detectable by the dye test.[95] These reports were from various countries throughout the world. A world-wide average figure of 36 per cent has been estimated.[56] No difference was found between sexes[55,56] although a significant difference was shown between young dogs (<6 months) and older dogs (>6 months):[55] the prevalence rate in older dogs being greater (28%) than those aged under 6 months (2%). These figures may be indicative of infection by the ingestion of tissue cysts in raw or undercooked meat. However dogs are known to be copraphagic implying that oocysts may also play a role in infection. The tachyzoite, too, can play a role in infection in dogs; congenital infection has been reported.[96] Tachyzoites have also been reported in dog's milk[97] and this is suspected to be a source of infection for young puppies. There have been comparatively few clinical reported cases of canine toxoplasmosis.[52] Neuromuscular, respiratory, and gastrointestinal signs have been associated with infection. When the disease involves the respiratory system death can follow.[96] Dogs concomitantly infected with canine distemper virus became clinically ill with toxoplasmosis.[96,98] It is possible that the toxoplasma infection becomes reactivated by the coincident infection with distemper virus. Similarly a dog, who had been receiving irradiation treatment and cortisone therapy died from toxoplasmosis.[96] The age of the dog, clinical signs, suspected concomitant infection(s) or vaccination history, and serology can help diagnosis. The cerebrospinal fluid may be parasitized in neurological cases. However, a parasite similar to toxoplasma, *Neospora caninum*, produces an illness that presents like toxoplasmosis[52] and it is believed that it is the cause of many of the neurological cases of suspected canine toxoplasmosis. Pyrimethamine, sulphonamides, and streptomycin have been used with some success in different combinations in the treatment of canine toxoplasmosis.[52] The dog is a very unlikely source of infection to humans. It is possible infection may be transmitted by tachyzoites if a person is bitten by an infected dog. However as the infective dose will be small and the parasite is fragile in the external environment, the chances of infection by this route are low. The simple measure of feeding dogs tinned pet food or well-cooked meat will minimize the risk of infection in dogs.

There is very little documented information concerning natural infections in other pets. However, there have been reports of outbreaks, and prevalence studies, in groups of animals that are commonly kept as pets. An outbreak of toxoplasmosis was reported in small passerine birds on a bird-farm in Italy.[99] A variety of birds were infected: canaries, greenfinches, goldfinches, siskins, bullfinches, and linnets. Initially 26 per cent (60/230) of the birds succumbed to infection and died. Four months after the acute phase of infection some birds suffered from ocular disease. The original sources of infection may have been contaminated food or water.

Exotics

Seroprevalence results of a sample of exotic animals are summarized in Table 2.4. The group of animals represents three different continents: Africa,[100] Asia,[101] and South America.[102] The South American animals were tested by the indirect haemagglutination test and the others by the dye test. One study[100] distinguished the prevalence rate between carnivores and herbivores of the African Savannah. Eleven of 14 carnivores (78%) had a positive dye test while 97 of 118 (82%) herbivores were positive. Obvious constraints meant that sample size was very small for the various animals. In the African Savannah grazing or drinking oocyst-contaminated water are the likely

Table 2.4   Seroprevalence in exotic animals (%)[100-102]

	Animal	Seroprevalence
S. America	Opossum	18/33 (55)
	Squirrel monkey	24/49 (49)
Africa	Spotted hyena	6/6 (100)
	Zebra	9/10 (90)
	Giraffe	5/10 (50)
	Silver-backed jackal	4/6 (66)
Asia	Cynomolgus monkey	11/50 (22)
	Rhesus monkey	0/64 (0)

routes of infection. Strict grazers, such as zebras, had the highest prevalence whilst giraffes, which feed on bushes and trees where oocysts will not be found, the lowest (Table 2.4). Their route of infection is probably by drinking contaminated water or whilst licking minerals from the earth. Thus free-living Felidae may play an important role in the spread of infection. Infection in carnivores is reliant on the capture of infected prey, scavenging of carcasses, or drinking oocyst-contaminated water.

The spectrum of toxoplasma hosts is well illustrated in zoos and animal parks. The kangaroo, spider monkey, polar bear, lion, cheetah, California sea lion, elephant, llama, yak, and impala are amongst some of the animals that have had antibodies detected in their serum.[103] Toxoplasma in many exotic animals is either asymptomatic or produces a vague, non-specific illness which rarely becomes serious. However there are reports of epizootic toxoplasma infections resulting in the death of animals.[104,105] During 1982, at Perth zoological gardens, Australia, 16 of 17 adult squirrel monkeys died within 4 days of each other.[104] The four young were thereafter hand reared. Histological examination of various organs of the adults revealed lesions, some with toxoplasma present. Bacteriological and virological tests were negative. The squirrel monkeys were fed on a diet of maize, fruit, vegetables, and a meat loaf containing raw, minced sheep hearts. At the same time as this outbreak four of a colony of nine red-tailed wambengers (an Australian marsupial) died. Investigation of an un-named specimen, possibly analysis of an infected organ, suggested toxoplasma infection. Both the wambengers and the squirrel monkeys were fed the same meat loaf. As the suspected source of infection the meat loaf was well cooked thereafter, with no further deaths. Oocyst infection was implicated in a fatal outbreak of toxoplasma in ring-tailed lemurs from Cincinnati zoo in the USA.[105] All infected lemurs died (4/4). Their cages were adjacent to those housing cats and it is suspected that the lemurs food was contaminated by oocysts during the cleaning of the cats' cages. The lemurs had not been fed meat. It is unusual that an infection of toxoplasma should produce such drastic results in the squirrel monkey and the lemur, when it is usually asymptomatic in most animals. It is possible that both animals have little natural resistance because of their limited exposure to toxoplasma; the squirrel monkey lives in trees and will not come in contact with oocysts while lemurs are from Madagascar where domestic cats are not native animals.

STRAINS

As toxoplasma can infect a wide range of animals it is not surprising that research workers have been reluctant to accept that only one toxoplasma species is involved. Yet this is generally agreed to be the case (Chapter 1). Nevertheless differences are apparent. A strain isolated from a six-year-old-boy[106] who succumbed to toxoplasmic encephalitis, was highly virulent in mice; while other strains isolated can be avirulent, producing only a latent infection in mice.[107] In this section the virulence of strains, the effects in various hosts and their influence upon the clinical features of infection will be discussed.

Virulence

One of the most common characteristics of the many toxoplasma strains is their variation in virulence. Virulence can be recorded as either the time lapsed before the animal(s) succumb to infection or the percentage of animals that do succumb.[108] Using these criteria a broad spectrum of strains have been characterized: the highly virulent RH strain (8 days), the moderately virulent S_5 strain (11 days), and the less virulent 113 CE strain (24 days)[108]. Generally a strain virulent in mice is virulent in other animals. Thus, the RH strain is highly pathogenic in guinea pigs, rabbits, chicks, and rhesus monkeys as well as mice.[106] Virulence can be enhanced by the continuous passage of the parasite in laboratory animals. When the RH strain was initially isolated, mice succumbed to infection 17–21 days in the first passage, 7–8 days in the second, and 3–5 days in the third passage and thereafter.[106] This increase in pathogenicity has been confirmed in other experiments using different host species.[107] Obviously the continuous passage of the parasite may select for the more rapidly multiplying forms. An increased multiplication rate of the parasite in cell cultures has also been correlated with increasingly virulent strains of toxoplasma.[108]

Many avirulent or less virulent strains have been isolated.[107] The Beverley strain was isolated from a healthy rabbit in 1959.[109] Commonly used in laboratories as a source of tissue cysts, the Beverley strain has a low virulence in mice, producing a latent toxoplasma infection. Although usually less or non-pathogenic, this and other similar strains can have serious results in both animals and humans. Breeding ewes may become infected with an apparently avirulent strain which may be passed transplacentally and cause abortion. Also a reactivated latent infection in an immunocompromised person can have grave results (Chapter 7). Thus the consequences of infection are not solely due to virulence. Unfortunately it is not fully understood what is the cause of the variation of virulence in toxoplasma strains. Few other biological differences between strains have been recognized; the mutant ts-4 strain is of low virulence in mice and is not able to produce tissue cysts or oocysts.[110] Therefore it is hoped to be used as a potential vaccine (Chapter 9). Antigenic differences exist between strains but these are minor (Chapter 8). The lengths of two virulent strains of toxoplasma were noticeably different, the more virulent strain being longer; however, those of similar virulence were indistinguishable.[107]

Clinical features

Strain differences may explain the spectrum of clinical disease produced by toxoplasma infection in animals and humans (Chapter 3). Strains isolated from severe cases of human toxoplasmosis are generally highly virulent in mice.[107] Thus, in the immunocompetent host it would be reasonable to partly attribute the non-specific symptoms or asymptomatic illness observed during infection (Chapter 3) to a less virulent or avirulent strain of toxoplasma. As the more virulent strains appear to multiply more rapidly *in vitro*,[108] this may indicate how they might cause a more obvious clinical picture. This characteristic may also account for the range of incubation periods observed in human cases of toxoplasmosis (Chapter 3). However an isolate from a severe case of human toxoplasmosis only produced an observable infection in mice 84 days after inoculation.[107] This lack of an obvious pathogenic response in mice to a suspected highly virulent strain may be due to host factors. Differing susceptibility to infection has also been noted within a host species; strains of mice have different susceptibilities to some degree.[52] However it is not known whether this also happens in breeds of large animals or in humans.[52]

The inoculation of sulphadiazine-treated animals with the RH strain of toxoplasma produces different effects. Tissue cysts developed primarily in the cardiac muscle of guinea-pigs but in the brains of mice and hamsters.[111] It is possible that toxoplasma strains preferentially infect specific organs in different animals and in humans. However the differentiation of toxoplasma strains by the characterization of their antigens (Chapter 8) may allow for a comparison of various clinical features and the responsible toxoplasma strains. Similarly, different strains may be associated with differences in drug sensitivity.[108] Thus, the identification of the strain may determine a more efficient drug regime. As with other infections strain variations are likely to be important in terms of clinical disease (Chapter 3), diagnosis (Chapter 8), and treatment (Chapter 5).

SUMMARY

Toxoplasma is an important world-wide infection in both animals and humans. Difficulties in the nomenclature of some of the stages of the parasite are disappearing. There are three infective stages: the oocyst, tachyzoite, and tissue cyst. The sporozoite is the product of the sexual cycle and sporulation. The sexual cycle is found only in cats and involves a two-phase development of schizogony and gametogony within the intestinal epithelium. The net result of the sexual cycle is the production of millions of unsporulated oocysts. Sporulation does not take place within a host but within the environment. Sporulation is the development of sporozoites within the shed oocysts. Oocysts are transmitted by the faecal–oral route which is one of the major routes of infection in animals and humans. They are very hardy and can survive extreme conditions. The tachyzoite and bradyzoite are part of the asexual cycle. The asexual cycle takes place in a wide range of animals. The proliferation of tachyzoites within pseudocysts results in dissemination of infection throughout the host. Tachyzoites are transmitted by: ingestion, inoculation, and transplacentally. Bradyzoites encyst to form tissue cysts and result in latent infection. Tissue cysts can be transmitted to another host by ingestion of infected meat or by organ transplantation.

As the complete host of toxoplasma, the cat plays a major role in toxoplasma infection. Infection rarely causes clinical disease, but results in the shedding of millions of oocysts. The careful disposal of the oocysts is important in the control of infection. Although clinical disease is rare in domestic animals such as sheep, pigs, goats, poultry, and horses seroprevalence rates can be high. Thus commonly eaten meats like pork and beef may have a high risk of infection. Cows' milk which is usually pasteurized has a low risk of infection whilst the drinking of raw goats' milk has been implicated in cases of human toxoplasmosis. The thorough cooking of meat and boiling of goats' milk are practical methods of controlling infection. Abortion in sheep and goats can cause severe economic loss to livestock owners, particularly in abortion 'storms' when a large proportion of the breeding flock may abort. Treatment and the possibility of vaccination may mimimize the extent of infection. Toxoplasma has been found in common household pets and in the exotic animals of the world. Infection is usually asymptomatic. Various strains of toxoplasma have been isolated from animals and humans and usually characterized according to virulence. The strategies for the control of toxoplasma infection in humans are dependent on the understanding of the sexual and asexual cycles of infection in animals.

REFERENCES

1 Buxton, D. (1990). Ovine toxoplasmosis: a review. *J. R. Soc. Med.*, **83**, 509–11.

2 Dubey, J. P. and Kirkbride, C. A. (1990). Toxoplasmosis and other causes of abortions in sheep from north central United States. *J. Am. Vet. Med. Assoc.*, **196**, 287–90.

3 Hutchison, W. M., Dunachie, J. F., Work, K., and Siim, J. C. (1971). The life cycle of the coccidian parasite, *Toxoplasma gondii*, in the domestic cat. *Trans. R. Soc. Trop. Med. Hyg.*, **65**, 380–99.

4 Miller, N. L., Frenkel, J. K., and Dubey, J. P. (1972). Oral infections with toxoplasma cysts and oocysts in felines, other mammals, and in birds. *J. Parasitol.*, **58**, 928–37.

5 Frenkel, J. K. (1973). Toxoplasmosis: parasite life cycle, pathology, and immunology. In *The Coccidia: Eimeria, Isospora, Toxoplasma and related genera.* (ed. D. M. Hammond and P. L. Long), pp. 343–410. University Park Press, Baltimore.

6 Goldman, M., Carver, R. K., and Sulzer, A. J. (1958). Reproduction of *Toxoplasma gondii* by internal budding. *J. Parasitol.*, **44**, 161–71.

7 Hoare, C. A. (1972). The developmental stages of toxoplasma. *J. Trop. Med. Hyg.*, **75**, 56–8.

8 Jacobs, L. (1973). New knowledge of toxoplasma and toxoplasmosis. *Adv. Parasitol.*, **11**, 631–69.

9 Sibley, L. D., Krahenbuhl, J. L., Adams, G. M. W., and Weidner, E. (1986). Toxoplasma modifies macrophage phagosomes by secretion of a vesicular network rich in surface proteins. *J. Cell. Biol.*, **103**, 867–74.

10 Jacobs, L. (1974). *Toxoplasma gondii*: parasitology and transmission. *Bull. N. Y. Acad. Med.*, **50**, 128–45.

11 Dubey, J. P. and Frenkel, J. K. (1976). Feline toxoplasmosis from acutely infected mice and the development of toxoplasma cysts. *J. Protozool.*, **23**, 537–46.

12 Jacobs, L., Remington, J. S., and Melton, M. L. (1960). The resistance of the encysted form of *Toxoplasma gondii*. *J. Parasitol.*, **46**, 11–21.

13 Hutchison, W. M. (1965). Experimental transmission of *Toxoplasma gondii*. *Nature*, **206**, 961–2.

14 Dubey, J. P. and Frenkel, J. K. (1972). Cyst-induced toxoplasmosis in cats. *J. Protozool.*, **19**, 155–77.

15 Freyre, A., Dubey, J. P., Smith, D. D., and Frenkel, J. K. (1989). Oocyst-induced *Toxoplasma gondii* infections in cats. *J. Parasitol.*, **75**, 750–5.

16 Overdulve, J. P. (1978). Studies on the life cycle of *Toxoplasma gondii* in germ free, gnotobiotic and conventional cats. *Proc. K. Ned. Akad. Wet., Ser. C*, **81**, 19–59.

17 Dubey, J. P. (1979). Direct development of enteroepithelial stages of toxoplasma in the intestines of cats fed cysts. *Am. J. Vet. Res.*, **40**, 1634–7.

18 Ferguson, D. J. P., Birch-Andersen, A., Siim, J. C., and Hutchison, W. M. (1979). Ultrastructural studies on the sporulation of oocysts of *Toxoplasma gondii* I. Development of the zygote and formation of the sporoblasts. *Acta Pathol. Microbiol. Scand. Ser. B*, **87**, 171–81.

19 Ferguson, D. J. P., Birch-Andersen, A., Siim, J. C., and Hutchison, W. M. (1979). Ultrastructural studies on the sporulation of oocysts of *Toxoplasma gondii* II. Formation of the sporocyst and structure of the sporocyst wall. *Acta Pathol. Microbiol. Scand. Ser. B*, **87**, 183–90.

20 Ferguson, D. J. P., Birch-Andersen, A., Siim, J. C., and Hutchison, W. M. (1979). Ultrastructural studies on the sporulation of oocysts of *Toxoplasma gondii* III. Formation of the sporozoites within the sporocysts. *Acta Pathol. Microbiol. Scand., Ser. B*, **87**, 253–60.

21 Dubey, J. P., Miller, N. L., and Frenkel, J. K. (1970). Characterization of the new fecal form of *Toxoplasma gondii*. *J. Parasitol.*, **56**, 447–56.

22 Jewell, M. L., Frenkel, J. K., Johnson, K. M., Reed, V., and Ruiz, A. (1972). Development of toxoplasma oocysts in neotropical Felidae. *Am. J. Trop. Med. Hyg.*, **21**, 512–7.

23   Werk, R. (1985). How does *Toxoplasma gondii* enter host cells? *Rev. Infect. Dis.*, **7**, 449–57.

24   Lycke, E., Carlberg, K., and Norrby, R. (1975). Interactions between *Toxoplasma gondii* and its host cells: function of the penetration–enhancing factor of toxoplasma. *Infect. Immun.*, **11**, 853–61.

25   Remington, J. S. and Desmonts, G. (1976). Toxoplasmosis. In *Infectious diseases of the fetus and newborn infant.* (ed. J. S. Remington and J. O. Klein), pp. 191–332. W. B. Saunders, Philadelphia.

26   Jacobs, L. (1967). Toxoplasma and toxoplasmosis. In *Advances in parasitology*, vol. 5. (ed. B. Dawes), pp. 1–45, Academic Press, London.

27   Garnham, P. C. C., Baker, J. R., and Bird, R. G. (1962). Fine structure of cystic form of *Toxoplasma gondii. Br. Med. J.*, **1**, 83–4.

28   Wolf, A. and Cowen, D. (1937). Granulomatous encephalomyelitis due to an encephalitozoon (encephalitozoic encephalomyelitis). A new protozoan disease of man. *Bull. Neurol. Inst. N. Y.*, **6**, 306–71.

29   Desmonts, G., Couvreur, J., Alison, F., Baudelot, J., Gerbeaux, J., and Lelong, M. (1965). Étude épidémiologique sur la toxoplamose: de l'influence de la cuisson des viandes de boucherie sur la fréquence de l'infection humaine. *Rev. Franc Études Clin. Biol.*, **10**, 952–8.

30   Kean, B. H., Kimball, A. C., and Christenson, W. N. (1969). An epidemic of acute toxoplasmosis. *JAMA*, **208**, 1002–4.

31   Ferguson, D. J. P., Birch-Andersen, A., Siim, J. C., and Hutchison, W. M. (1979). An ultrastructural study on the excystation of the sporozoites of *Toxoplasma gondii. Acta Pathol. Microbiol. Scand., Ser. B*, **87**, 277–83.

32   Yilmaz, S. M. and Hopkins, S. H. (1972). Effects of different conditions on duration of infectivity of *Toxoplasma gondii* oocysts. *J. Parasitol.*, **58**, 938–9.

33   Frenkel, J. K., Ruiz, A., and Chinchilla, M. (1975). Soil survival of toxoplasma oocysts in Kansas and Costa Rica. *Am. J. Trop. Med. Hyg.*, **24**, 439–43.

34   Benenson, M. W., Takafuji, E. T., Lemon, S. M., Greenup, R. L., and Sulzer, A. J. (1982). Oocyst-transmitted toxoplasmosis associated with ingestion of contaminated water. *N. Engl. J. Med.*, **307**, 666–9.

35   Saari, M. and Raisanen, S. (1974). Transmission of acute toxoplasma infection. The survival of trophozoites in human tears, saliva, and urine and in cow's milk. *Acta Ophthalmol.*, **52**, 847–52.

36   Spence, J. B., Beattie, C. P., Faulkner, J., Henry, L., and Watson, W. A. (1978). *Toxoplasma gondii* in the semen of rams. *Vet. Rec.* **102**, 38–9.

37   Siegel, S. E., Lunde, M. N., Gelderman, A. H., Halterman, R. H., Brown, J. A., Levine, A. S. *et al.* (1971). Transmission of toxoplasmosis by leukocyte transfusion. *Blood*, **37**, 388–94.

38   Nelson, J. C., Kauffmann, D. J. H., Ciavarella, D., and Senisi, W. J. (1989). Acquired toxoplasmic retinochoroiditis after platelet transfusions. *Ann. Ophthalmol.*, **21**, 253–4.

39   Skinner, L. J., Timperley, A. C., Wightman, D., Chatterton, J. M. W., and Ho-Yen, D. O. (1990). Simultaneous diagnosis of toxoplasmosis in goats and goatowner's family. *Scand. J. Infect. Dis.*, **22**, 359–61.

40   Beverley, J. K. A. and Beattie, C. P. (1958). Glandular toxoplasmosis. A survey of 30 cases. *Lancet*, **ii**, 379–84.

41   Price, J. H. (1969). Toxoplasma infection in an urban community. *Br. Med. J.*, **4**, 141–3.

42   Riemann, H. P., Meyer, M. E., Theis, J. H., Kelso, G., and Behymer, D. E. (1975). Toxoplasmosis in an infant fed unpasteurized goat milk. *J. Pediatr.*, **87**, 573–6.

43   Dubey, J. P. (1974). Effect of freezing on the infectivity of toxoplasma cysts to cats. *J. Am. Vet. Med. Assoc.*, **165**, 534–6.

44   Dubey, J. P., Kotula, A. W., Sharar, A., Andrews, C. D., and Lindsay, D. S. (1990). Effect of high temperature on infectivity of *Toxoplasma gondii* tissue cysts in pork. *J. Parasitol.*, **76**, 201–4.

45 Neville, P. (1984). *Pest Control News*, **8**, 6.

46 Deeb, B. J., Sufan, M. M., and DiGiacomo, R. F. (1985). *Toxoplasma gondii* infection of cats in Beirut, Lebanon. *J. Trop. Med. Hyg.*, **88**, 301–6.

47 Wallace, G. D. (1973). The role of the cat in the natural history of *Toxoplasma gondii*. *Am. J. Trop. Med. Hyg.*, **22**, 313–22.

48 Wallace, G. D. (1975). Observations on a feline coccidium with some characteristics of toxoplasma and sarcocystis. *Z. Parasitenk.*, **46**, 167–78.

49 Dubey, J. P., Christie, E., and Pappas, P. W. (1977). Characterization of *Toxoplasma gondii* from the feces of naturally infected cats. *J. Infect. Dis.*, **136**, 432–5.

50 Frenkel, J. K. and Smith D. D. (1982). Immunization of cats against shedding of toxoplasma oocysts. *J. Parasitol.*, **68**, 744–8.

51 Dubey, J. P. (1976). Reshedding of toxoplasma oocysts by chronically infected cats. *Nature*, **262**, 213–14.

52 Dubey, J. P. and Beattie, C. P. (1988). *Toxoplasmosis of animals and man*. CRC Press, Florida.

53 Dubey, J. P. (1987). Toxoplasmosis. *Vet. Clin. North Am. Small Anim. Pract.*, **17**, 1389–404.

54 Dubey, J. P. (1973). Feline toxoplasmosis and coccidiosis: a survey of domiciled and stray cats. *J. Am. Vet. Med. Assoc.*, **162**, 873–7.

55 Jackson, M. H., Hutchison, W. M., and Siim, J. C. (1987). Prevalence of *Toxoplasma gondii* in meat animals, cats and dogs in central Scotland. *Br. Vet. J.*, **143**, 159–65.

56 Childs, J. E. and Seegar, W. S. (1986). Epidemiologic observations on infection with *Toxoplasma gondii* in three species of urban mammals from Baltimore, Maryland, USA. *Int. J. Zoonoses*, **13**, 249–61.

57 Hay, J. and Hutchison, W. M. (1983). *Toxoplasma gondii* — an environmental contaminant. *Ecol. Dis.*, **2**, 33–43.

58 Jackson, M. H., Hutchison, W. M., and Siim, J. C. (1986). Toxoplasmosis in a wild rodent population of central Scotland and a possible explanation of the mode of transmission. *J. Zool.*, **209**, 549–57.

59 Wallace, G. D. (1973). Intermediate and transport hosts in the natural history of *Toxoplasma gondii. Am. J. Trop. Med. Hyg.*, **22**, 456–64.

60 Beaver, P. C. (1975). Biology of soil-transmitted helminths: the massive infection. *Health Lab. Sci.*, **12**, 116–25.

61 Tobias, S. (1986). Diagnosing the cause of feline pyrexia and anorexia. *Vet. Med.*, **81**, 930–3.

62 Lappin, M. R., Greene, C. E., Winston, S., Toll, S. L., and Epstein, M. E. (1989). Clinical feline toxoplasmosis. Serologic diagnosis and therapeutic management of 15 cases. *J. Vet. Intern. Med.*, **3**, 139–43.

63 Meier, H., Holzworth, J., and Griffiths, R. C. (1957). Toxoplasmosis in the cat — fourteen cases. *J. Am. Vet. Med. Assoc.*, **131**, 395–414.

64 Piper, R. C., Cole, C. R., and Shadduck, J. A. (1970). Natural and experimental ocular toxoplasmosis in animals. *Am. J. Ophthalmol.*, **69**, 662–8.

65 Dubey, J. P. (1986). Toxoplasmosis. *J. Am. Vet. Med. Assoc.*, **189**, 166–70.

66 Deeb, B. J., Sufan, M. M., and DiGiacomo, R. F. (1986). *Toxoplasma gondii* antibodies in cats: detection by indirect haemagglutination and indirect fluorescent antibody tests. *J. Parasitol.*, **72**, 355–7.

67 Lappin, M. R., Greene, C. E., Prestwood, A. K., Dawe, D. L., and Tarleton, R. L. (1989). Diagnosis of recent *Toxoplasma gondii* infection in cats by use of an enzyme-linked immunosorbent assay for immunoglobulin M. *Am. J. Vet. Res.*, **50**, 1580–5.

68 Rawal, B. D. (1959). Toxoplasmosis. A dye-test survey on sera from vegetarians and meat eaters in Bombay. *Trans. R. Soc. Trop. Med. Hyg.*, **53**, 61–3.

69 Wallace, G. D. (1969). Serologic and epidemiologic observations on toxoplasmosis on three pacific atolls. *Am. J. Epidemiol.*, **90**, 103–11.

70   Huldt, G., Lagercrantz, R., and Sheehe, P.R. (1979). On the epidemiology of human toxoplasmosis in Scandinavia especially in children. *Acta Paediatr. Scand.*, **68**, 745–9.

71   Woodruff, A. W., de Savigny, D. H., and Hendy-Ibbs, P. M. (1982). Toxocaral and toxoplasmal antibodies in cat breeders and in Icelanders exposed to cats but not to dogs. *Br. Med. J.*, **284**, 309–10.

72   Meat and Livestock Commission with the Health Education Authority. *Meat, diet and health. A report on red meat today.* Booklet, obtainable from: HEA, Snowdon Drive, Winterhill, Milton Keynes, MK6 1HQ.

73   Dubey, J. P. (1990). Status of toxoplasmosis in sheep and goats in the United States. *J. Am. Vet. Med. Assoc.*, **196**, 259–62.

74   Malik, M. A., Dreesen, D. W., and de la Cruz, A. (1990). Toxoplasmosis in sheep in north eastern United States. *J. Am. Vet. Med. Assoc.*, **196**, 263–5.

75   Beverley, J. K. A. and Watson, W. A. (1961). Ovine abortion and toxoplasmosis in Yorkshire. *Vet. Rec.*, **73**, 6–11.

76   Dubey, J. P., Murrell, K. D., Fayer, R., and Schad, G. A. (1986). Distribution of *Toxoplasma gondii* tissue cysts in commercial cuts of pork. *J. Am. Vet. Med. Assoc.*, **188**, 1035–7.

77   Beverley, J. K. A., Henry, L., and Hunter, D. (1978). Experimental toxoplasmosis in young piglets. *Res. Vet. Sci.*, **24**, 139–46.

78   Dubey, J. P. (1990). Status of toxoplasmosis in cattle in the United States. *J. Am. Vet. Med. Assoc.*, **196**, 257–9.

79   Dubey, J. P. (1986). A review of toxoplasmosis in cattle. *Vet. Parasitol.*, **22**, 177–202.

80   Ruppanner, R., Riemann, H. P., Farver, T. B., West, G., Behymer, D. E., and Wijayasinghe, C. (1978). Prevalence of *Coxiella burnetii* (Q fever) and *Toxoplasma gondii* among dairy goats in California. *Am. J. Vet. Res.*, **39**, 867–70.

81   Dubey, J. P. and Adams, D. S. (1990). Prevalence of *Toxoplasma gondii* antibodies in dairy goats from 1982 to 1984. *J. Am. Vet. Med. Assoc.*, **196**, 295–6.

82   Dubey, J. P., Sharma, S. P., Lopes, C. W. G., Williams, J. F., Williams, C. S. F., and Weisbrode, S. E. (1980). Caprine toxoplasmosis: abortion, clinical signs, and distribution of toxoplasma in tissues of goats fed *Toxoplasma gondii* oocysts. *Am. J. Vet. Res.*, **41**, 1072–6.

83   Mehdi, N. A. Q., Kazacos, K. R., and Carlton, W. W. (1983). Fatal disseminated toxoplasmosis in a goat. *J. Am. Vet. Med. Assoc.*, **183**, 115–17.

84   Munday, B. L. and Mason, R. W. (1979). Toxoplasmosis as a cause of perinatal death in goats. *Aust. Vet. J.*, **55**, 485–7.

85   Dubey, J. P. (1988). Lesions in transplacentally induced toxoplasmosis in goats. *Am. J. Vet. Res.*, **49**, 905–9.

86   Dubey, J. P. (1981). Toxoplasma-induced abortion in dairy goats. *J. Am. Vet. Med. Assoc.*, **178**, 671–4.

87   Dubey, J. P., Sundberg, J. P., and Matiuck, S. W. (1981). Toxoplasmosis associated with abortion in goats and sheep in Connecticut. *Am. J. Vet. Res.*, **42**, 1624–6.

88   Dubey, J. P. (1982). Repeat transplacental transfer of *Toxoplasma gondii* in dairy goats. *J. Am. Vet. Med. Assoc.*, **180**, 1220–1.

89   Biancifiori, F., Rondini, C., Grelloni, V., and Frescura, T. (1986). Avian toxoplasmosis: experimental infection of chicken and pigeon. *Comp. Immunol. Microbiol. Infect. Dis.*, **9**, 337–46.

90   Erichsen, S. and Harboe, A. (1953). Toxoplasmosis in chickens. I. An epidemic outbreak of toxoplasmosis in a chicken flock in south eastern Norway. *Acta Pathol.*, **XXXIII**, 56–71.

91   Al-Khalidi, N. W. and Dubey, J. P. (1979). Prevalence of *Toxoplasma gondii* infection in horses. *J. Parasitol.*, **65**, 331–4.

92   Dubey, J. P. (1985). Persistence of encysted *Toxoplasma gondii* in tissues of equids fed oocysts. *Am. J. Vet. Res.*, **46**, 1753–4.

93   Blewett, D. A. and Trees, A. J. (1987). The epidemiology of ovine toxoplasmosis with especial respect to control. *Br. Vet. J.*, **143**, 128–35.

94 Faull, W. B., Clarkson, M. J., and Winter, A. C. (1986). Toxoplasmosis in a flock of sheep: some investigations into its source and control. *Vet. Rec.*, **119**, 491–3.
95 Dubey, J. P. (1985). Toxoplasmosis in dogs. *Canine Pract.*, **12**, 7–26.
96 Capen, C. C. and Cole, C. R. (1966). Pulmonary lesions in dogs with experimental and naturally occurring toxoplasmosis. *Pathol. Vet.*, **3**, 40–63.
97 Chamberlain, D. M., Docton, F. L., and Cole, C. R. (1953). Toxoplasmosis. II. Intra-uterine infection in dogs, premature birth and presence of organisms in milk. *Proc. Soc. Exp. Biol. Med.*, **82**, 198–200.
98 Campbell, R. S. F., Martin, W. B., and Gordon, E. D. (1955). Toxoplasmosis as a complication of canine distemper. *Vet. Rec.*, **67**, 708–16.
99 Parenti, E., Cerruti Sola, S., Turilli, C., and Corazzola, S. (1986). Spontaneous toxoplasmosis in canaries (*Serinus canaria*) and other small passerine cage birds. *Avian Pathol.*, **15**, 183–97.
100 Mas Bakal, P., Karstad, L., and In't Veld, N. (1980). Serologic evidence of toxoplasmosis in captive and free-living wild mammals in Kenya. *J. Wildl. Dis.*, **16**, 559–64.
101 Remington, J. S., Soave, O. A., and Davis, J. (1965). A serological survey in three species of monkeys for antibodies to toxoplasma. *Am. J. Trop. Med. Hyg.*, **14**, 724–6.
102 Ferraroni, J. J., Reed, S. G., and Speer, C. A. (1980). Prevalence of toxoplasma antibodies in humans and various animals in the Amazon. *J. Parasitol.*, **47**, 148–50.
103 Riemann, H. P., Behymer, D. E., Fowler, M. E., Schulz, T., Lock, A., Orthoefer, J. G. *et al.* (1974). Prevalence of antibodies to *Toxoplasma gondii* in captive exotic mammals. *J. Am. Vet. Med. Assoc.*, **165**, 798–800.
104 Dickson, J., Fry, J., Fairfax, R., and Spence, T. (1983). Epidemic toxoplasmosis in captive squirrel monkeys (*Saimiri sciureus*). *Vet. Rec.*, **112**, 302.
105 Dubey, J. P., Kramer, L. W., and Weisbrode, S. E. (1985). Acute death associated with *Toxoplasma gondii* in ring-tailed lemurs. *J. Am. Vet. Med. Assoc.*, **187**, 1272–3.
106 Sabin, A. B. (1941). Toxoplasmic encephalitis in children. *JAMA*, **116**, 801–7.
107 Jacobs, L. (1956). Propagation, morphology, and biology of toxoplasma. *Ann. N. Y. Acad. Sci.*, **64**, 154–79.
108 Kaufman, H. E., Remington, J. S., and Jacobs, L. (1958). Toxoplasmosis: the nature of virulence. *Am. J. Ophthalmol.*, **46**, 255–61.
109 Beverley, J. K. A., Henry, L. Hunter, D., and Brown, M. E. (1977). Experimental toxoplasmosis in calves. *Res. Vet. Sci.*, **23**, 33–7.
110 Frenkel, J. K. (1990). Transmission of toxoplasmosis and the role of immunity in limiting transmission and illness. *J. Am. Vet. Med. Assoc.*, **196**, 233–40.
111 Frenkel, J. K. (1967). Toxoplasmosis. In *Comparative aspects of reproductive failure.* (ed. K. Bernirschke), pp. 296–321. Springer Verlag, FRG.

3

Clinical features

DARREL O. HO-YEN

In the 1970s, there was a complaint that medical practitioners failed to appreciate toxoplasmosis.[1] Many looked upon this infection as a tropical disease and most stated that they had not seen a case. Yet, toxoplasmosis is a common, world-wide infection (Chapter 1). Like many such infections, the clinical features of *Toxoplasma gondii* in humans range a wide spectrum from asymptomatic infection to severe illness with high mortality, and the vast majority of illnesses are not diagnosed.[1] In the 1990s, there is a greater awareness of this infection, but there is still not a complete appreciation of its importance.[2]

The spectrum of the clinical features of toxoplasma infection is as in Table 3.1. In the late 1980s there has been much media attention to congenital toxoplasmosis.[2] We believe that this is justified and that an antenatal screening programme combined with health promotion is likely to obtain the best results.[3] Clinical features in pregnant women are similar, but slightly different (Chapter 6) when compared to the effects of infection in non-pregnant women. The effects of infection in the fetus, neonate, child, and young adult are also complex and warrant separate consideration (Chapter 6). It is important for the medical practitioner to realize that toxoplasmosis is not an uncommon infection and its consequences in pregnancy may be devastating.

Another area in which it is critical to recognize toxoplasmosis is in immunocompromised patients.[1,2,4] In these individuals, toxoplasmosis may easily become disseminated and produce an illness with high morbidity and mortality. Immunocompromised patients, either as a result of infections such as human immunodeficiency virus or medical treatment, are likely to increase.[2] The number and types of organ transplants are also increasing and these patients are being subjected to quite complex immunosuppression. Such treatment may easily result in unforeseen reactivation of toxoplasmosis[5] which may have disastrous consequences. Clinical features of infection in immunocompromised individuals are considered separately (Chapter 7).

As most attention has been in the areas of infection in pregnancy and those with immunocompromise, there is least appreciation of the other manifestations of toxoplasmosis. This chapter is specifically concerned with acquired infection in the immunocompetent individual (Table 3.1). Infection is usually 'primary' (in individuals who have not previously had toxoplasmosis). However, toxoplasma has the ability to encyst after the acute infection and remain dormant for many years (Chapter 2). If the body's immune

Table 3.1   Toxoplasma infection in humans

Congenital infection (Chapter 6)
  Asymptomatic at birth (primary)
  Clinically affected at birth (primary)

Acquired infection
  Immunocompetent
    Asymptomatic (primary/reactivation)
    Symptomatic (primary/reactivation)
      non-specific illness (primary/reactivation)
      lymphadenopathy (primary/reactivation)
      myalgic encephalomyelitis (primary)
      ocular disease (primary/reactivation)
      other manifestations (primary/reactivation)

  Immunocompromise (Chapter 7)
    Asymptomatic (primary/reactivation)
    Non-specific illness (primary/reactivation)
    Disseminated infection (primary/reactivation)

---

system is disturbed there may be 'reactivation' of dormant infection. It can be difficult to distinguish the clinical features of a primary infection from a reactivated infection (Table 3.1) and the distinction is often made on the presence or absence of serological evidence of past infection (Chapter 4). The diagnosis of acquired infection is considered separately (Chapter 4), as is appropriate treatment (Chapter 5).

INCUBATION PERIOD

After an individual has been exposed to toxoplasma, there is an incubation period before the signs and symptoms of infection develop. Sources of toxoplasma infection have been described in Chapter 2. Unfortunately precise information on the incubation period is not available, but there is some indication from outbreaks of infection. These suggest that the incubation period is 4–21 days, with most illness in the second week after exposure.[6-8] Extrapolation of information from outbreaks may be inappropriate as more virulent strains may be involved in outbreaks. However, as with all infections, infection is a result of several factors within the host and the parasite. Thus, it is noticeable that in outbreaks there is a large range of incubation periods.[6-8] One important consideration is the infecting dose. A report of an outbreak from oocyst-contaminated water showed a shorter incubation period related to the use of one water source which may have been more contaminated than others.[7] The stage of the parasite may also be important, for example as a tachyzoite in goats' milk, the parasite may have a longer incubation period with a milder illness.[9] Nevertheless, tachyzoites

Table 3.2   Frequency of symptoms (%) in outbreaks of acquired toxoplasma infection[6-9,12,16]

Symptoms	New York	California	Alabama	California	Panama	Atlanta	Total
Pyrexia	5/6	1/10	6/10	5/18	28/31	33/37	78/112
	(83)	(10)	(60)	(28)	(90)	(89)	(70)
Lymphadenopathy	2/6	0/10	6/10	9/18	24/31	31/37	72/112
	(33)	(0)	(60)	(50)	(77)	(84)	(64)
Malaise	4/6	1/10	4/10	3/18	23/31	25/37	60/112
	(67)	(10)	(40)	(17)	(74)	(68)	(54)
Myalgia	2/6	0/10	0/10	3/18	21/31	22/37	48/112
	(33)	(0)	(0)	(17)	(68)	(60)	(43)
Asymptomatic	1/6	9/10	3/10	9/18	3/31	2/37	27/112
	(17)	(90)	(30)	(50)	(10)	(5)	(24)

involved in laboratory accidents (such as needle-sticks or splashing of the organism onto face and eyes) are associated with short incubation periods of 3–10 days[10] and may be due to greater virulence of these strains as well as the route of infection.

ASYMPTOMATIC ILLNESS

Asymptomatic illness with infections is usually regarded as being more common in children. Unfortunately, there is very little information on toxoplasma infection in children. One study suggested that toxoplasma infection was very rare in symptomatic children under 10 years admitted to hospital, occurring in 3/187 (1.6%).[11] This is a surprising result as children can be regarded as being more exposed to infection by having more intimate contact with cats or soil which may be contaminated. When these conditions occur, there can be a high rate of infection (7/11, 65%) with many children symptomatic (6/7, 86%).[12] Although the previous study[11] considered symptomatic illness, the technical methods more accurately detect prevalence of toxoplasma antibody. Thus, the results suggest that toxoplasma infection, whether symptomatic or asymptomatic, is rare in children. In Britain, this conclusion is also consistent with the finding of toxoplasma antibody in 15/180 (8%) of normal children aged under 10 years.[13]

In adults, the likelihood of asymptomatic illness is unknown, but reputable medical textbooks suggest that most infection is asymptomatic.[14,15] Although this seems a reasonable statement, it is probably based on a high prevalence of antibody in the general population. Thus, another explanation may be that symptoms of toxoplasma infection are non-specific in many patients. In an attempt to decide on these two possible explanations, it is sensible to look at the information from outbreaks of toxoplasma infection[6-9,12,16] (Table 3.2). In one of these studies,[16] only four of nine families have been included in Table 3.2 as one family had an infection in pregnancy and the other four families had only one member with a current infection. This table also does not include one of the first documented food-related outbreaks[17] as few individuals were involved and it was unclear if the sample was biased towards symptomatic individuals. Similarly, a large outbreak of 110 individuals in Brazil[18] was excluded because the information was very difficult to interpret.

The six studies in Table 3.2 show an average of asymptomatic infection in 24 per cent of individuals. Both tropical and temperate countries were considered with the results showing a wide range (5–90%) of asymptomatic infection. Thus, it is likely that strains of different virulence were considered. The two larger studies[7,8] are particularly interesting as they show: dependable reporting of results; results that were easy to interpret; a large number of individuals; a tropical and a temperate country; and the lowest level of asymptomatic infection. It is possible that this level of 5–10 per cent is more accurate when individuals are carefully questioned on the presence of non-specific symptoms such as fever and malaise. Nevertheless, one must remember that the virulence of different strains is likely to vary (Chapter 2). The outbreaks in Table 3.2 are probably representative of toxoplasma infection and these results suggest that it may be more accurate to say that 'a quarter' of patients have asymptomatic infection instead of 'the majority'.

Reactivation of toxoplasma infection can be clinically obvious, serious, and disseminated in individuals who have been severely immunocompromised.[2] Much less severe changes in the immune system or mechanical trauma may result in localized release of parasites without symptoms from previously dormant tissue cysts. For this reason toxoplasma infection has been described as 'chronic'.[4] However, it is probably more accurate to describe toxoplasma infection in dormant tissue cysts as 'latent' and release of the parasite, locally or systemically, as 'reactivation'. Such release may explain the persistence of toxoplasma antibody for many years after the acute infection. In immunocompetent individuals, reactivation is likely to be mostly asymptomatic. As there are no tests to identify reactivation, this cannot be confirmed. Indirect support for the hypothesis is the seriousness of the infection in the immunocompromised individual (Chapter 7). The theory is that toxoplasmosis is a latent infection which requires the immune system to keep it dormant. Changes in the immune system result in reactivation which is usually asymptomatic; if immunocompromise is severe, reactivation becomes serious.[2,4]

SYMPTOMATIC ILLNESS

In the past, it was believed that acquired toxoplasmosis generally resulted in infection rather than disease.[19] This was probably because of the paucity of characteristic clinical signs. In this chapter, symptomatic illness is considered under the sections shown in Table 3.1.

Non-specific illness

The analysis of outbreaks (Table 3.2) shows that the most common clinical symptom was pyrexia, and it was present in 70 per cent of those infected. As low-grade pyrexia is a frequent occurrence of many infectious and other medical conditions, it is not surprising that it may not be considered to be due to toxoplasma infection. It is also of little help that pyrexia rarely occurs alone, as the other symptoms which usually accompany it (lymphadenopathy, malaise, and myalgia, Table 3.2) are all equally non-specific. Myalgia is an interesting symptom. It can be very memorable,[6] common,[7,8] or absent.[9,12] In these reports, myalgia is present in the more severe clinical outbreaks, and so may be associated with the more virulent strains of toxoplasma. In the immunocompetent, it is unlikely that severe myalgia is commonly present in reactivated infections. Such infections may be likely to manifest as mild, pyrexial illnesses. At present it would be difficult to diagnose such illnesses as specific IgM or high dye test titres (Chapter 4) are not usually produced.

Lymphadenopathy

Lymphadenopathy is an important clinical sign of acquired primary toxoplasmosis in the immunocompetent (Fig. 3.1). Although it was first described in the early 1950s in pregnancy,[20] it is more common in non-pregnant individuals.[14,15,21–27] From these studies, a scheme depicting the relationships between lymphadenopathy, pyrexia, and lymphocytosis can be postulated (Fig. 3.2). Nevertheless, it must be remembered that other epidemiological factors, including strain variation, may influence the relative importance of any of these clinical syndromes.

Age and sex

Variations in the age–sex distribution of patients with toxoplasma lymphadenopathy has long been of interest to researchers.[22,24] Large studies show that more boys than girls under 15 years of age have toxoplasma lymphadenopathy.[22–25] In one study[23] this result was surprising as three-quarters of the patients were female. Explanations for the preponderance in boys are that boys attract more dirt and therefore more infection (Chapter 1); and that care of boys is more solicitous so that they are seen more by doctors and are more likely to be investigated.[22] It is also possible that investigations are more common in boys as their lymph nodes are larger.[22] In adults, women are far more affected than men.[22–25] Although this may be a real finding, it is more likely that it is a result of sampling bias. Lymphadenopathy is more common in the neck[23,24] and women, because of

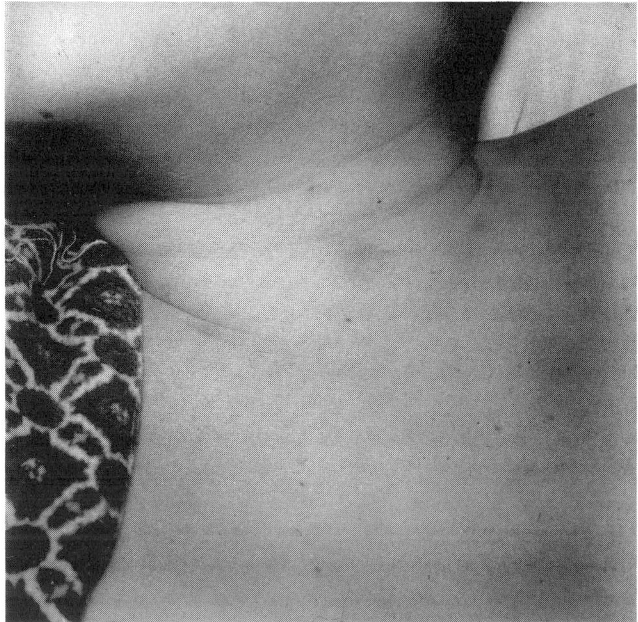

Fig. 3.1 Lymphadenopathy in the neck of an adolescent with acquired toxoplasmosis.

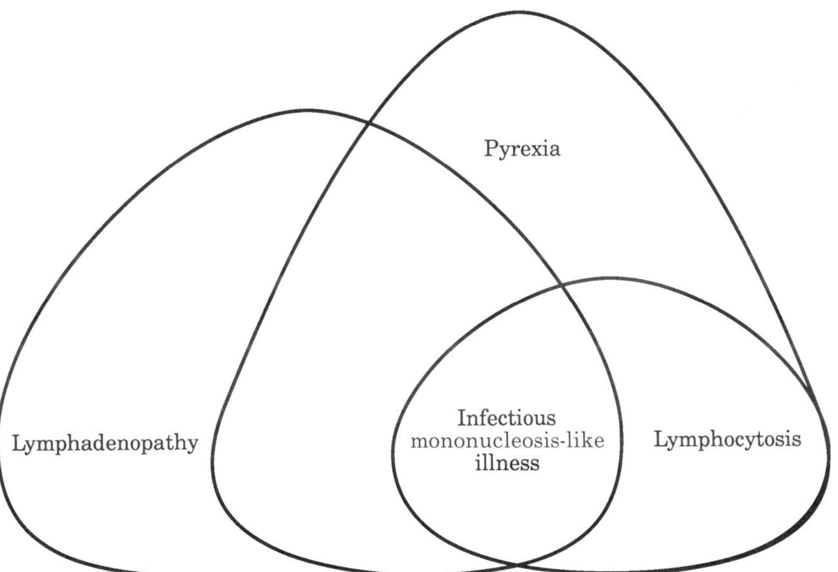

Fig. 3.2 The relationships between lymphadenopathy, pyrexia, lymphocytosis, and infectious mononucleosis-like illness in toxoplasma infection.

clothing, hair, and make-up examine their necks more than men.[22] It is also possible that they are more 'lump conscious'[22] and consult their medical practitioners more often.[28] Nevertheless, women are probably more exposed to infection as they more often prepare food, own cats, and clean cat litter trays. These reasons appear strong and there is no need to suggest immunological reasons why women appear to have lymphadenopathy.[22] In both sexes, the 21–25 year age group is more often affected,[23] and is probably most adventurous in travel and willingness to experiment with food.

Sites

Large studies often have incomplete information on the sites of toxoplasma lymphadenopathy[26,27] but one study is exceptional in having detailed information in 231 of 234 cases.[23] When the available information from the three largest studies[23,26,27] are amalgamated, a total of 654 cases are available for consideration. These show that lymphadenopathy is more common in the neck (65%), followed by the axillae (24%) and then the groin (11%). Although anterior neck lymphadenopathy is more common (54% of cases),[24] it has been suggested that posterior lymph node enlargement is characteristic of toxoplasmosis.[29] This latter study examined only 18 posterior cervical nodes and four from other sites; it shows sampling bias and there is no information on whether other nodes were enlarged in the patients who were studied. In sporadic infection, posterior cervical lymphadenopathy can be as common as 27 per cent.[23] Nevertheless, in one outbreak of 31 cases, lymphadenopathy was most prominent in the posterior cervical region (58%), axillary (48%), and anterior cervical region (19%).[8] In this study, infection was believed to be due to oocyst-contaminated water. Thus, differences in the presence of anterior or posterior cervical lymphadenopathy may be due to strain variation or mode of infection. Lymphadenopathy is much more frequent at single sites (62–75%)[23,25] in adults; but, in children, multiple sites (51%) may be more common.[27] Obviously any group of lymph nodes in the body may be affected and of the other sites, anterior chest wall and mammary nodes are most often involved.[23,25] In addition, as with other causes of lymphadenopathy, hilar lymphadenopathy may result in the collapse of part of the lung and mesenteric lymphadenopathy may cause abdominal complaints.[14]

Characteristics

Enlarged lymph nodes are usually firm, unattached to the overlying skin and initially tender. There may be pain in the vicinity of the affected node but this quickly subsides. Most nodes are noticed by the patients themselves.[23] The size can vary up to 6 cm,[24] but most patients (50%) have nodes of 1–2 cm.[23] As a group, patients with toxoplasma lymphadenopathy have larger lymph nodes than patients with non-specific hyperplastic lymphadenopathy.[23] It is possible that studies of excised lymph nodes are biased towards larger

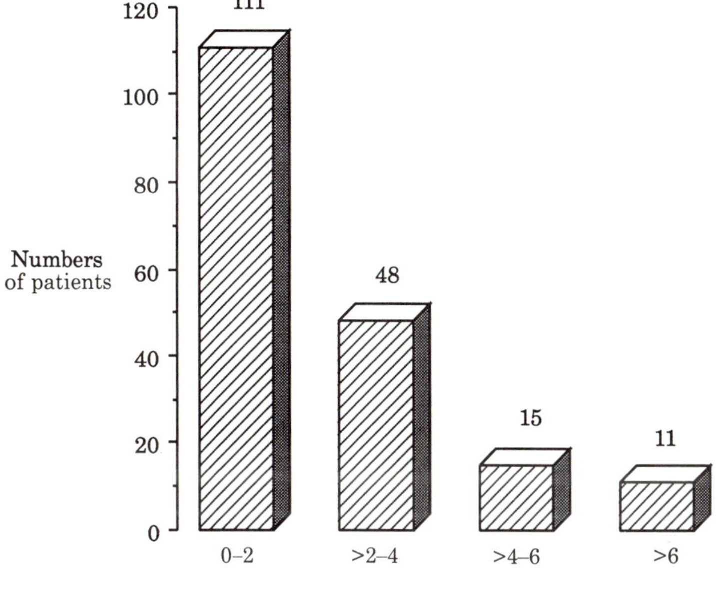

Fig. 3.3 Persistence of lymphadenopathy.

nodes which may be more suitable for biopsy. However, a more likely explanation is that the sizes of these nodes are representative of the range with toxoplasmosis, especially as 18 per cent were less than 1 cm in diameter.[23] The more likely reason for biopsy was the persistence of the nodes in a concerned patient, often female, who has found the abnormality herself. Several workers have emphasized that lymphadenopathy may persist for very many months.[15,23–26] In three studies[23,24,26] the information on persistence of lymphadenopathy was good and when this information was collated, the results on 185 patients were obtained (Fig. 3.3). The vast majority (60%) recovered before 2 months, and most of these within one month.[23] However, a quarter of patients take 2–4 months to return to normal and 8 per cent take 4–6 months. A substantial number of patients (6%) do not return to normal until much later (Fig. 3.3). These findings suggest that toxoplasma infection has a significant immune response in humans and that it should not be surprising if symptoms persist.

Infectious mononucleosis-like illness

Infectious mononucleosis is a disease of young adults with definite clinical (lymphadenopathy, sore throat, pyrexia), and haematological (atypical lymphocytes) findings.[30] The illness is now known to be caused by the Epstein–Barr virus and the diagnosis is usually made by the presence of specific IgM for Epstein–Barr viral capsid antigen.[30] A very similar illness can be produced by toxoplasma infection and it has been estimated that in the UK, toxoplasma accounts for 5 per cent of illness that appear to be infectious mononucleosis.[24,31] This finding is consistent with that in USA.[32] However, infectious mononucleosis is a well-defined clinical disease[30] and toxoplasma may not always produce a clinical picture which fulfills the criteria for 'infectious mononucleosis'. It is appropriate here to consider the term 'glandular fever' which has been used as being synonymous with 'infectious mononucleosis'; in reality, the latter is an illness with distinct clinical and laboratory findings, whereas the former is a descriptive term for illnesses with enlarged glands and fever.[30] It is likely that toxoplasma infection is a common cause of glandular fever and an uncommon cause of infectious mononucleosis-like illness (Fig. 3.2).

Systemic symptoms vary in different groups of patients with toxoplasma lymphadenopathy.[23–25] There is no doubt that many patients can have asymptomatic lymphadenopathy (Fig. 3.2). In some studies, pyrexia was uncommon (16–26%);[23,25] whereas in another the majority of patients had pyrexia (59%).[24] These findings are similar to those found in outbreaks (Table 3.2) where the prevalence of pyrexia varied between 10–90 per cent and may be mostly explained by strain variations. Thus, generally most patients with toxoplasma lymphadenopathy have other symptoms. In addition, systemic symptoms are not necessarily related to clinically detectable enlargement of other reticulo-endothelial organs such as the liver or spleen.[24,33] Splenomegaly is probably more common than hepatomegaly, but both are rare in toxoplasmosis.

The clinical features of toxoplasma lymphadenopathy are usually quite different from those of infectious mononucleosis. In the latter disease, there is often persistence of pyrexia; prominent sore throat; more generalized lymphadenopathy, and more common hepatosplenomegaly.[34] Perhaps the most useful differentiating factor is that sore throat is uncommon in toxoplasma lymphadenopathy.[31] Laboratory results easily differentiate the two illnesses. Liver function tests are abnormal in most patients with infectious mononucleosis, and atypical lymphocytosis is a prerequisite for diagnosis.[30,31] Hepatitis in acquired toxoplasmosis is infrequent; it may present as part of the initial illness or it may be a late complication.[35] The haematological results in toxoplasma lymphadenopathy also differ from infectious mononucleosis. A normal total white cell count appears to be the rule,[23,33] however, leucocytosis was present in 20 per cent of patients in one study.[24] There is agreement that lymphocytosis is common (50–84%),[23,24] but eosinophilia is absent (0/77 cases)[23] or very rare (1/30 cases).[24] For the diagnosis of infectious mononucleosis, atypical lymphocytosis has to be present.[30,31] In patients with toxoplasma lymphadenopathy, atypical lymphocytosis may be uncommon (6/77, 8%),[23] more common (5/30, 17%)[24] or very common (11/28, 39%).[8] It is likely that these differences in atypical lymphocytes may reflect strain variations of toxoplasma or sampling bias. As Epstein–Barr virus is the accepted cause of infectious mononucleosis,[30] when toxoplasma infection produces a similar illness, it is appropriate to use the label 'infectious mononucleosis-like illness'. Although toxoplasmosis is a rare cause of an infectious mononucleosis-like illness in the UK, this is not the case in other countries. Malaysia is an excellent example that the epidemiology of infections varies with the different conditions in other countries. In many tropical countries such as Malaysia, Epstein–Barr virus infection occurs in early childhood in the vast majority of the population, so infectious mononucleosis is rare in adolescents. Such individuals suspected of having infectious mononucleosis are thus more likely to have toxoplasmosis.[36]

Diagnosis

The differential diagnosis of lymphadenopathy, with or without systemic symptoms, is extensive: infectious mononucleosis, cytomegalovirus, cat scratch disease, tuberculosis, other viruses (for example herpes simplex, adenovirus), lymphoma, leukaemia, other malignancies and autoimmune disease.[1,14] Among the infections, Epstein–Barr virus and cytomegalovirus infections are probably the most common in the UK among patients who are investigated. Among those patients who are biopsied, the most common malignancies are probably Hodgkin's disease and non-Hodgkin's lymphoma.[37] Most patients with toxoplasma lymphadenopathy are not investigated by biopsy; serological diagnosis is easier. The histological appearance of the lymph node during toxoplasma infection is considered in Chapter 4. Of patients with lymphadenopathy who have a biopsy, the diagnosis of toxoplasmosis, often with the aid of serology is made in 85 per cent.[23] This study also concluded that patients with lymph node hyperplasia have a longer history and relative lymphocytosis is less common; patients with Hodgkin's disease had more systemic symptoms, especially weight loss and anaemia.[23] Fortunately, most patients do not require treatment (Chapter 5).

Ocular disease

The first description of toxoplasma ocular disease was a good indication of its future. Probably the first description was in 1923 by a Czech ophthalmologist, Dr J. Jankû, in the retina of an infant with congenital toxoplasmosis;[38] but as the manuscript was in Czech, it was not noticed by most of the world. Today, nearly 70 years later, toxoplasma eye disease remains an enigma. The dilemma is simple: is toxoplasma ocular disease a result of congenital or acquired infection?

Congenital or acquired infection

Although the early description showed that the organism was in the retina, the clinical picture is usually of uveitis[39] and consequently pathologists have often unsuccessfully concentrated on investigations of the uveal tract. The realization that retinitis is the principal finding has resulted in 'retinochoroiditis' replacing the term 'chorioretinitis' when the aetiogical agent is believed to be toxoplasma. Unfortunately, this is of little help in deciding if ocular disease is congenital or acquired. Some researchers have concluded that 2/100, 2 per cent of cases are congenital;[40] whereas, others have felt that 61/63, 97 per cent of cases were congenital.[41] These disparate findings are difficult to reconcile. In addition, the absence of a retinochoroidal scar does not exclude congenital toxoplasmosis, as toxoplasma cysts have been seen in the retina when tissue reaction was minimal and not recognizable clinically.[39]

The question is not resolved by consideration of epidemiological information. In a South Pacific island, it was found that 90 per cent of the population over 22 years had dye test titres of more than 1:16 and yet no case of congenital or ocular toxoplasmosis was found.[42] It was concluded that mothers had been infected in childhood and thus did not transmit infection to their children. This seems like a good story, but totally different epidemiological information is found in Africa. In west Africa, it appears that there is a high prevalence of antibody and a high incidence of the ocular form of the disease.[43,44] One important consideration is that these last two studies were done on patients from west Africa who were in the USA and UK. Thus, these studies may be biased. Nevertheless, there is strong serological evidence of past (possibly congenital) infection in patients with toxoplasma uveitis compared with controls.[39,43] Even in a study from west Africa (Nigeria) 95.4 per cent patients with uveitis had toxoplasma antibodies compared with 72.2 per cent of controls.[45] Obviously in such a situation, the serological result for the individual patient is of limited diagnostic value.

The main difficulty is in diagnosis. Acquired ocular infection can be easily diagnosed when there has been a seroconversion or high dye test, no evidence of old retinal scars, and when the illness is accompanied by other systemic symptoms of toxoplasmosis.[6,39] Unfortunately there are many cases without these findings. Firstly, patients are usually not seen early enough for a seroconversion to be demonstrated, and acquired infection does not always result in a high dye test. Secondly, the retina may be infected without the presence of scars;[39] and an old scar could have resulted from acquired infection earlier in life, but after birth or from another infection such as cytomegalovirus. Lastly in some patients, retinochoroiditis may present many months after the systemic infection.[6] Thus it is probable that many cases are wrongly attributable to congenital infection. A possible explanation may be found in the argument that as toxoplasma antibodies increase with age, ocular toxoplasmosis should also increase if it is due to acquired infection.[6] As the peak for ocular disease is 10–35 years, it is more likely to be due to congenital infection.[6] This argument does not consider all the facts. Large studies of ocular disease show that new cases continue to arise up to the age of 80 years.[43] When this latter finding is coupled with difficulties in

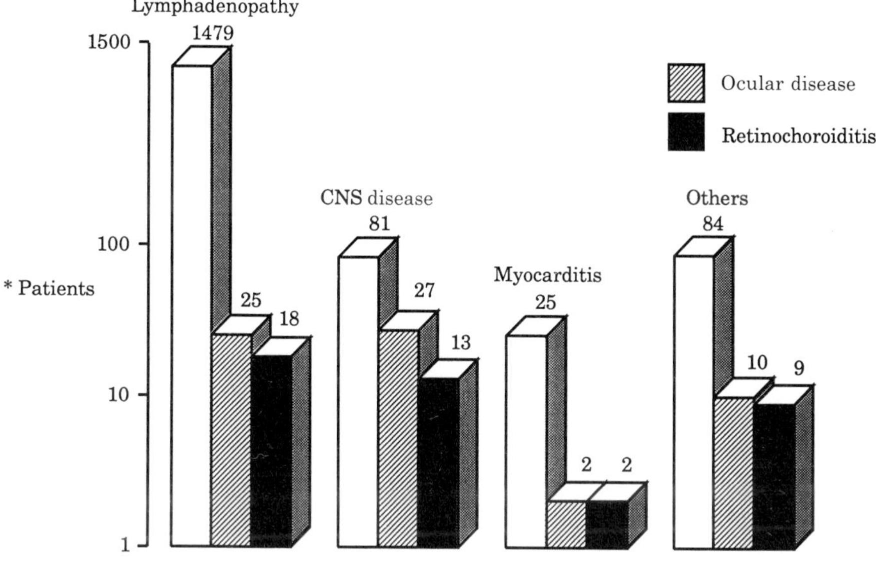

Fig. 3.4   Incidence of ocular disease and retinochoroiditis in patients with acquired toxoplasmosis (adapted from Perkins[39]).

diagnosis, a possible explanation of the aetiology of ocular toxoplasmosis is that there is a relatively constant acquired ocular disease throughout life, but a peak of congenital ocular disease in the first half of life.

Acquired infection

The review of the literature by Perkins[39] was meticulous, extensive and informative. He clearly showed that ocular disease was not a common manifestation of acquired toxoplasma infection. Information was obtained by considering 1669 cases of acquired systemic toxoplasmosis in the literature up to 1973 and determining the relative frequency of ocular disease and retinochoroiditis. His results are highly likely to be a true representation of acquired infection and they have been converted to a graphic presentation in Fig. 3.4. It shows that the most diagnosed toxoplasma manifestation, lymphadenopathy, is rarely associated with ocular disease (1.7%) or retino-choroiditis (1.2%). Ocular disease and retinochoroiditis were most commonly found in patients with central nervous system manifestations, however these patients were a rare presentation (81/1669, 4.9%) of acquired systemic toxoplasmosis. Perkins extends his observations to say that as ocular disease was rare in patients with systemic toxoplasmosis, it is likely that such complications were even less common in subclinical toxoplasmosis.[39]

Ocular disease as the sole manifestation of acquired toxoplasmosis is difficult to prove. Isolation of the organism and lack of previous retinochoroidal scars may also be associated with congenital infection.[39] Perhaps the presence of specific IgM for toxoplasma and a high dye test titre are the best findings in favour of acquired disease. With our current knowledge diagnosis of acquired disease is not possible in the absence of these findings. Thus, one should be cautious about being definite that such cases are either congenital or acquired infection.

Clinical features

With most illnesses, it is usual to separate the clinical presentation into congenital and acquired disease. With current knowledge (as shown above), to say that the vast majority of cases with ocular disease are of congenital origin,[43] is probably premature. Hopefully, as diagnostic techniques improve, it will be possible to determine those cases which are acquired. In the meantime, some cases are obviously congenital (as a result of diagnosis *in utero* or in the neonate), and the clinical features are considered in Chapter 6. For others, it is pragmatic to consider them together without differentiation into acquired or congenital.

Ocular toxoplasmosis principally involves the uveal tract[39,43] which consists of the choroid (providing the vascular supply to the outer layers of the retina and the retinal pigment epithelium), ciliary body (muscle of accommodation and site of formation of aqueous humour) and the iris (which form the pupillary aperture). The relative positions of these structures are shown in Fig. 3.5. As the uveal tract is a continuous layer, inflammation can easily affect the whole uveal tract (panuveitis). Nevertheless, clinically the distinction is often made between anterior uveitis (iritis or iridocyclitis) and posterior uveitis (choroiditis). The relative frequency of all manifestations of uveitis are: anterior uveitis (63%), posterior uveitis (19%) and panuveitis (18%), with acute infections being more common except in panuveitis where chronic infection predominates.[46] When ocular toxoplasmosis is considered a different pattern emerges: with posterior uveitis (64%) being most common; accompanied by panuveitis (12%), papillitis producing optic atrophy (12%) , conjunctivitis (3%) and others (10%).[39] Focal retinochoroiditis is the major lesion (Fig. 3.6). This may be seen as a healed lesion, or as a new area of activity often at the edge of a previous scar.[43] Although lesions are variable in size, they are usually posterior to the equator, unilateral and single.[43] A disseminated retinochoroiditis may be due to toxoplasma, but this is very unusual. Most lesions appear along a main vessel and this is consistent with haematogenous spread. These findings are consistent with the teaching that lesions are congenital when bilateral and acquired when unilateral, however, this theory may reflect only the severity of the infection. The condition of juxtapapillary retinochoroiditis (Jensen's choroiditis) probably does not warrant separate consideration and is best regarded as a variant of ocular

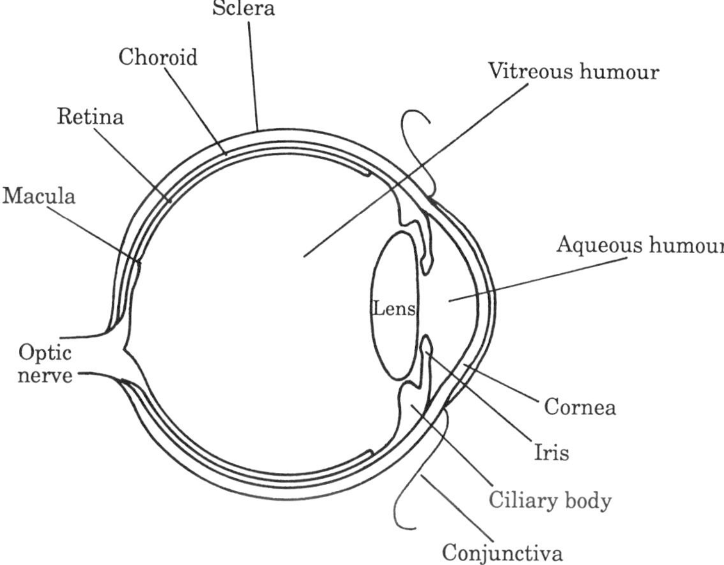

Fig. 3.5 Diagrammatic representation of the eye.

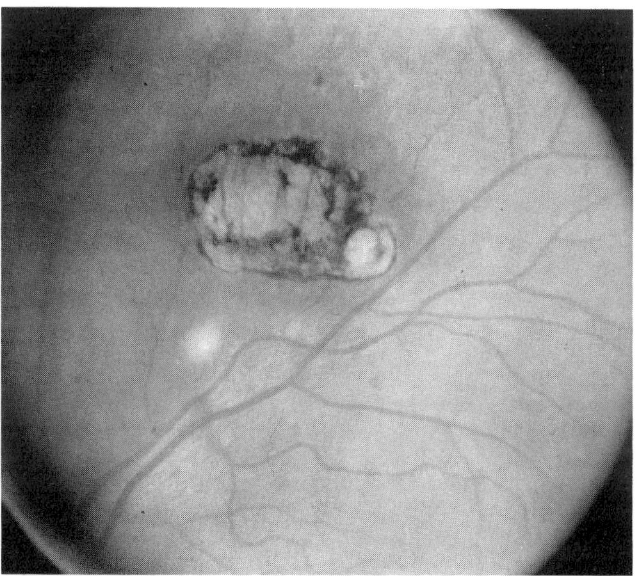

Fig. 3.6 Toxoplasma retinochoroidal scar. There is a sharply circumscribed area of choroidal and retinal pigment epithelial defect with scattered pigment clumping due to damaged pigment epithelial cells.

toxoplasmosis.[43,46] Observations of congenital toxoplasmosis have led to the belief that the macula is nearly always involved.[43,46] Obviously, macular involvement leads to early patient referral because of symptoms, so studies may be biased in favour of such lesions. Nevertheless one good study of uveitis cases suggests that macular lesions occur alone in 30 per cent of cases and in combination with other lesions in 41 per cent of cases.[47]

Inflammation of the optic nerve (papillitis) with the production of optic atrophy are the next important conditions. One problem is that clinicians tend to describe such inflammation of the optic nerve as papillary oedema as there is no loss of vision and a swelling is observed, whereas pathologists label the same process as papillitis as inflammatory cells are seen.[43] Usually papillitis producing optic atrophy is found in patients with disease of the central nervous system.[39] Vitreous opacities as a result of inflammatory cells in the vitreous humour are seen in about two thirds of cases of ocular toxoplasmosis, and their absence in the remaining one-third probably suggests old, inactive disease.[47] The commonest presentation of uveitis is anterior uveitis (iridocyclitis),[46] however, this condition is uncommon in ocular toxoplasmosis. The organism has not been isolated from the human iris or ciliary body; nevertheless, it is customary to regard agranulomatous iridocyclitis as due to toxoplasma if there is retinochoroiditis and serological evidence of toxoplasma infection.[43] There seems little good evidence to link Eales's disease or geographic choroiditis to toxoplasma infection.[39]

The outlook for patients with ocular toxoplasmosis is not straightforward. On average, toxoplasmic retinochoroiditis lasts 4 months, but when there is visual loss the duration of illness is 7 months.[41] However, there are considerable variations with attacks lasting as little as 2 weeks or as long as 2 years. Although each attack settles with time, the effect of retinochoroidal scarring may leave the patient with a visual defect which may be particularly significant if lesions are central. In one series, 38 per cent of patients had one attack, but the average was 2.7 episodes.[41] The difficulty with ocular toxoplasmosis is that the organism is likely to encyst after an attack and can be reactivated at a later date if there is immunocompromise. This is further discussed in Chapter 7.

Myalgic encephalomyelitis

There has been much public interest in this illness which is also known as post viral fatigue syndrome, chronic fatigue syndrome and Royal Free disease.[48] Cardinal features of the syndrome are: generalized, relapsing fatigue for at least 3 months; complaints of prominent disturbance of concentration and/or short term memory impairment; and the exclusion of any obvious organic cause for a similar illness.[49] The patients are usually previously well, most remember an initiating pyrexial illness and muscle pain is not invariably present.[49] This syndrome is a frequent complication of infectious mononucleosis.[48] However the association of toxoplasma infection with myalgic encephalomyelitis has only recently been made, probably because many cases of infectious mononucleosis are diagnosed clinically (and believed to be due to Epstein–Barr virus infection), and toxoplasma is only a rare cause of an infectious mononucleosis-like illness (Fig. 3.2). Although, both malaise (54%) and myalgia (43%) can be common symptoms in toxoplasma infection (Table 3.2), the diagnosis of myalgic encephalomyelitis also requires persistence of symptoms for more than 3 months and disturbance in concentration and/or memory. In patients with toxoplasma lymphadenopathy persisting symptoms were present in 5/30 (17%) cases,[24] whilst in a study of patients with myalgic encephalomyelitis, 2/50 (4%) were believed to have the illness as a result of toxoplasma infection.[2] Therefore toxoplasmosis can be regarded as an uncommon cause of myalgic encephalomyelitis. When it does happen, it is a result of a primary infection in which myalgic encephalomyelitis is a complication. Fortunately, such patients appear to have a better prognosis than if the illness were due to a viral infection.[2] Nevertheless, specific anti-toxoplasma therapy is not indicated and patients should be managed as for myalgic encephalomyelitis.[49]

Other manifestations

Other manifestations of toxoplasma infection may present as part of a disseminated toxoplasmosis, or, more rarely, occur alone. These latter descriptions vary in frequency and their precise incidence is difficult to ascertain.

Encephalitis

The first confirmation that toxoplasma produced encephalitis was when the organism was propagated in animals from post-mortem material.[50] However, this and the previously described cases[50] were in the very young and probably represented congenital infection. Encephalitis as a complication of congenital infection is described in Chapter 6. Similarly, encephalitis as a result of immunocompromise,[51] and especially associated with the acquired immune-deficiency syndrome[52] (AIDS) are considered in Chapter 7. There is no doubt that in the last decade there has been a great increase in toxoplasma encephalitis, principally as a result of immunosuppression, either as part of medical treatment or related to human immunodeficiency virus infection.[52]

A superb review by Dr Townsend and colleagues allows acquired toxoplasma encephalitis to be put into perspective.[53] These researchers reported the nervous system manifestations of six cases and reviewed 39 cases from the literature. They found that primary acquired toxoplasma infection accounted for nearly half of the cases (22/45, 49%). There appeared to be three separate clinical syndromes: diffuse encephalitis, with and without seizures (9/45, 20%); meningoencephalitis (12/45, 27%); and single or multiple progressive mass lesions (24/45, 53%). Individuals with acquired infection were equally likely to develop any of the three syndromes. An interesting observation was that the two patients with laboratory-acquired infection both had a diffuse encephalitis. Diagnosis was very difficult and was not made before death in 2/23 immunocompromised patients. These findings are consistent with those showing difficulties in diagnosis among AIDS patients.[52] Nevertheless there was a cerebrospinal fluid pleocytosis in 27/34 cases and brain biopsy allowed identification of the organism in 4/5 cases.[53] The outlook is dramatically changed when appropriate treatment is given and 13/14 of those who received such treatment did well.

Hepatitis

The frequency of this complication is unknown. In one outbreak, hepatitis was found in 4/37 (11%);[7] whilst in another, abnormalities of liver function were detected in 31/35 (89%).[8] Strain variations may explain these discrepancies, and although liver function may be tested in an outbreak, it may not be investigated in other patients unless they have jaundice. One clinical entity appears to be the patient without lymphadenopathy, but with fever and jaundice and when hepatitis A is the suspected diagnosis.[35,54] In such patients it is possible for lymphadenopathy to develop some weeks later. In cases where hepatitis and lymphadenopathy present simultaneously, the diagnosis of infectious mononucleosis is usually suspected. The last variant is where the patient initially has lymphadenopathy but develops hepatitis in the course of a protracted illness.[35] In all patients with acquired toxoplasma hepatitis, the value of specific treatment is not established and some patients may recover quickly without treatment.[35]

Cardiac complications

Cardiac toxoplasma complications of congenital infection or immunocompromise are well recognized.[55] Nevertheless, with acquired toxoplasmosis in the immunocompetent, cardiac complications can be important to establish. The usual lesions are myocarditis and pericarditis[55] but endocarditis (as part of a pancarditis) has also been described.[56] Obviously, generalized toxoplasmosis may result in localized cysts in the heart and other tissues of the body. The problem is that such cysts have also been found in healthy individuals following accidental death.[55] When specific toxoplasma IgM is present and there is a response to therapy,[56] the diagnosis is acceptable. However, when there are only normal or moderate dye test titres,[55] the diagnosis may be difficult to make. An interesting observation is that toxoplasma may produce a huge pericardial effusion lasting several years.[57] In these two cases, toxoplasma organisms were seen in the pericardial fluid and isolated in mice in one case. These authors[57] suggest that toxoplasma may be an important causative agent in chronic pericardial effusion where almost half the cases are without an aetiological diagnosis. In cases with familiar cardiac lesions, the possibility of a strain of toxoplasma with cardiac affinity should be considered.[56] Although there are not many cases in the literature, prognosis appears good when the appropriate treatment is given.[55-57]

Muscle and skin

With an acute toxoplasma infection, it is not too surprising that there may be widespread involvement of muscle and skin after the systemic distribution of tachyzoites.[58,59] Of greater interest, is the association of toxoplasma infection with polymyositis, a rare disease of unknown aetiology.[60] The problem is that in most cases of polymyositis diagnosis is difficult as there is no evidence of acute infection. In addition, as polymyositis is associated with immunocompromise and many patients have immunosuppressive therapy, it is suggested that most cases are due to reactivated infection.[60] This review concluded that toxoplasmosis was a complication of the disease and that appropriate systemic anti-toxoplasma therapy may result in amelioration of the disease, but not usually a cure. The sequential development of myositis with acute infection, a fall of antibodies with treatment and the progression to polymyositis is instructive.[58] It suggests that in some cases there may be altered antigenic stimulation (as a result of toxoplasma infection) as well as a disturbed immune response in this disease. Thus, most cases may result from reactivated infection but some may be due to a late complication of acute infection.

Cutaneous manifestations may occur separately or with muscle involvement as in dermatomyositis.[59] As with most infections the demonstration of toxoplasma infection at the same time as skin disease does not prove a causal relationship. With severely ill patients, especially with systemic disease, the parasite may be found in tissues as a result of haematogenous spread, and may thus be an innocent bystander. Nevertheless, when there is a local inflammatory response and necrotic foci containing free toxoplasma tachyzoites,[38] the evidence for a causal role for toxoplasma appears strong. Toxoplasma infection is associated with a wide range of common (maculopapular rash, urticaria, pruritus, erythema multiforme, purpura, ulcers) and rare skin conditions (lichen spinulosus, dermohypodermitis, lichen parapsoriasis).[59] Other associations, such as with erythema chronicum migrans (now known to be caused by *Borrelia burgdorferi*) and Behcet disease may be fortuitous. Thus, the available evidence suggests that skin manifestations as a result of toxoplasma infection are uncommon, but may be more likely with systemic disease and in the immunocompromised (Chapter 7). Among the immunocompetent, it is critical that there is a distinction between coincident toxoplasma infection with a skin disease and a causative role of toxoplasma.

Miscellaneous conditions

Lung disease can be an important complication of toxoplasmosis. In 1963 it was suggested that toxoplasma may be an important cause of atypical pneumonia, but toxoplasma was forgotten with the demonstration of *Myco-plasma pneumoniae* as a major cause of this condition.[61] Twenty-five years later, the prominence of the immunocompromised has raised the awareness of toxoplasma infection of the lung as the second most affected organ at autopsy.[62] In the immunocompetent, toxoplasma lung disease is rare and probably underdiagnosed. Most patients may be breathless and pyrexial but cough and sputum are usually absent, and radiological evidence of disease is usually non-specific.[62] Diffuse, bilateral infiltrates are usual but there may occasionally be one or more lobes involved.

Kidney disease is rare in acquired toxoplasma infection. The association of acute nephritis and nephrotic syndrome with congenital toxoplasmosis[63] might suggest that renal disease should be more commonly found. One difficulty may be linking the renal disease with toxoplasma infection. Although it was possible to link a case of nephritis with acute toxoplasmosis and demonstrate immune-complexes in the renal biopsy, it was only possible to demonstrate toxoplasma antigen in the first biopsy.[64] As in immune-complex nephritis the offending antigen may only be found in early disease and as toxoplasma is not usually looked for, the true incidence of the infection is unknown.

A variety of other conditions have been incidentally found with toxoplasma infection. To what extent toxoplasma is a rare cause of these conditions, as opposed to a chance occurrence, is difficult to ascertain. Thus, thrombocytopenic purpura, a condition which is usually idiopathic, has been associated with toxoplasmosis.[65] Similarly, other conditions such as Guillain–Barré Syndrome, polyneuritis, facial palsy, and haemolytic anaemia may be fortuitous associations. Nevertheless, some associations are convincing, as in the case of polyarthritis which did not respond to aspirin but was cured with anti-toxoplasma therapy.[66]

SUMMARY

Acquired toxoplasma infection is often unrecognized as many medical practitioners are not aware of this infection. The incubation period probably varies with the infectious dose, the toxoplasma strain, the stage of the parasite, the route of infection, and the individual infected. For sporadic cases, the usual incubation period is 10–14 days. Although many textbooks state that most infections are asymptomatic, this is probably untrue. An analysis of outbreaks of toxoplasmosis shows that most infections are symptomatic, but that the symptoms are frequently non-specific. Symptoms such as pyrexia, malaise, and myalgia have many causes and a toxoplasma aetiology is rarely considered. Nevertheless, asymptomatic illness may be present in a quarter of infections.

Lymphadenopathy is possibly the most easily recognizable manifestation of acquired toxoplasmosis. This manifestation appears to be more common in boys and young women, but in both sexes the 21–25 year age group is most affected. The neck lymph nodes are firm, unattached to overlying skin, and initially tender. Although most become normal within one month, 14 per cent of patients take more than 4 months. Toxoplasma may produce an infectious mononucleosis-like illness and causes about 5 per cent of such cases. The most important differential diagnosis is Epstein–Barr virus infection, but difficulty in making the diagnosis is not common.

Ocular toxoplasmosis is perhaps the most common manifestation of toxoplasma infection, however, the question of whether this manifestation is due to congenital or acquired infection remains to be answered. It is possible to show that many cases are congenital and some are acquired; however, for the majority of patients, current diagnostic tests cannot differentiate congenital from acquired infection. The future will probably show that there is relatively constant acquired ocular disease throughout life, but a peak of congenital ocular disease in the first half of life. Ocular disease principally involves the uveal tract, especially posterior uveitis. Retinochoroiditis is the most common lesion and lasts an average of 4 months (range 2 weeks to 2 years), and patients have an average of 2.7 attacks.

Myalgic encephalomyelitis is a recently recognized complication of toxoplasmosis and appears to have a better prognosis than if this illness is associated with other infections. Other manifestations of toxoplasma infection are uncommon and their precise frequency is difficult to determine. Encephalitis is particularly common among the immunocompromised. Hepatitis is probably more frequent than is recognized. Cardiac, muscle, and skin involvement have important consequences and test the acumen of the clinician. Lung, kidney, and other manifestations are clinical curiosities. Sadly, the clinical features of acquired toxoplasmosis are not diagnostic and diagnosis depends mainly on the awareness of the condition by the clinician.

REFERENCES

1 Remington, J. S. (1974). Toxoplasmosis in adults. *Bull. N. Y. Acad. Med.*, **50**, 211–27.
2 Ho-Yen, D. O. (1990). Toxoplasmosis in humans: discussion paper. *J. R. Soc. Med.*, **83**, 571–2.
3 Ho-Yen, D. O. and Chatterton, J. M. W. (1990). Congenital toxoplasmosis — why and how to screen. *Rev. Med. Microbiol.*, **1**, 229–35.
4 Krick, J. A. and Remington, J. S. (1978). Toxoplasmosis in the adult — an overview. *N. Engl. J. Med.*, **298**, 550–3.
5 Pendry, K., Tait, R. C., McLay, A., Ho-Yen, D. O., Baird, D., and Burnett, A. K. (1990). Toxoplasmosis after BMT for CML. *Bone Marrow Transplant.*, **5**, 65–7.
6 Masur, H., Jones, J. C., Lempert, J. A., and Cherubini, T. D. (1978). Outbreak of toxoplasmosis in a family and documentation of acquired retinochoroiditis. *Am. J. Med.*, **64**, 396–402.
7 Teutsch, S. M., Juranek, D. D., Sulzer, A., Dubey, J. P., and Sikes, R. K. (1979). Epidemic toxoplasmosis associated with infected cats. *N. Engl. J. Med.*, **300**, 695–9.
8 Benenson, M. W., Takafuji, E. T., Lemon, S. M., Greenup, R. L., and Sulzer, A. J. (1982). Oocyst-transmitted toxoplasmosis associated with ingestion of contaminated water. *N. Engl. J. Med.*, **307**, 666–9.
9 Sacks, J. J., Roberto, R. R., and Brooks, N. F. (1982). Toxoplasmosis infection associated with raw goat's milk. *JAMA*, **248**, 1728–32.
10 Rawal, B. D. (1959). Laboratory infection with toxoplasmosis. *J. Clin. Pathol.*, **12**, 59–61.
11 Edwards, J. M. B., Kessel, I., Gardner, S. D., Eaton, B. R., Pollock, T. M., Fleck, D. G. *et al.* (1981). A search for a characteristic illness in children with serological evidence of viral or toxoplasma infection. *J. Infect.*, **3**, 316–23.
12 Stagno, S., Dykes, A. C., Amos, C. S., Head, R. A., Juranek, D. D., and Walls, K. (1980). An outbreak of toxoplasmosis linked to cats. *Paediatrics*, **65**, 706–11.
13 Fleck, D. G. (1963). Epidemiology of toxoplasmosis. *J. Hyg.*, **61**, 61–5.
14 Christie, A. B. (1987). *Infectious diseases*. Churchill Livingstone, Edinburgh.
15 Kwantes, W. (1987). Toxoplasmosis. In *Oxford textbook of medicine*. (ed. D. J. Weatherall and J. G. G. Ledingham), pp. 5.507–9. Oxford University Press, Oxford.
16 Luft, B. J. and Remington, J. S. (1984). Acute toxoplasma infection among family members of patients with acute lymphadenopathic toxoplasmosis. *Arch. Intern. Med.*, **144**, 53–6.
17 Kean, B. H., Kimball, A. C., and Christenson, W. N. (1969). An epidemic of acute toxoplasmosis. *JAMA*, **208**, 1002–4.
18 Magaldi, C., Elkis, H., Pattoli, D., and Coscina, A. L. (1969). Epidemic of toxoplasmosis at a university in Sao-Jose-des-Campos, S. P.-Brazil. *Rev. Lat-Am. Microbiol. Parasitol.*, **11**, 5–13.
19 Feldman, H. A. (1974). Toxoplasmosis: an overview. *Bull. N. Y. Acad. Med.*, **50**, 110–27.
20 Gard, S. and Magnusson, J. H. (1951). Glandular form of toxoplasmosis in connection with pregnancy. *Acta Med. Scand.*, **141**, 59–64.
21 Siim, J. C. (1956). Toxoplasmosis acquisita lymphonodosa: clinical and pathological aspects. *Ann. N. Y. Acad. Sci.*, **64**, 185–206.
22 Beverley, J. K. A., Fleck, D. G., Kwantes, W., and Ludlam, G. B. (1976). Age–sex distribution of various diseases with particular reference to toxoplasmic lymphadenopathy. *J. Hyg.*, **76**, 215–28.
23 Miettinen, M., Saxen, L., and Saxen, E. (1980). Lymph node toxoplasmosis. *Acta Med. Scand.*, **208**, 431–6.
24 Beverley, J. K. A. and Beattie, C. P. (1958). Glandular toxoplasmosis. A survey of 30 cases. *Lancet*, **ii**, 379–84.
25 Rafaty, F. M. (1977). Cervical adenopathy secondary to toxoplasmosis. *Arch. Otolaryngol.*, **103**, 547–9.

26 Tentunen, A. (1964). Glandular toxoplasmosis: occurrence of the disease in Finland. *Acta Pathol. Microbiol. Scand., Suppl.* **172**.

27 Lelong, M., Bernard, J., Desmonts, G., and Couvreur, J. (1960). La toxoplasmose acquise. Etude de 227 observations. *Arch. Fr Pédiatr.*, **17**, 281–331.

28 Fry, J. (1983). Present state and future needs in general practice. *Royal College General Practitioners*, London.

29 Gray Jr, G. F., Kimball, A. C., and Kean, B. H., (1972). The posterior cervical lymph node in toxoplasmosis. *Am. J. Pathol.*, **69**, 349–58.

30 Ho-Yen, D. O., Pinkerton, I. W., and Reid, D. (1984). Infectious mononucleosis. *Update*, **28**, 184–92.

31 Ho-Yen, D. O. and Martin, K. W. (1981). The relationship between atypical lymphocytosis and serological tests in infectious mononucleosis. *J. Infect.*, **3**, 324–31.

32 Remington, J. S., Barnett, C. G., Meikel, M., and Lunde, M. N. (1962). Toxoplasmosis and infectious mononucleosis. *Arch. Intern. Med.*, **110**, 744–53.

33 Thomaidis, T., Anastassea-Vlachou, K., Mandalenaki-Lambrou, C., Theodoridis, C., and Vrahnou, E. (1977). Chronic lymphoglandular enlargement and toxoplasmosis in children. *Arch. Dis. Child.*, **52**, 403–7.

34 Chang, R. S. (1980). *Infectious mononucleosis*. Hall, Boston.

35 Weitberg, A. B., Alper, J. C., Diamond, I., and Fligiel, Z. (1979). Acute granulomatous hepatitis in the course of acquired toxoplasmosis. *N. Engl. J. Med.*, **300**, 1093–6.

36 Tan, D. S. K., Zaman, V., and Lopes, M. (1978). Infectious mononucleosis or toxoplasmosis? *Med. J. Malays.*, **33**, 23–5.

37 Miettinen, M. and Franssila, K. (1982). Malignant lymphoma simulating lymph node toxoplasmosis. *Histopathology*, **6**, 129–40.

38 Pinkerton, H. and Weinman, D. (1940). Toxoplasma infection in man. *Arch. Pathol.*, **30**, 374–92.

39 Perkins, E. S. (1973). Ocular toxoplasmosis. *Br. J. Ophthalmol.*, **57**, 1–17.

40 Chodos, J. B. and Habegger-Chodos, H. E. (1963). An epidemiologic study of 100 cases of toxoplasmic uveitis: implications for diagnosis and prevention. *Can. Med. Assoc. J.*, **88**, 505–11.

41 Friedman, C. T. and Knox, D. L. (1969). Variations in recurrent active toxoplasmic retinochoroiditis. *Arch. Ophthalmol.*, **81**, 481–93.

42 Darrell, R. W., Pieper, S., Kurland, L. T., and Jacobs, L. (1964). Chorioretinopathy and toxoplasmosis. An epidemiologic study on a south Pacific island. *Arch. Ophthalmol.*, **71**, 63–8.

43 Schlaegel, T. F. (1978). *Ocular toxoplasmosis and pars planitis*. Grune and Stratton, New York.

44 Perkins, E. S. (1976). Epidemiology of uveitis. *Trans. Ophthalmol. Soc. UK*, **96**, 105–7.

45 Olurin, O., Fleck, D. G., and Osuntokun, B. (1972). Toxoplasmosis and chorioretinitis in Nigeria. *Trop. Geogr. Med.*, **24**, 240–5.

46 Perkins, E. S. (1961). *Uveitis and toxoplasmosis*. Churchill, London.

47 Hogan, M. J., Kimura, S. J., and O'Connor, G. R. (1964). Ocular toxoplasmosis. *Arch. Ophthalmol.*, **72**, 592–600.

48 Ho-Yen, D. O. (1987). *Better recovery from viral illnesses*. Dodona Books, Inverness.

49 Ho-Yen, D. O. (1990). Patient management of post-viral fatigue syndrome. *Br. J. Gen. Pract.*, **40**, 37–9.

50 Wolf, A., Cowen, D., and Paige, B. H. (1939). Toxoplasmic encephalomyelitis III. A new case of granulomatous encephalomyelitis due to a protozoan. *Am. J. Pathol.*, **15**, 657–95.

51 Lancet. (1984). Toxoplasmosis diagnosis and immunodeficiency. **i**, 605–6.

52 Holliman, R. E. (1988). Toxoplasmosis and the acquired immune deficiency syndrome. *J. Infect.*, **16**, 121–8.

53 Townsend, J. J., Wolinsky, J. S., Baringer, J. R., and Johnson, P. C. (1975). Acquired toxoplasmosis. A neglected cause of treatable nervous system disease. *Arch. Neurol.*, **32**, 335–43.

54 Vethanyagam, A. and Bryceson, A. D. M. (1976). Acquired toxoplasmosis presenting as hepatitis. *Trans. R. Soc. Trop. Med. Hyg.*, **70**, 524–5.

55 Theologides, A. and Kennedy, B. J. (1969). Toxoplasmic myocarditis and pericarditis. *Am. J. Med.*, **47**, 169–74.

56 Cunningham, T. (1982). Pancarditis in acute toxoplasmosis. *Am. J. Clin. Pathol.*, **78**, 403–5.

57 Sagrista-Sauleda, J., Permanyer-Miralda, G., Juste-Sanchez, C., Buen-Sanchez, M. L., Pujadas-Capmany, R., Arcalis-Arce, L. *et al.* (1982). Huge chronic pericardial effusion caused by *Toxoplasma gondii. Circulation*, **66**, 895–7.

58 Adams, E. M., Hafez, G. R., Carnes, M., Weisner, J. K., and Grazians, F. M. (1984). The development of polymyositis in a patient with toxoplasmosis: clinical and pathological findings and a review of the literature. *Clin. Exp. Rheumatol.*, **2**, 205–8.

59 Binazzi, M. (1986). Historical aspects of cutaneous toxoplasmosis. *Int. J. Dermatol.*, **25**, 401–4.

60 Behan, N. M. H., Behan, P. O., Draper, I. T., and Williams, H. (1983). Does toxoplasma cause polymyositis? *Acta Neuropathol.*, **61**, 246–52.

61 Ludlam, G. B. and Beattie, C. P. (1963). Pulmonary toxoplasmosis? *Lancet*, **ii**, 1136–8.

62 Caterall, J. R., Hofflin, J. M., and Remington, J. S. (1986). Pulmonary toxoplasmosis. *Am. Rev. Respir. Dis.*, **133**, 704–5.

63 Wickborn, B. and Winberg, J. (1972). Coincidence of congenital toxoplasmosis and acute nephritis with nephrotic syndrome. *Acta Paediatr. Scand.*, **61**, 470–2.

64 Ginsburg, B. E., Wasserman, J., Huldt, G., and Bergstrand, A. (1974). Case of glomerulonephritis associated with acute toxoplasmosis. *Br. Med. J.*, **3**, 664–5.

65 Diamant, S. and Spirer, Z. (1980). Chronic thrombocytopenic purpura associated with toxoplasmosis. *Br. Med. J.*, **280**, 1505–6.

66 Gernou, V., Messanitakis, J., Karpathios, T., and Kingo, A. (1983). Chronic polyarthritis of toxoplasmic etiology. *Helv. Paediatr. Acta*, **38**, 295–6.

4

Diagnosis

ALEX W. L. JOSS

Diagnosis of toxoplasmosis is not straightforward. There are no obvious clinical characteristics which are unique or predominantly specific to the disease. Indeed, many infections are asymptomatic, or the symptoms so mundane that they are overlooked.[1] This has serious implications for the pregnant woman because it does not allow an easy way to diagnose potential congenital infections. Diagnosis of newly acquired infection relies on good laboratory tests. Laboratory tests are less helpful when disease recurs, usually by reactivation of latent infection. For such patients, whether immunocompromised with toxoplasmosis affecting vital organs or immuno-competent with ophthalmic toxoplasmosis, the clinical diagnosis assumes greater significance (Fig. 4.1).

The mainstay of laboratory diagnosis is serological testing for evidence of specific antibody. With an appropriate combination of tests it is usually possible to distinguish between current and past infection. In healthy patients, acute, fulminant toxoplasmosis is very rare, so diagnosis is only usually considered some time after the initial illness when there is an antibody response. Demonstrating the presence of the parasite itself in various body tissues is possible. It is more invasive and the presence of *Toxoplasma gondii* cysts would not necessarily relate to current infection.[2] It is therefore not often attempted except during reactivated infection in the immunocompromised when serological tests are largely inadequate. This chapter will assess the various diagnostic tests, and consider how appropriate each is in different clinical situations. It will identify the areas in which there are shortcomings in laboratory diagnosis and it will point to new techniques which may improve diagnosis.

In healthy adults, lymphadenopathy is the commonest indication for suspecting toxoplasmosis. Symptoms are of an infectious mononucleosis-like illness[3] (Chapter 3). Blood samples for serological testing provide the means of distinguishing it from infections which produce similar symptoms. Toxoplasmosis in some instances can also be confused with malignancy, such as lymphoma,[4] especially as lymphadenopathy may persist for months (Chapter 3). Other persisting symptoms are fatigue and muscle pain.[3] Toxoplasma may be considered in the diagnosis of retinochoroiditis,[5] hepatitis,[6] and encephalitis[7] if they occur without more obvious causes. An early diagnosis becomes more essential in pregnant or immunocompromised patients. Unfortunately, the signs of infection are not any clearer. The features of toxoplasmosis in pregnancy and the neonate and the necessary

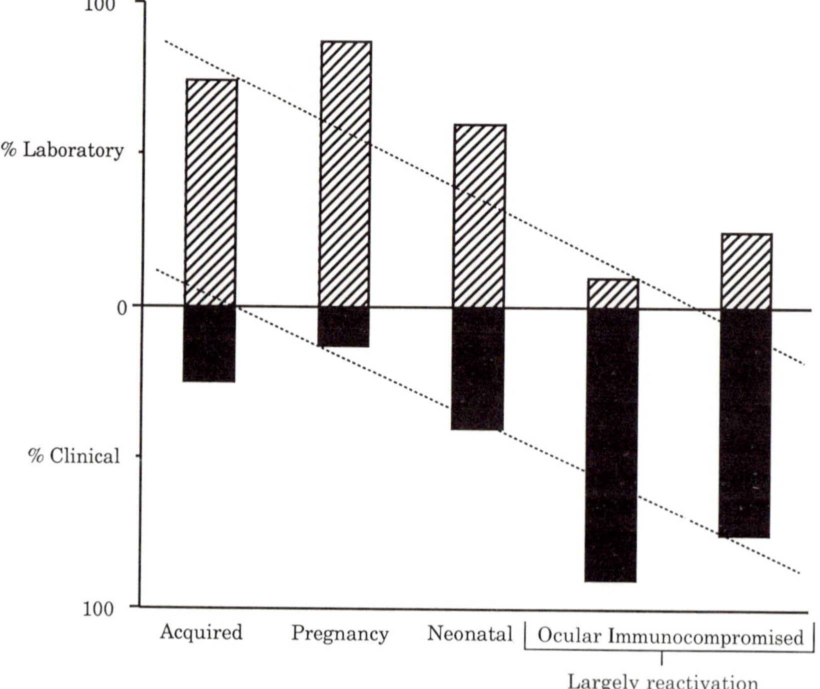

Fig. 4.1 Types of infection currently amenable to laboratory diagnosis, with percentage of laboratory diagnosis above the line, and percentage clinical diagnosis below the line.

investigations to confirm it are described in Chapter 6. In patients with acquired immune-deficiency syndrome (AIDS) or patients subjected to immunosuppression during transplant or cancer treatment, infection may disseminate affecting a wider range of organs. The more unusual clinical manifestations such as encephalitis or pneumonitis then become common (Chapter 7). The appropriate specimens to submit for laboratory diagnosis in each of these conditions are indicated in Table 4.1.

PRELIMINARY TESTS

The definitive diagnosis of toxoplasmosis is still a relatively rare occurrence and requires specialist expertise. Most local bacteriology or virology laboratories will be able to carry out preliminary tests but positive diagnoses will often eventually be made in a Reference Laboratory. The initial local testing will usually entail one, occasionally two, simple tests to screen out samples

Table 4.1 Specimens required in different clinical conditions

Clinical	Type and choice of specimen						
	Blood (clotted)	Blood (EDTA)[¶]	CSF	Amniotic fluid	Aqueous humour	Biopsy	Placenta
Normal acquired	1					2 (node)	
Pregnant	1						
Fetal	2	1		3			
Fetal death	1+					2 (organ)	3
Neonatal	1+		3				2
Ophthalmic	1				2		
Immunocompromised transplant/AIDS	1	2					3

1, 2, 3 = choice of specimen.

+ = maternal blood sample also required.

¶ = for detection of parasitaemia, or non-specific tests.

Table 4.2 Characteristics of serological tests

Test	Antigen template	Detection	Reading
Latex	Latex	Agglutination	Macroscopic
HA	Tanned red cells	Agglutination	Macroscopic
ELISA	Polystyrene well, bead	Colour	Automatic
IF	Glass slide (multiwell)	Fluorescence	Microscopic
Dye test	Live suspension	Killing	Microscopic
DA	Dead suspension	Agglutination	Macroscopic
CF	Solution	Red cell lysis	Macroscopic
FIAX	Plastic paddle	Fluorescence	Automatic

HA = indirect haemagglutination assay.
ELISA = enzyme-linked immunosorbent assay.
IF = indirect immunofluorescence.
DA = direct agglutination.
CF = complement fixation.
FIAX = (commercial) fluoroimmunoassay.

which do not merit further investigation. The tests which are currently used are latex agglutination,[8] indirect haemagglutination (HA),[8,9] and immunoassay, usually enzyme-linked immunosorbent assays (ELISA).[10] Another immunoassay, less commonly used in this country, is indirect immunofluorescence (IF).[11] Which one is used will depend on local policy, which format fits in best with other routine test commitments, and availability at low costs. As a general principle, latex and HA tests are more suited to laboratories with a low throughput and immunoassays to those with large numbers.

The specialist expertise sought when submitting specimens to a Reference Laboratory derives from a number of factors. They deal with a larger number of specimens, and more positive specimens. They perform a greater range of tests, in particular the Sabin–Feldman dye test (DT)[12] and tests for different antibody classes. The immune response to any organism in a patient does not consist of a single antibody. Usually there is a mixture of many unique antibodies and there are also five general classes of antibody (IgG, IgM, IgA, IgE, and IgD). The serum concentration of specific antibody of some of these classes can give an indication of how recently infection has occurred. Following primary infection, specific IgM appears early in the serum;[13] IgG is synthesized slightly later and remains detectable much longer. IgM usually ceases to be detectable after 4 to 6 months, so its presence is taken to indicate current or recent infection. A function of the Reference Laboratory is to interpret the results from this range of tests on one or more specimens from each patient and comment on their relevance to the patient's current symptoms. The main tests used in diagnosis of toxoplasmosis and the principles involved are summarized in Table 4.2.

Agglutination tests

In all serological tests, a positive result depends on being able to demonstrate an interaction between antibody present in the patient's serum and parasite antigen(s). In the latex and HA tests (Table 4.2), the antigen is a soluble extract of toxoplasma tachyzoites covalently bound to latex particles[8] or chemically to sheep[9] or turkey erythrocytes.[8] Antibodies are multivalent, so that one molecule can bind simultaneously with more than one antigen-coated particle. Interaction therefore leads to agglutination and large aggregates of the particles are less able to slide down the slopes of a U-bottom or V-bottom plastic well than free particles. Tests are performed by introducing a constant volume of coated particles to dilutions of patient's serum in saline. Antibody negative sera show only tight buttons of settled erythrocytes or particles at all dilutions; positive sera produce carpets of agglutination up to a dilution at which 50 per cent agglutination, 50 per cent settling occurs. This end-point dilution (e.g. 1/64) is quoted as a measure of antibody titre; the higher the dilution, the higher the titre. With both these assays, it is possible to screen larger numbers of specimens by testing at a single serum dilution (e.g. 1/16 or 1/32) and only investigate further when the results are positive. Some modified latex tests are available which produce qualitative results rapidly (in a few minutes).[14] One of these involves enzymatic detection of latex-bound specific antibodies on a membrane filter.[15]

ELISA

In ELISA, the soluble toxoplasma extract is hydrostatically adsorbed to plastic (usually polystyrene) in a variety of formats[16-18] (Table 4.2) but two are more commonly used. Usually the wells of 8 × 12 flat bottom microtitre trays[10,17] or 8- or 12-well (flat or round bottom) strips,[17] are coated, but polystyrene spheres are an alternative.[16] The latter are easier to manufacture to a uniform standard so in theory there should be less variation between duplicate sample results. However, as the washing equipment for spheres is more specialized, wells form the basis of more commercial assays. Apart from random well-to-well variation, edge effects have been reported[10] where higher readings are consistently observed in wells at the end of rows or around the perimeter of plates. Irreversible drying of reagents such as conjugate on to the plastic is one possible way of generating false positive results. These can be eliminated or substantially reduced by firmly sealing the wells with parafilm to avoid evaporation during incubations. Once coated, wells or spheres are stable for months at 4 °C or −20 °C and these templates form the basis of a plethora of commercially available kits. They are also simple to produce for in-house assays.[17]

In the test, patients' sera are incubated with the antigen-coated plastic for up to an hour, at room temperature or 37 °C (Fig. 4.2). Unbound proteins

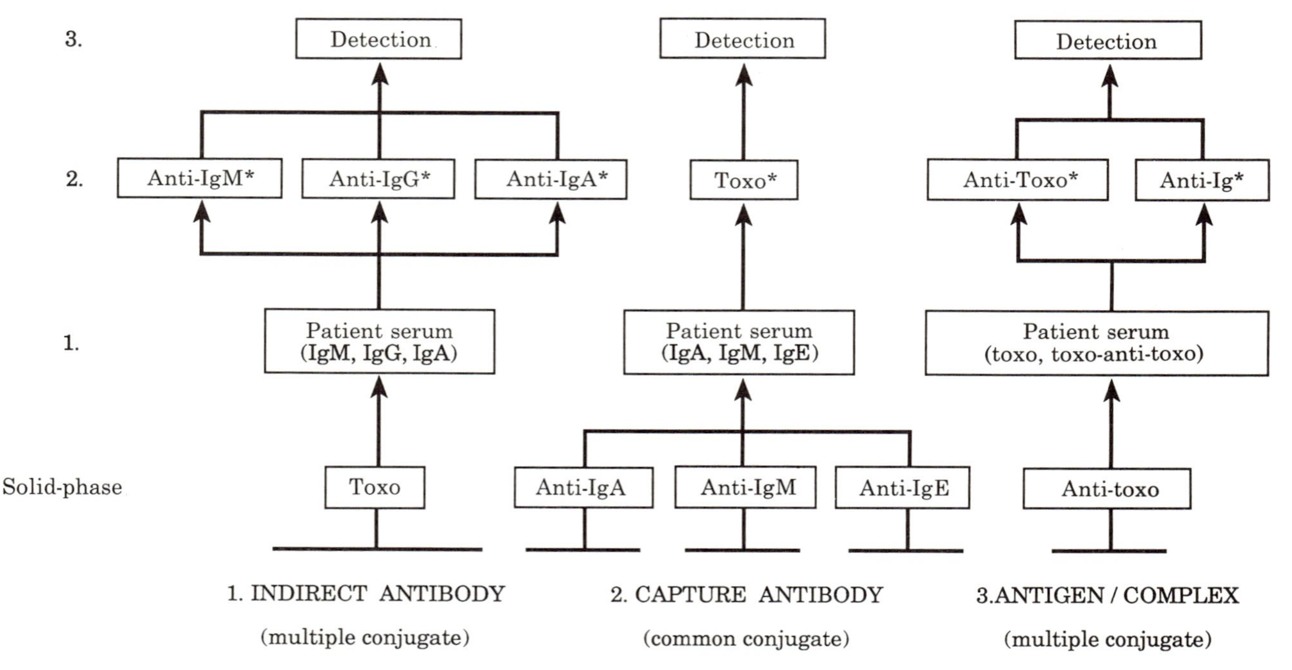

Fig. 4.2   Component stages of three types of immunoassay.

are washed off, usually with automatic ELISA washer equipment, and bound antibodies detected by addition of an animal antibody (usually goat or rabbit) with specificity for human immunoglobulin (antibody). The animal antibody has been covalently coupled to an enzyme (usually alkaline phosphatase or peroxidase) to form a conjugate, and when the appropriate substrate is added the amount of coloured product is directly proportional to the amount of human antibody bound from patient's serum. ELISAs can be carried out on several dilutions of serum, and titred to an end-point as in the previous tests,[10] but it is more usual to assay on a single serum dilution and equate antibody concentration to optical density, using an ELISA reader. It is generally accepted that ELISAs are qualitative or semi-quantitative with results graded relative to a positive/negative threshold.[17]

Interpretation: diagnostic

Preliminary tests serve different functions. For diagnosis high sensitivity is important,[17] so that cases which might merit further investigation are not dismissed at the first examination. However, a balance has to be struck between high sensitivity and poor specificity. The latter leads to excessive referral of false positive specimens to the Reference Laboratory, which is wasteful of the time and resources of both laboratories. Testing for immunity, either before or during pregnancy, is not at present systematically undertaken, although our laboratory, in common with others in the UK, has noted a marked increase in the demand for it in the last few years. In this situation, the more important consideration is not to falsely reassure women that they are protected, so specificity has to be guaranteed.

The HA and latex tests theoretically have the advantage of being able to detect both IgG and IgM antibody. They should therefore be more sensitive than IgG–ELISA in the early stages of infection.[19] However, the HA test has been shown to produce negative results in a few patients with high DT titres and symptoms characteristic of active infection.[20] It was not any more sensitive than IgG–ELISA in the early stages of a case of infection in pregnancy which resulted in a congenitally damaged baby.[17] The latex test can also produce false negative results in relation to the DT.[8] Both HA and latex also produce significant numbers of false positive results.[8,9,20,21] In the case of HA, these can be controlled by parallel testing against unsensitized erythrocytes, i.e. cells lacking in toxoplasma antigen. As they are read by eye, interpretation is subjective and sometimes the end-points in the HA are not clearcut.[22]

The most comprehensive procedure for preliminary diagnostic testing is probably a combination of IgG– and IgM–ELISA tests.[17,19] Many laboratories, however, might find this too expensive in terms of commercial reagents, particularly if their workload is small. If limited to one of these tests, IgM–ELISA might appear the more appropriate for diagnosis of active infection. It will detect acute infection (usually more effectively than IgG–ELISA) but it may give little indication of prolonged or recurrent infection. IgG–ELISA therefore is the preliminary test used by many laboratories. This is permissable because the undramatic nature of the clinical disease means that a diagnosis is very seldom sought before IgG antibodies are detectable.[3] The test threshold must be set low, but in practice the limit is around eight international units, otherwise too many false positives are generated.[17]

Whichever preliminary test is chosen, the important point is to interpret the results according to the clinical information and the limitations of the test. Positive or borderline results of single tests should be confirmed in a Reference Laboratory. Negative results may be directly reported as evidence against current infection, but not in all circumstances. It may be too soon after the onset of illness for antibody to be detectable, in which case a follow up specimen has to be requested. In ophthalmic toxoplasmosis[5] or toxoplasmic encephalitis in immunocompromised patients,[23] what is sought is any evidence of previous infection which might support the clinical diagnosis. In these circumstances the dye test provides the most reliable combination of sensitivity and specificity, especially when the threshold is lowered to two international units.

Interpretation: immunity testing

For immunity screening, before or during pregnancy, IgG–ELISA is the most appropriate test both in terms of its capacity for dealing with large specimen numbers[24] and its specificity.[19] To avoid false reassurance of immunity, a higher detection threshold, equivalent to approximately 15 international units in the dye test, would be advisable. IgG–ELISA readings above this level in asymptomatic women can be interpreted as evidence of immunity. It is preferable also to demonstrate absence of specific IgM antibody, to rule out asymptomatic current infection. If the latter is not done locally, these specimens along with those which yield equivocal screen results, might benefit from referral to a laboratory that can perform additional tests. Women with negative IgG–ELISA readings should be regarded as non-immune and advised of risks of infection. They may also be re-tested later in pregnancy on request or as part of a formal surveillance programme if it exists (Chapter 6).

DEFINITIVE TESTS

Dye test

The dye test measures the total amount of specific antibody in a serum which is capable of participating in antibody-mediated killing of tachyzoites by complement. Originally, as the name implies, the dye methylene blue was used to stain parasites which survived exposure to patient serum and complement.[12] Now, the dye test has been modified for reading by phase contrast microscopy[25] in microtitre plates.[26,27] Patient sera are diluted in saline and exposed at 37 °C to a constant concentration of live tachyzoites and human complement. Results are read through the flat bottom wells of the plate using a phase objective on an inverted microscope. The titre is the dilution at which 50 per cent of parasites are phase-dark, i.e. dead; negative sera leave phase-bright parasites at all dilutions.

Live parasites are derived from infected mouse[26] or rat[27] peritoneal exudates, although tissue culture[28] could be used provided the constancy of supply was reliable. The important prerequisite is that the suspension should be fully viable, and uncontaminated with host cells or bacteria. The human complement is obtained by screening human blood donations for antibody-negative samples with suitable complement levels for the test to work. Serum is preferable to citrated plasma,[29] but it is essential that non-specific killing does not occur and large volumes are required. Commercial sources of animal complement are possible substitutes. The dye test is highly sensitive and fully quantitative. Results are usually expressed in international units by direct (or more often indirect, via an in-house standard) comparison with a standard reference 1000 units/ml serum supplied without charge by the Standard Serum Institut, Denmark. Positive titres can range from 2 to several 100 000 units/ml, but the normal range is 2–125 units/ml.

Given the complexities involved in maintaining a test such as the DT which requires live parasites, it is not surprising that alternatives have been sought which are capable of matching its performance. Direct agglutination (DA)[30] and IF[11,22] are the more likely candidates, although HA, latex and ELISA have all been claimed to show satisfactory correlation with the DT. Direct agglutination is very similar in principle to the HA test, with formol-fixed tachyzoites in high concentrations substituted for sensitized erythrocytes (Table 4.2). IF is similar to ELISA: the antigen consists of undisrupted tachyzoites, formol-fixed[22] or dried unfixed[31] on a glass microscope slide (multiwell); the conjugate is fluorescein labelled. Results can be expressed as an end-point dilution titre, namely that at which 50 per cent of parasites fluoresce when observed at magnification ×400 on a fluorescence microscope.

Specific IgM tests

Early methods for measurement of specific IgM were aimed at physically separating IgM from IgG. Techniques such as ultracentrifugation[9,32] on a preformed gradient of increasing sucrose density (10–40 per cent) or column chromatography on columns of agarose gel,[33] which may be purchased prepacked,[34] took advantage of the fact that IgM is a much bigger molecule

(5×) than IgG. The separated IgM fraction has to be tested by one of the previously described tests, e.g. HA, DT,[9] to determine whether it has specificity for toxoplasma. Fractionation methods require relatively large volumes of serum (>0.3 ml). They tend not to be used now because they are relatively labour intensive, but there is a basic simplicity in the principle involved which ought not to be entirely ignored.

Present day tests for specific IgM work on unfractionated serum, and on smaller volumes (10–50 µl). There is a multiplicity of different types of immunoassay but the basic format is common to all. Specific IgM in specimen serum interacts with toxoplasma antigen on a solid surface. The solid surface is usually plastic in the case of ELISA, or glass in the case of IF; a positive reaction is visualized by enzymatic,[35–37] fluorescence,[27,38] or agglutination methods.[39,40] There are two general types of assays; IgM capture, in which captured specific IgM then binds toxoplasma antigen; and indirect IgM, in which antigen coated matrices then bind specific IgM (Fig. 4.2).

IgM capture tests

An IgM capture assay is the current method of choice for determining the presence of toxoplasma-specific IgM.[35,36,41] Patient IgM is the first interactant to be bound, by capture onto anti-human IgM coated plastic microtitre wells or beads. The capture antibody is usually the globulin fraction of a rabbit, sheep, or goat serum. The binding and subsequent washing away of the non-bound serum proteins, including the other immunoglobulin classes, is in effect serum fractionation on a smaller scale. If toxoplasma-specific IgM is present it will then bind toxoplasma antigen. The binding of antigen is demonstrated by different procedures according to how it is labelled and presented. Capture assays are subject to competition from non-specific IgM.

Capture ELISA methods use an enzyme-labelled conjugate to detect binding of toxoplasma antigen by specific IgM-positive specimens (Fig. 4.2). There are a considerable number of in-house[35,36,42] or commercial assays available,[43,44] with different ways of labelling the detection stage. In some the antigen itself is directly labelled either with the enzyme peroxidase[42] or with biotin[35] which is then complexed to peroxidase-labelled avidin. In others the antigen is indirectly labelled by complexing with a peroxidase-labelled anti-toxoplasma antibody which is either polyclonal[41] or monoclonal.[36] The complex can even be a triple complex of unlabelled antigen, biotin-labelled anti-toxoplasma and avidin–peroxidase. Peroxidase is preferred to alkaline phosphatase, for no apparent reason other than its lower cost. Results are measured as absorbance readings and interpreted as positive or negative relative to a threshold defined by the control sera in the kit. Readings within 10 per cent of the threshold are generally regarded as equivocal, but quantitation is not usually attempted.

IgM–immunosorbent agglutination assays (IgM–ISAGA)[39,40] combine the features of IgM-capture and DA assays. The patient's IgM is captured on anti-IgM coated wells of U-bottom microtitre strips, and specificity for toxoplasma detected by addition of formalin-fixed tachyzoites. An IgM positive serum prevents the parasites from settling, so they form a carpet on the base of the well; with negative sera cells settle to form a button. Sera are tested at a single dilution. The degree of positivity is scored subjectively on a scale of 0–4, but accuracy is improved by testing more than once against different amounts of antigen. The principle of agglutination at the detection stages also applies to the IgM capture haemadsorption assay,[45] when toxoplasma coated sheep erythrocytes are substituted for the DA antigen.

Indirect IgM tests

Indirect IgM assays parallel the IF and ELISA assays described earlier for total antibody. An anti-human IgM conjugate is substituted at the detection stage for a whole globulin or anti-IgG conjugate; the conjugate is fluorescein-[27,38,46] or enzyme-labelled[37] (Fig. 4.2). This type of assay is subject to inhibition by high concentration specific IgG[33] in test specimens or false positive results when rheumatoid (IgM auto anti-IgG)[40] or anti-nuclear factors[40, 47] are present. To avoid these problems, it is necessary to preadsorb sera with insolubilized human IgG aggregate,[46] or anti-human IgG.[43]

There is some evidence that IgM-capture ELISA correlates better than indirect IF with the results of sucrose gradient fractionation,[41] which may be regarded as the reference standard. Indirect assays are also generally regarded as less sensitive than capture assays.[37,40,41] This may appear somewhat paradoxical given that capture assays measure the fraction of the patient's IgM which is specific to toxoplasma, whereas indirect assays should measure the absolute specific IgM content.[48] The reason may stem from the pentavalent nature of IgM. In indirect assays five binding sites on the plastic could bind one molecule of IgM and eventually one molecule of conjugate whereas in capture assays one binding site could amplify to five molecules of antigen bound. Competition from non-specific IgM in capture assays may thus have less effect on the result than competition from specific IgG in indirect assays. Removal of IgG increases the sensitivity of indirect IF tests.[33,34]

Interpretation

The prime purpose of tests such as the dye test is to confirm the results of preliminary tests. The actual level of antibody in international units (i.u.) per ml can bear some relationship to active infection,[3,41] but results are usually now interpreted in conjunction with results of specific IgM tests. Obviously rising DT titres (by >4-fold), on sera collected one or two weeks apart and tested at the same time, provide evidence of active infection. Seroconversion or a rising titre plus specific IgM suggest primary infection; but rising titres, with or without detectable specific IgM, can also be found in reactivated infection. DT titres higher than 250 i.u./ml are unusual in healthy individuals, so levels of this order suggest infection.[3] However, as titres do not fall rapidly following infection, high titres may be detectable many months later, when specific IgM has become undetectable. Persistence of such titres may be associated with persisting symptoms,[41] such as malaise, myalgia, and general tiredness. Titres greater than 125 i.u./ml have been considered significant in patients with ophthalmic disease, but it is accepted that any level of DT confirms past infection and can be consistent with a clinical diagnosis of ocular toxoplasmosis.[5] Testing to the limits of sensitivity (2 i.u./ml) is therefore advisable. Similar flexibility in interpretation has to be applied in immunocompromised patients whether due to AIDS, transplant, or malignancy. Primary infection in these patients may not always produce high DT titres.[49,50] Reactivation is common, and severe, usually cerebral, damage may be caused by the parasite without significantly raising the DT titre.[23,51,52]

Apart from the absolute test levels, DT titre relative to other tests can help to determine the stage of infection (Table 4.3). Because it detects either IgG or IgM antibody it will yield higher titres than IgG–ELISA in the early stage of infection, for reasons which are self explanatory. Less obvious, however, is the fact that a similar phenomenon is evident in our experience when HA and DT are compared in acute specimens, and similar experiences are reported for the DA and DT.[30] In sera from latently infected patients, the titres of all these tests reach equivalence with the DT. Part of the reason may be that early stage antibodies probably react mainly with antigens from the surface membrane of the parasite. As the principle of the DT is antibody induced damage of live parasite membranes, these are the antibodies it detects. Membrane antigens are not so important in the other tests. In the IgG–ELISA or HA tests, which use disrupted parasite extracts, they may form a minor constituent of the antigen present. The DA is designed to detect specific IgG, not IgM.[30]

It is possible, but not universally agreed that the IF is less sensitive than the DT,[11,53,54] although there is good overall correlation. Like the DA[30,55] it is susceptible to false positive reactions due to 'natural antibodies'.[53] These are antibodies of IgM class which are found in (some) non-infected individuals. In the DA interference is avoided by pretreating sera with 2-mercaptoethanol which destroys the activity of IgM antibody.[30] This effectively restricts the test to detection of specific IgG antibody. In the IF tests, natural antibodies produce polar fluorescence,[53] so operator skill is essential to distinguish false positive from genuine results. The insensitivity of the DT that is sometimes observed during episodes of reactivated infection in AIDS or other immunocompromised patients[23,51] is probably mainly due to immunosuppression, but it is also possible that the antibodies produced at this time largely fail to recognize normal membrane antigens. Modification of membrane antigens by fixation improves sensitivity,[56] so that an elevated ratio of DA to DT in AIDS patients could be diagnostic of

Table 4.3 Common examples of serological results and interpretation in immunocompetent patients

	Example	Test results				Interpretation (stage of infection)
		Preliminary		Definitive		
		IgG	IgM	DT	IgM	
Good clinical evidence	1	–	–	–	ND	No significant titres yet*
of acute toxoplasmosis	2	–	+	–	–	No significant titres yet*
	3	±	+	+	–	No significant titres yet* (possible early acute; possible false +ve IgM)
	4	+	++	++	++	Current**
	5	++	+	++	±	Current/recent**
	6	++	–	+	–	Probable past
Poor clinical evidence of	7	–	–	ND	ND	No evidence of current
acute toxoplasmosis	8	+	–	–	ND	No evidence of current (possible false +ve IgG)
	9	–	+	–	–	No evidence of current (possible false +ve IgM)
	10	++	+	++	–	Possible recent* (possible false +ve IgM)
	11	+	+	+	–	Probable past (possible false +ve IgM)
	12	+	–	+	ND	Past

ND = test not done.
* = request second specimen.
** = request second specimen to confirm.

reactivated infection,[57] particularly if the method of DA antigen fixation uses acetone instead of the standard formaldehyde.[58,59] The difference in ratios comparing patients with active to those with latent infection, however, may be too marginal for reliable routine diagnoses.

False negative DT results are very rare. Because the DT is the accepted 'gold standard' test, discrepancies between it and positive preliminary test results are likely to be attributed to non-specificity of the latter. However, there are documented reports of negative DT results in the face of convincing histological[30,59,60] or serological[60] evidence of active toxoplasmosis. These occurred in immunocompromised patients, with AIDS and biopsy-proven cerebral toxoplasmosis[59], or cardiac involvement following transplantation.[60] With increasing numbers of patients in these categories this problem is bound to become more common.

If tests for toxoplasma-specific IgM were fully sensitive and specific and only produced positive results for a short period (<3 months) after infection, they would be definitive of current or recent infection on their own. No single test, however, meets all these criteria. Capture assays are designed to give improved sensitivity and specificity, but are not perfect in either respect. In addition, the considerable variation in the length of time specific IgM is produced by different patients, sometimes for over a year[42] (Table 4.4), means that caution is required when interpreting the precise time of onset. Positive results should be confirmed by a second IgM test and interpretation should take account of DT and specific IgG results (Table 4.3). A repeat serum after 10 to 14 days can help give a more accurate assessment of the infection stage, and this is essential when diagnosing infection in pregnancy. In pregnancy, deciding the date of infection is the most important consideration; in other clinical situations, maximum sensitivity may be more important, so the IgM tests which are appropriate differ according to these requirements.

In regard to specificity, IgM capture assays are designed to avoid the problems of interference from IgM anti-IgG rheumatoid factors or anti-nuclear antibodies which plague indirect immunoassays,[42,47,61] although false positive results are still possible when these auto-antibodies occur together.[62] Occasional false positive results also occur with some heterophile-antibody positive sera,[35] jaundiced or post-mortem sera, and tests on haemolysed sera are not advised for some commercial kits. Capture assay design may render some more susceptible to non specificity. Those in which the detection stage involves directly labelled antigen,[35,42] or an unlabelled antigen plus poly-clonal antiserum conjugate, may be vulnerable if impurities in the antigen (e.g. bacterial contaminants) are recognized by some patients' IgM. Antigens used in these assays should therefore be pure. Tests which use monoclonal anti-toxoplasma conjugate are theoretically the least susceptible to this potential false reaction. Finally, patient treatment may influence IgM results. Transplant patients immunosuppressed with rabbit anti-thymocyte globulin

Table 4.4 Persistence of toxoplasma specific IgM

Patient	Clinical	Months post infection	Dye test (i.u./ml)	IgM–ELISA		Sucrose gradient IgM	IFM
				Screen	Confirmatory		
1.	Thrombocytopenia; abdominal pain	3*	1000	+	+	+	+
	at 7 months; premature delivery;	5	500	+	+	+	+
	neonatal death	8 (birth)	125	+	+	+	±
	Thrombocytopenia;	21	125	+	+	+	+
	healthy baby	23	250	+	+	+	+
		27.5 (birth)	125	+	+	NT	+
2.	Gland biopsy suggestive;	4	4000	+	+	+	+
	persisting symptoms	7.5	1000	+	±	+	+
		9.5	500	+	−	NT	−
		13.5	250	±	−	NT	+
3.	Gland biopsy suggestive;	X	2000	+	+	±	+
	hepatomegaly;	7.5	1000	−	−	NT	±
	rheumatoid factor +ve;	14.5	1000	NT	−	+	+
	persisting symptoms	36	4000	−	−	NT	−
4.	Uveitis; then lymphoma;	X	30	NT	NT	NT	−
	rheumatoid factor	48	500	−	+	−	+
	seroconversion;	49	500	+	+	±	+
	reactivation	52.5	4000	+	+	NT	+
5.	Hodgkins lymphoma;	X	500	−	−	NT	−
	reactivation	17.5	1000	+	+	NT	+
		19.5	500	+	±	NT	+

X = time of infection unknown; serial samples timed from this specimen.
NT = not tested.
* = months pregnant when first tested.

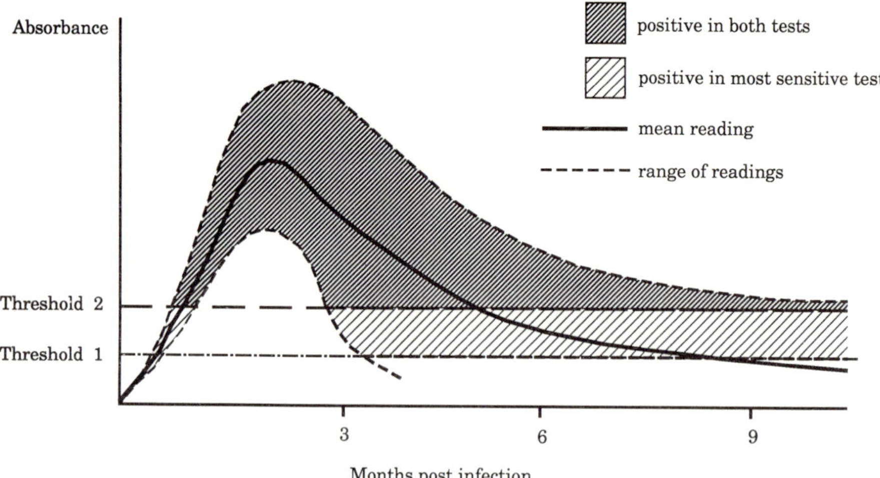

Fig. 4.3 Effect of immunoassay sensitivity on duration of specific IgM detection. Threshold 1 = absorbance above which test 1 produces positive results. Threshold 2 = absorbance above which test 2 produces positive results.

may produce false positive readings via reaction of anti-rabbit antibodies with rabbit anti-toxoplasma conjugates.[63]

Several studies have examined the relative sensitivity of some of the many toxoplasma IgM kits available.[43,44] The sensitivity is to some extent arbitary, as the threshold can be adjusted to restrict positive results to a desirable time span. In Fig. 4.3, test number 1 is designed for high sensitivity but will produce positive results for longer than test number 2, which has a higher threshold. The latter may limit positive results to a more meaningful time post infection, but at the expense of a decrease in sensitivity very early in infection. Capture ELISAs are generally more sensitive than indirect immunoassays[41,42] and are claimed to discriminate better between early and late infection.[37] The relative sensitivity is best illustrated in sera of congenitally infected neonates in which a capture ELISA detected specific IgM in 75 per cent, compared to only 25 per cent with IgM–IF.[64] IgM–ISAGA appears to be even more sensitive.[39,65] The problem posed by IgM persistence is demonstrated in Table 4.4. Patient 1 had strongly positive IgM titres in all tests throughout two pregnancies. She was thought to be recently infected when she was tested in the first pregnancy, but had she presented for the first time during the second pregnancy only careful analysis of serial specimens would have ruled out a recent infection then. The first pregnancy resulted in neonatal death at 2 days, with toxoplasma infection not confirmed; the second pregnancy in a healthy baby. Patient 2 shows the relative persistence in two ELISA tests; patient 3 shows persisting elevation of the dye test titre; and patients 4 and 5 show positive IgM results during reactivated infection.

It is up to individual laboratories to choose the test(s) which best meet their requirements. Our policy is to use an in-house capture ELISA with a threshold set deliberately low to screen all sera referred to the laboratory and to confirm positive results with a less sensitive commercial capture ELISA. Strong positive results in both tests indicate current or recent infection (possibly 1–5 months post infection) (Fig. 4.3). A positive screen IgM with a weakly positive or negative confirmatory result either indicates early acute infection (0–1 month), infection in the recent past (5–12 months), or a false positive screen result. Interpretation in conjunction with the preliminary IgG and DT results usually discriminates between these alternatives (Table 4.3). Early in infection the DT titre can be significantly higher in relative terms than the IgG titre; late in infection they are more likely to both be high; and they are often both negative or low in the third instance. The above strategy is suitable for most situations and particularly so when the priority is to determine the onset of infection. When sensitivity is paramount, e.g. for diagnosis of congenital infection, provided the neonate has a positive DT titre we test by IgM–ISAGA regardless of screen IgM–ELISA results. Similar sensitivity is sought when testing immunocompromised patients, especially AIDS patients with encephalitis, for reactivated infection. Although IgM is detectable in some reactivated infections[23,49] (Table 4.4), in the majority of cases[23,58,59] (including ocular infections), IgM results are not diagnostic, even with IgM–ISAGA.[58] It is therefore assumed that reactivation does not always stimulate specific IgM synthesis.

OTHER ANTIBODY CLASSES

The presence of specific IgM is almost synonymous with acute or recent infection, but useful diagnostic information may be gained from studying other antibody classes. IgG synthesis increases early in infection and a seroconversion or significant rise in total antibody has of course, always provided reliable evidence of active infection. By the time diagnosis is sought it may be too late to demonstrate such a rise.[66] Apart from a rise in titre, the strength of the binding interaction between IgG antibody and toxoplasma[67] can indicate the stage of infection. Early in an infection, antibody binds loosely to the antigen, i.e. antibody of low avidity. Subsequently binding is improved when antibody of higher avidity is produced. When IgG–ELISA plates are washed with 6 M urea after the serum incubation step, IgG antibodies with low avidity are washed off, reducing the final absorbance reading. Avidity can be expressed in terms of a ratio (of the end-point titre with a urea wash as a percentage of the titre with a normal wash). Acute phase sera produce ratios of less than 20 per cent, compared to 25–97 per cent (mean 50%) in those from latently infected individuals.

IgG antibody can be subdivided into four subclasses, IgG_1, IgG_2, IgG_3, and

IgG$_4$. Their relative contribution to the immune response to toxoplasma can be ascertained by substituting mouse monoclonal antibodies at the stage when anti-human IgG is used in a standard IgG-ELISA. The unlabelled monoclonals bind to IgG$_1$, IgG$_2$, IgG$_3$ and IgG$_4$ anti-toxoplasma antibody in four separate ELISAs and are detected using enzyme-labelled anti-mouse antibody conjugate.[68,69] The assay only has practical application if the predominating IgG subclass(es) in acute, latent, or reactivated infection are different. There is little evidence that they are. The contributions of toxoplasma-specific IgA[70,71] and IgE[72] antibody at different stages of infection are measured in modified capture assays, simply by substituting anti-IgA or anti-IgE for anti-IgM on the capture matrix (Fig. 4.2). The specificity of the assay can be refined to detect antibody against a purified toxoplasma surface protein,[70] as opposed to measuring the response to the type of crude total antigen extract that is normally employed in IgM-capture assays. Specific IgA and IgE are only found in active infection, but the time scales for these antibodies are different from that of IgM. All of these antibody classes are measured in a more complex assay, the enzyme-linked immunofiltration assay (ELIFA),[73] described in Chapter 8, and claimed to be the most sensitive method of detecting congenital infection in neonates.

ANTIGEN DETECTION

Serological tests for antibody, with their relative ease and speed compared to demonstrating the presence of the parasite itself, are the standard approach to the diagnosis of active toxoplasma infection. However, serological tests have also been developed to measure soluble toxoplasma antigens circulating in the blood.[74,75] The techniques are again ELISA-based, involving purified rabbit antibodies to whole parasites, both as the capture and the detector[74] (i.e. a sandwich ELISA, analogous to methods routinely used to detect hepatitis B surface antigen) (Fig. 4.2). Modification of the capture antibody by digestion with pepsin prevents binding of IgM anti-IgG rheumatoid factors and reduces the incidence of false positive results.[75] An alternative method involves binding of small volumes of patient serum (4 µl) directly to nitrocellulose and demonstrating the presence of toxoplasma antigen by adding successively, with intervening washes, rabbit anti-toxoplasma and enzyme-labelled goat anti-rabbit antibody[76] and a substrate which produces a coloured precipitate on the nitrocellulose when enzyme conjugate has been bound to the reaction spots. The sandwich ELISAs have the advantage of generating coloured solutions, which allow objective readings on single dilutions of sera; the dot-immunobinding assays are read subjectively.

A surprising feature of these assays is that positive results are obtainable in patient specimens with high antibody levels despite the fact that detection of control toxoplasma lysate is inhibited by addition of human serum with a similar high concentration anti-toxoplasma antibody.[75] This has led to the development of tests for circulating immune complexes. To test for complexes which consist specifically of toxoplasma antigen and bound antibody, a modification of the sandwich ELISA is used.[77] Antigen is captured by pepsin-digested rabbit anti-toxoplasma antibodies as before and, because patient antibody in the complex is simultaneously bound, it can be detected with an anti-human immunoglobulin enzyme conjugate. Alternatively, non-specific measurements of total complex can be measured by radioprecipitation methods.[78] This is only appropriate for patients who do not have underlying autoimmune disease. If necessary, the precipitates can subsequently be analysed for toxoplasma antigen and antibody content.

Circulating antigens and complexes are apparently only detected during active infection, both in acute primary infection and in reactivated infection.[75,79] The tests therefore could have a role in diagnosis of the disease. They have been developed as in-house tests, with the prime requirement being a good, specific anti-toxoplasma antibody. Antigen detection, with only one variable involved, is inherently more straightforward than measurement of complexes. However the initial capture step in the assay for toxoplasma-specific complexes[77] uses the same type of anti-toxoplasma antibody coated on plastic as in antigen detection tests;[74,75] so there is potential for measuring both antigen and complex in parallel, simply by adding labelled anti-toxoplasma and labelled anti-human IgG to different replicates of the same test specimen (Fig. 4.2).

Interpretation

Toxoplasma antigenaemia might provide a more accurate indication of current or recent infection than the presence of toxoplasma specific IgM, given the variable persistence of the latter. Published evidence would appear to support this premise,[74,75] although while one report suggests a short period of circulation,[74] the second finds antigen present in lymphadenopathic toxoplasmosis for 4 months in most cases and occasionally up to 8 months, or even as long as 20 months.[75] Unfortunately, diagnosis could not depend on the test, as only 20–65 per cent of confirmed cases of acute toxoplasmosis had detectable circulating antigen.[75,76] This apparent failure in sensitivity is not explained by sequestration of antigen within immune complexes when increasing amounts of antibody are synthesized because antigenaemia is frequently evident in the presence of high antibody titres.[74,75] It is more likely that the transition from antigenaemia to circulating antibody is diverse, with a restricted range of antigens present in the circulation.[79] Tests for circulating antigen are therefore subject to at least as much, and probably more, patient-to-patient variability than specific IgM tests. Moreover they are not universally available (in commercial kit form) so their usefulness is limited to laboratories with in-house tests. The parallel investigation for immune complexes might not improve the accuracy of diagnosing current infection, because complexes have been detected in more than 10 per cent of blood donors.[77]

Antigen detection has greater potential in reactivated infection, especially with immunocompromise. Antigenaemia, however is reported to be detectable in a minority (33 per cent) of transplant patients with serological or clinical evidence of reactivation[80] and even less (17–26%) in AIDS patients with severe symptoms.[79,81] In one of the latter studies,[79] it was not stated whether severe AIDS always involved toxoplasmosis. Rather it is implied that 26 per cent with detectable antigen is close to the percentage (30%) expected to contract toxoplasmosis. The patients in the other (preliminary) study[81] all had toxoplasmic encephalitis, so a 17 per cent detection rate is not impressive. Despite the apparently disappointing sensitivity of antigen detection, it was an improvement on IgM detection in the same specimens.[79,80] In a continuation of one of these studies,[79] 10 of 11 AIDS patients with diagnosed toxoplasmosis were transiently positive for circulating antigen, compared to 0 of 7 with no evidence of current infection.[82] Because of the difficulties in finding any serological test to assist diagnosis in the immunocompromised, development of a standard assay for circulating antigen should prove useful. There is anecdotal evidence that it may help confirm the diagnosis of ocular toxoplasmosis.[74]

MISCELLANEOUS TESTS

Inhibition tests

The complement fixation (CF) test is still widely used in diagnostic virology, and to a lesser extent elsewhere in microbiology. Its use in toxoplasma diagnosis has largely been superceded by immunoassays. It has been employed as a screening test, with confirmation of positive results by dye test.[27] The antigen used to fix guinea-pig complement when toxoplasma antibody is present in test specimens is the same as the clarified, soluble antigen used to coat HA cells. Positive fixation results are identified by inhibition of erythrocyte lysis when haemolysin-sensitized erythrocytes (e.g. human group 'O') are added as an indicator. The test could still have a role in laboratories with a small demand for toxoplasma tests, but where CF tests are routinely used for other diagnostic serology. Using mouse or rat peritoneal exudate antigen, the test has lower sensitivity than the DT but acceptable specificity.[27]

Haemagglutination inhibition (HI), another test associated with viral diagnostic serology, can also be applied to toxoplasmosis diagnosis.[83] Given that this test is now less frequently used in virology, and the indicator red cells tend to be unique to a particular antigen (for toxoplasma they are mouse, with toxoplasma exotoxin), it is not important.

Skin tests

Immunity testing, as opposed to diagnostic testing, can dramatically increase the laboratory's workload. Multipuncture skins tests, using tissue culture derived excretory–secretory antigen, have been proposed as a means of reducing the workload of prepregnancy screening.[84] Although the antigen meets current safety standards and the test does not yield false positive results, which could lead to false reassurance of immunity, the case for its application remains unconvincing. Follow-up checks on the patient merely replace laboratory time with clinician (and patient) time, which is unlikely to be acceptable.

Other immunoassays

Apart from IF and ELISA, which have already been described, a variety of other formats of immunoassay are possible. Radioimmunoassay, which has been widely used in diagnostic viral serology, has been developed for toxoplasmosis.[85] It was designed to measure total antibody, using antigen-coated plastic beads and a ^{125}I-labelled conjugate. The wish to avoid radioactivity and the short shelf-life of the reagents is probably the reason this type of assay has failed to compete with ELISAs. A variant of the bead-based ELISA uses polycarbonate-coated iron instead of polystyrene beads to measure IgG antibody.[86] Used in conjunction with a magnetic transfer device, improved synchrony in the incubation and wash stages is achieved, which is claimed to improve the coefficients of variation between replicate assays. This assay has been commercially marketed and performs well in qualitative comparisons with a reference IF test.[22] A commercially produced fluoroimmunoassay, the FIAX test,[87] gave even better correlation with the IF, quantitatively as well as qualitatively.[22] It uses soluble-antigen-coated plastic paddles and fluorescein–anti-human IgG conjugate. Results are read objectively in a fluorometer (Table 4.2), but the expense of this specialized reading equipment is probably the reason the test is not widely used.

An immunoassay which is able to measure toxoplasma specific IgG and IgM antibody simultaneously and rapidly in an automated system would be the sought after ideal of most workers. This is apparently achieved in a new automated microparticle enzyme immunoassay (Toxo IgM MEIA) which can measure both IgG and IgM antibody in 24 samples in 35 minutes,[88] with 98 and 97 per cent agreement respectively with reference ELISA tests. The solid matrix in the test is provided by toxoplasma-antigen-coated polystyrene microparticles which bind specific IgG or IgM, with unbound antibodies removed by filtration. The detection stages involve alkaline phosphatase–anti-globulin conjugate and optical density reading in Abbott-IMX automated equipment. As both assays are equivalent to indirect ELISAs, all

IgM positive samples have to be re-tested following absorption to remove rheumatoid factors. Apart from this disadvantage, the other major disincentive to widespread use, for example in screening, will be the cost of the equipment and kits.

TESTS IN OTHER FLUIDS

Many of the tests used on sera; dye test,[89] IF,[89] ELISA,[90] ELISA for antigen,[75,76] have been successfully used on fluids such as aqueous humour,[89-91] CSF,[75,76] or amniotic fluid[75,92] to seek the presence of antigen or antibody from a site near the active infection. Detection of antigen in these fluids is on its own significant, but antibody titres have to be related to serum titres to distinguish between local synthesis, which would be significant, and mere diffusion from the vascular system which is not. It is therefore standard practice to measure toxoplasma specific antibody in both fluid and serum, and to compare the ratio to some reference which will not be affected by active toxoplasmosis. This can be total IgG or antibody to a common virus of childhood, such as measles, to which the patient might be expected to have immunity. Collection of CSF, amniotic fluids, or aqueous humour are all less readily undertaken than blood sampling. CSF collection is a fairly routine procedure, but not entirely without risk; aqueous humour is taken by oblique corneal puncture in an operating room.[90] Samples tend to be small, yet serological tests have to be carried out at lower dilutions simply because the antibody concentration is expected to be markedly less than in serum. They have to be free of blood contamination. All these factors conspire against frequent testing of these specimens, and limit laboratory experience of them.

Contrasting success rates for the detection of toxoplasma antigen are reported. One group failed to find antigen in CSF of AIDS patients even when they had toxoplasmic lesions.[79] Another group failed to find any in amniotic fluids from cases of congenital toxoplasmosis.[92] As 44 per cent of the amniotic fluids were positive by tissue culture and 77 per cent by mouse inoculation, the results suggest the antigen detection test lacked sensitivity. On the other hand, antigen has been detected in four CSFs from six congenitally infected infants by two different methods,[75,76] but in none of nine controls.[75] Amniotic fluids from two of the six cases were also positive but none of 13 controls.[75] The disparity between these studies is either due to different test sensitivity or because the ELISA possibly only detects antigen if it is in soluble or microparticulate form.[79] Brain cysts, even if they are expanding in size or number to the point where they affect the patient, perhaps only release minute amounts of antigen in this form into the CSF and then perhaps only transiently. More sensitive assays may aid diagnosis in samples like CSF (Chapter 8). Local synthesis of antibody in CSF (as measured by DT) is reported to be detectable in 50 per cent of cerebral toxoplasmosis cases.[93] Antibody (usually IgG) in aqueous humour is diagnostic of toxoplasmic retinitis in 70–90 per cent of cases.[94,95]

DIRECT TOXOPLASMA DETECTION

It has long been recognized that toxoplasma can be isolated[2,3] from various tissues of infected individuals. It can be directly identified[96,97] by a variety of staining techniques. Excised pieces of lymph node,[3,98] muscle,[98] brain,[57,97,99] lung,[96] placenta,[100] or autopsy samples from these[2,100] and other vital organs provide the study material. Parasites can also be isolated from blood[100,101] and other body fluids such as bronchial fluid,[96] CSF,[100] and amniotic fluids.[92] On rare occasions, tachyzoites can be seen in blood films, or cerebrospinal ventricular fluid.[102]

The question of whether or not to look for evidence of toxoplasmosis in tissue biopsies depends principally on three considerations: what is reasonable in terms of risk and inconvenience to the patient; how relevant the result is likely to be to the diagnosis; and the absence of reliable less invasive alternatives. Biopsies would not usually be undertaken by general practitioners, who deal with most patients. Lymph node biopsy is justified in the pursuit of a differential diagnosis, but its application and reliability in the diagnosis of toxoplasmosis *per se* will vary depending on the skill and experience of the histopathologist. Tissue cysts are easier to see by staining methods than tachyzoites (Table 4.5), but their presence in small numbers may be due to latent rather than active infection.[103] Confirmatory serological evidence is essential. Brain biopsy is only undertaken for the diagnosis of cerebral toxoplasmosis[57] as a last resort, when other evidence is not forthcoming (Fig. 4.4). It is avoided if the lesion is not in a favourable location for biopsy.[99] Muscle biopsies are rarely useful,[104] and investigation of lung material[96] is seldom necessary. Serial cardiac muscle biopsies are taken after heart transplant in some centres.[49] The search for the parasite is most actively and extensively engaged when proof of congenital infection is required.[105] In this situation, its presence is investigated at one or more of four stages: antenatally, in fetal blood or amniotic fluid;[92,105] at autopsy, in foetal tissues of a stillbirth or neonatal death;[106] at delivery, in the placenta;[100,106] and post-partum, in neonatal blood and CSF.[100]

In the case of solid tissue, parasites are not likely to be uniformly distributed. Biopsies are therefore taken from a site which is likely to be infected. This should be obvious in the case of a swollen lymph node, but brain biopsies require the assistance of computed tomograph (CT) scanning to identify areas of necrosis.[99,107] Craniotomy with excision, or burr hole needle biopsy, are alternative ways of obtaining tissue.[99] The latter might be less invasive but the larger quantity of tissue obtainable by open biopsy offers a better opportunity of finding the parasite. Sample size is also

Table 4.5 Reliability of methods for proving presence of parasites[92,96,97,99,102,109,110,115]

Method	Sensitivity	Specificity	Result available (no. of days)	Comments
Microscopy:				
Phase contrast	±	++	<1	Low probability of finding parasites
EM	+	++++	1–2	Low probability of finding parasites
Histology stains	++	++	2	Cysts easier than tachyzoites
IF	+++	+++	1	Background specificity problem
PAP	++++	++++	2–3	Most reliable stain for tachyzoites
Culture:				
Tissue + stains	++	+++	2–4	Intracellular parasite possibly not detected
Tissue + IF	+++	++++	2–4	Intracellular parasite possibly not detected
Animal	++++	++++	20–40	Latent cyst +ve ?relevant

EM = electron miscroscopy.
IF = immunofluorescence.
PAP = peroxidase–anti-peroxidase.
Sensitivity and specificity expressed in relative terms ± → ++++.

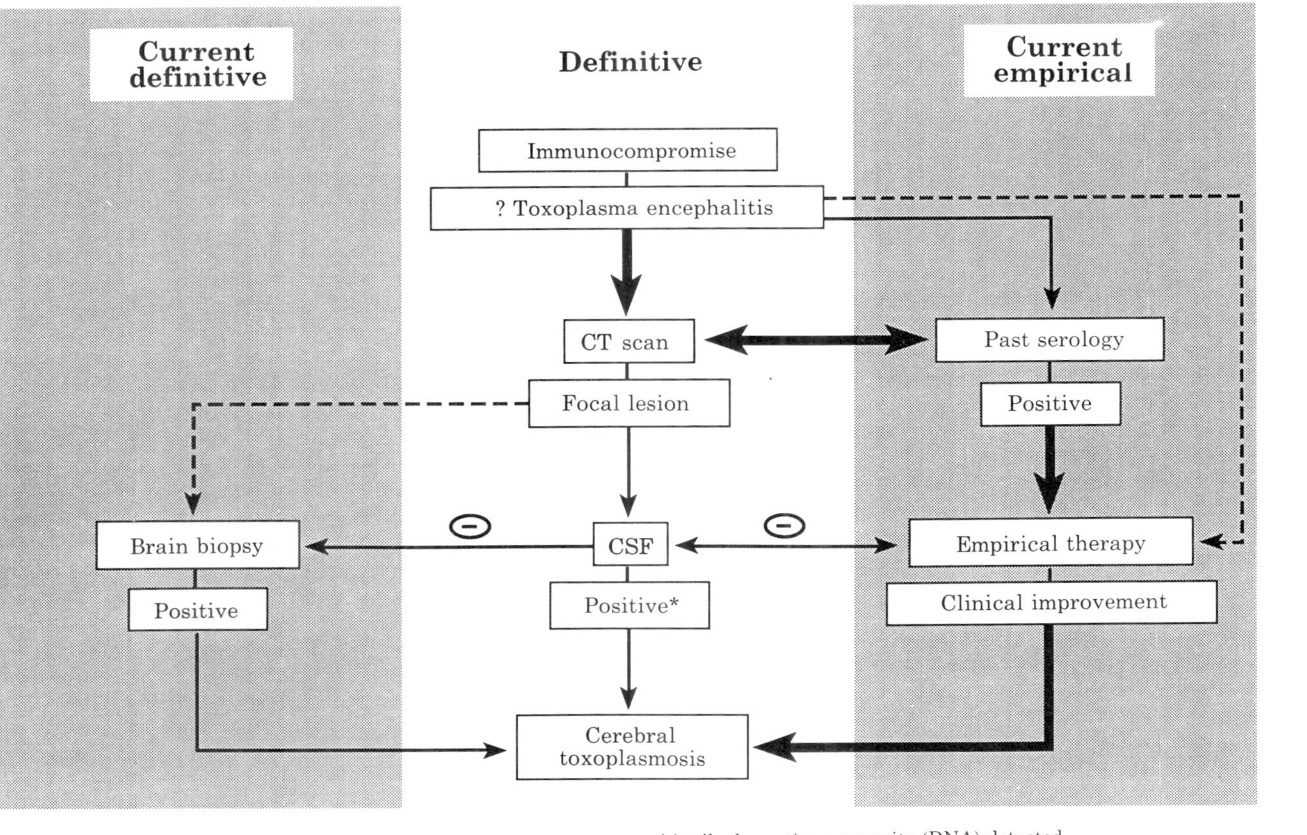

Fig. 4.4 Diagnosis of cerebral toxoplasmosis.

important with tissues such as the placenta where the location of the organism will not be apparent by visual inspection. In body fluids, toxoplasma, if present, will be uniformly distributed. The concentration may be low, however, so as large a sample as possible is advisable. Fairly large volumes of amniotic fluid (up to 36 ml) can be collected, by amniocentesis between the 15th and 20th weeks of gestation, but only 1–8 ml of fetal blood.[108] Cordocentesis is presently the preferred, and safest, method,[108] usually undertaken between the 17th and 24th weeks of gestation. According to these authors[108] it is also possible as early as the 12th week, which if confirmed, has important implications for the subsequent management of infected mothers.

Direct staining

Three sorts of evidence may suggest or confirm active toxoplasmosis in stained preparations: the presence of tachyzoites, tissue cysts, or characteristic histology. Standard histological stains, haematoxylin–eosin,[97,99] Giemsa,[97,99] eosin–methylene blue fast,[96] or periodic acid–Schiff[97] have been used to demonstrate tachyzoites or cysts, but the former are often more reliably identified by immunologically specific staining. The peroxidase–anti-peroxidase technique[97] is the method of choice (Table 4.5). Parasites in the section are stained by the action of a peroxidase enzyme (bound as a peroxidase rabbit anti-peroxidase complex to rabbit anti-toxoplasma via a divalent swine anti-rabbit IgG antibody, or preferably now as avidin peroxidase via biotinylated anti-rabbit IgG). The method is exquisitely sensitive and, with appropriate steps to reduce background staining, quite specific. Background staining can be a problem with indirect immunofluorescence staining.[109] Another advantage of the peroxidase–anti-peroxidase method is that it can be used on the standard formalin-fixed paraffin-embedded sections used for routine histology, whereas immunofluorescence requires unfixed material. The characteristic ultrastructure of toxoplasma is another way it can be identified. Electron microscopy on fixed or unfixed tissue has been used in diagnosis,[110] but its general use is probably limited by expense and the specialized nature of the skills and equipment required.[111]

In lymph node sections, tissue cysts or tachyzoites are rarely evident.[3] A presumptive diagnosis is made on the basis of a combination of characteristic histological features: reactive hyperplasia with scattered foci of histiocytic or epithelioid cells, focal medullary reticulosis, pseudogranulomas[3,112] etc (Figs 4.5, 4.6). Stains used are haematoxylin–eosin and periodic acid–Schiff,[112] with the latter possibly staining intracellular toxoplasma. Although the evidence is circumstantial rather than specific more than 90 per cent of 72 cases who fulfilled the criteria in one study[112] and 100 per cent in another[113] were subsequently found to have significant dye test titres (>250 units). Typical histology on positive brain sections shows well-demarcated necrotic areas,

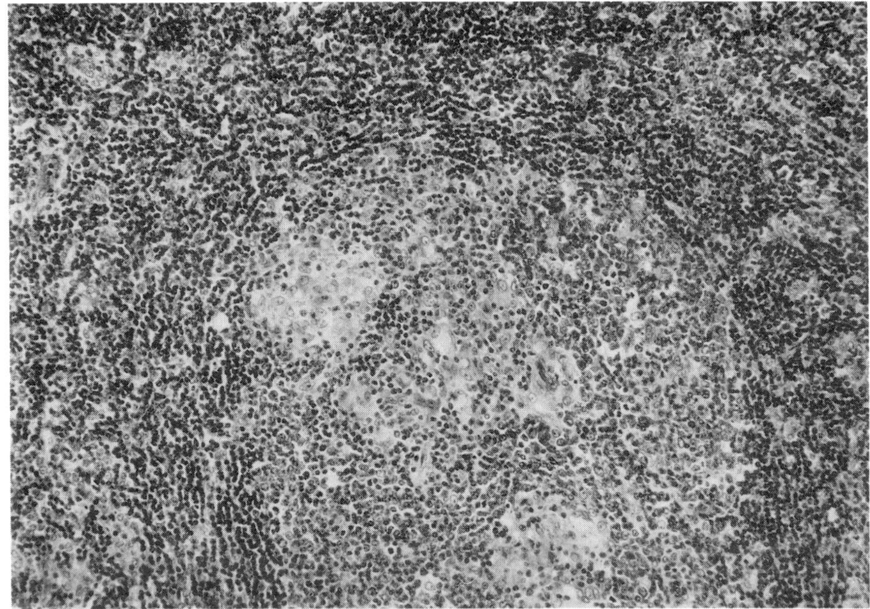

Fig. 4.5 Toxoplasmic lymphadenitis showing pale epithelioid cell clusters within a hyperplastic germinal centre. (H & E × 400).

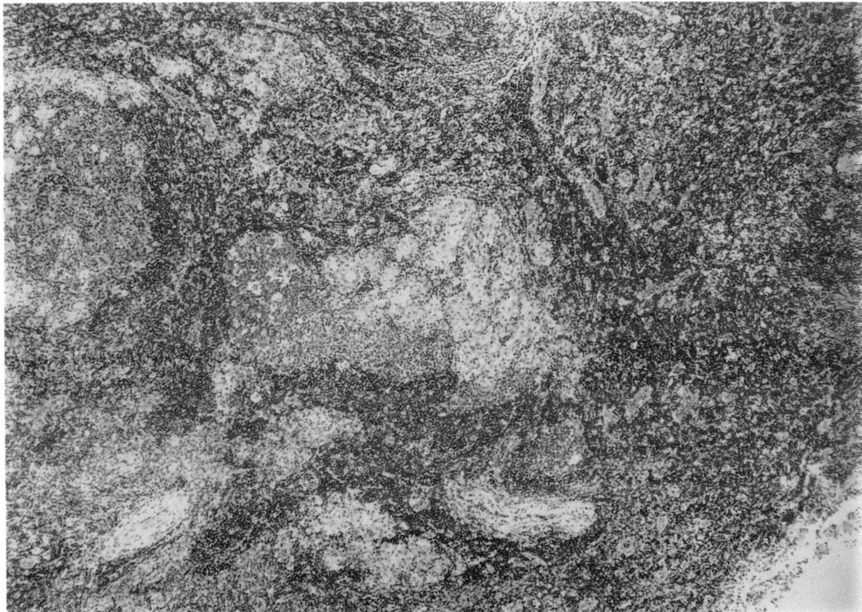

Fig. 4.6 Toxoplasmic lymphadenitis showing confluent pale epithelioid cell clusters and follicular hyperplasia. (H & E × 160).

with inflammatory infiltrates and arteritis at the edges, and oedema plus mild astrocytosis in adjacent brain sections.[57] This is not diagnostic on its own. In wet smears or in body fluids, such as CSF,[102] tachyzoites may be demonstrated by eosin–methylene blue fast,[96] immunofluorescence,[96] or Giemsa.[102] It is also possible, but difficult, to see tachyzoites in blood films, or tissue cysts in homogenised brain, by phase contrast microscopy.

Culture

The ability to grow toxoplasma in the laboratory, in animals or tissue culture, expands the number of parasites present in the original clinical specimen to a level at which they can be detected. The need to pretreat the specimen before inoculation depends on how accessible the parasites are. During infection body fluids such as CSF,[100] respiratory fluids,[96] amniotic fluid,[114] or blood[114] may contain tachyzoites. They can be inoculated directly, after centrifugal concentration. In the case of blood, parasites may be intracellular, within leucocytes, so the buffy coat[96] or clot[101] is usually separated as source inoculum. Solid tissues, such as placenta,[100] brain,[2] or muscle[2,100] in which cysts are likely, are predigested with pepsin[2] or preferably trypsin[100] (0.25%) for 1–2 h at 37 °C before inoculation. Lymph nodes are merely ground up[3] or teased apart. Toxoplasma can be grown in a variety of laboratory animals, mice, cotton rats, gerbils, but white mice are the standard for clinical specimens. Up to 1 ml of concentrated suspension is inoculated intraperitoneally. Results are slow and 3–6 weeks are normally allowed to elapse before testing for evidence of an antibody response. The brains of antibody-positive mice are homogenized, with glass beads in saline, and examined by phase contrast microscopy for the presence of tissue cysts. It is sometimes necessary to re-pass homogenized brain (up to 1 ml intraperitoneally) in more mice to attain a level of infection at which confirmation by microscopy is possible. This procedure is often advisable even when mice in the first passage are seronegative.

The possibility of using *in vitro* methods of cultivation of toxoplasma has been known for some time; its application in practice to isolate the parasite has only recently been introduced. The organism can multiply in a variety of cell lines: established continuous lines such as L-cells,[101] VERO,[28] or Hep-2;[28] finite passage lines[28] such as MRC-5; and primary monkey kidney.[28] The human embryonic fibroblast line, MRC-5, has been used to isolate toxoplasma from clinical specimens.[92,96] Fibroblasts on cover slips are exposed to a suspension of specimen concentrate in culture medium for 24 h before replacement of the medium and continuing culture. Infected cells can be identified within 4 days by direct microscopy with Giemsa staining,[115] but more rapidly by indirect immunofluorescence on acetone fixed cultures[115] using rabbit anti-toxoplasma antibody, then fluorescent anti-rabbit IgG. Positive isolation results have been obtained from amniotic fluids,[92] blood,[96] lung biopsy,[96] and bronchiolar lavage fluid,[96] sometimes as early as 2 days after inoculation.[96]

Interpretation

In the majority of instances when a laboratory diagnosis of toxoplasmosis is required, serological methods are still more useful than detection of the parasite directly or by isolation. There are also doubts about the sensitivity and specificity of these latter procedures. Non-homogeneous distribution or low concentration of the parasite in infected tissue leads to false negative results with some staining procedures;[96,99] the presence of tissue cysts in latently infected individuals means that a positive result is not necessarily equivalent to active infection.[99] In practice, in the two most serious situations when this sort of investigation is necessary, co-incidence of a cyst at the site of some other infection is unlikely. In cerebral toxoplasmosis, the most useful finding is the presence of tachyzoites, by peroxidase–anti-peroxidase[99] (PAP). It suggests active dissemination of the parasite. Cysts are rare in necrotic tissue.[97] In congenital infection, the fetus or neonate cannot be latently infected, so identification or isolation of parasites from any site must be diagnostic. Even from the placenta, a positive isolate is almost 90 per cent likely to be obtained when the fetus has been infected.[116] It may not prove that the fetus is infected, but there is strong evidence that it does.[116]

Ranking these procedures in order of relative sensitivity and specificity is difficult (Table 4.5). Animal inoculation is more sensitive than tissue culture,[92] possibly because intracellular parasites are not released in the 24 hr absorption period before the inoculum is removed and replaced by fresh medium. It is also more sensitive than direct staining.[99] Marginally more cases of cerebral toxoplasmosis were identified in one study by mouse inoculation of brain biopsies than by any staining procedure. The latter is seldom even considered as a method of proving congenital infection. Recognition of parasite morphology is important and the use of immunofluorescent-labelled antibody in tissue cultures enhances specificity. Tissue cysts are relatively easily identified even if rather faintly stained by some histological methods. Tachyzoites can also be seen using these stains,[99] but more easily with immunospecific methods such as PAP.[97] Reservations have been expressed concerning the reliability of immunofluorescence on sections without adequate controls or absorption of reagents.[109] It is probably necessary to use direct detection methods in combination to increase the chances of a positive diagnosis.[99] This is obviously more demanding on staff time per specimen, but, as speed is an important consideration, the following protocols are suggested. For brain biopsies, unfixed tissue should be examined by immunofluorescence while sections are prepared for routine histology and PAP.[97] Tissue culture is a reasonably rapid alternative if facilities and expertise are available, e.g. in an associated virology laboratory. Mouse inoculation should still be undertaken as the most sensitive technique in current use, but decisions on treatment are likely to have been made long before a result is available (Chapter 5). For diagnosis of fetal or neonatal infection, tissue culture can provide a rapid result,[92] but again mouse inoculation is essential to maximize detection. There is clearly a need for a technique which is both rapid and sensitive, and the polymerase chain reaction may fulfil it (Chapter 8).

OTHER INDICATORS OF INFECTION

In some conditions, circumstantial evidence of active toxoplasmosis can be deduced from radiological, haematological, or biochemical signs of damage. Clinical encephalitis in immunocompromised patients is investigated by CT scanning (Fig. 4.4), and cerebral toxoplasmosis features prominently in the differential diagnosis. AIDS is the commonest underlying cause for cerebral toxoplasmosis,[99,107,117] but it can occur in other immunocompromised patients.[118] Single or multiple nodular or annular lesions usually in the cerebral hemispheres appear hypo- or isodense and are enhanced by contrast medium.[97,107,117] They are more likely to be multiple and be associated with oedema than the lesions of cerebral lymphoma.[117] (Fig 4.7) Excision or needle biopsy is guided by CT scanning[99] and scans are also used to monitor reduction in lesion size during therapy.[117,118] A more sensitive, but less commonly used, alternative to CT is magnetic resonance imaging.[119] It is recommended as a backup to CT.[120] CT scanning can assist the diagnosis of congenital toxoplasmosis in neonates by identifying calcifications and hydrocephalus.[121] It could also help confirm whether a ventricular shunt is operating satisfactorily.[121] Another type of scanning technique, ultrasound, is regularly used during pregnancy to help diagnose fetal infection.[105,114] Ventricular dilation, presence of ascites, intracranial calcification, hepatomegaly, or placental width can indicate infection. Like CT scanning, it is used to guide sample collection, this time of fetal blood, for evidence of infection.

Various abnormal haematological or biochemical measurements are useful in some toxoplasma infections. Mild to moderate elevation of CSF protein concentration is the commonest, though not universal, abnormality in cerebral toxoplasmosis.[99,107] Non-specific signs which are useful in the diagnosis of fetal infection are elevated γ-glutamyltransferase, elevated total IgM, elevated white cell count, reduced platelet count, and to a lesser extent, eosinophilia or elevated lactic dehydrogenase.[105,114] None of these measurements on their own, or even in combination, provide proof of toxoplasma infection (they could be abnormal from a variety of other causes) but they support specific evidence which is no more than suggestive. One highly promising indicator of fetal infection is an elevated γ-interferon level in fetal

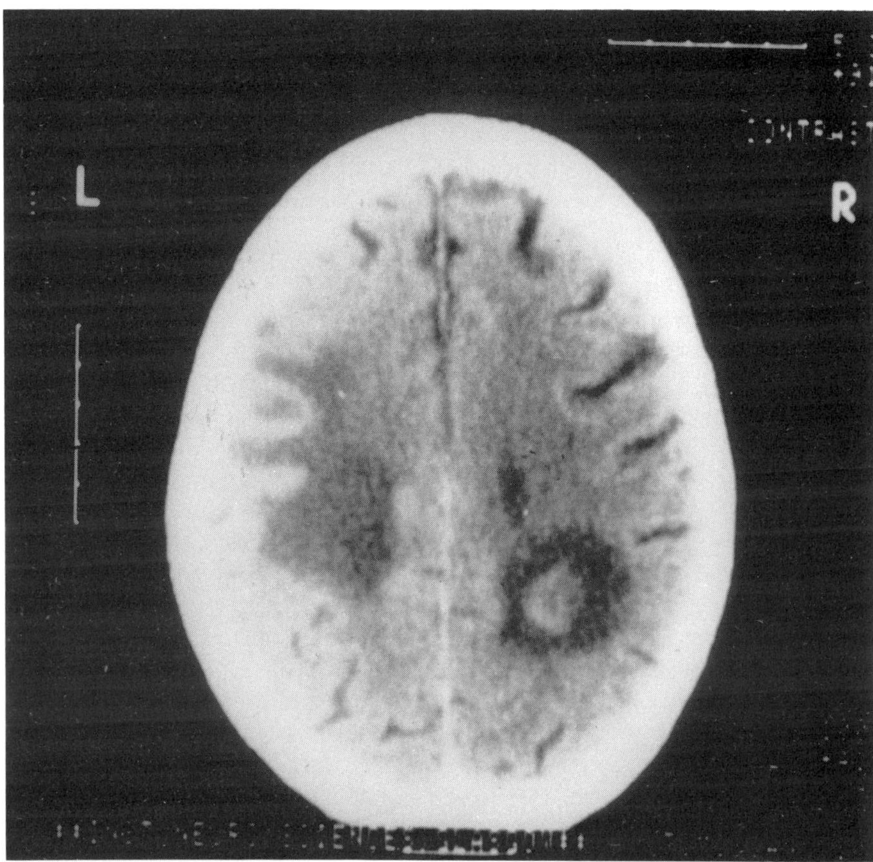

Fig. 4.7 Toxoplasma encephalitis CT scan: two regions with white matter oedema and an enhancing central focus on the right.

blood.[122] Measured by radioimmunoassay with monoclonal anti-γ-interferon, it offers a rapid sensitive diagnosis. Finally, response to anti-toxoplasma therapy itself is currently part of diagnosis of cerebral toxoplasmosis[123] (Fig. 4.4). Obviously, it is always desirable to have a diagnosis before therapy, but the urgency of the illness, the poor sensitivity of non-invasive procedures (serology) and the invasiveness of the currently most sensitive definitive procedure (brain biopsy) overrides this objection. More sensitive techniques (Chapter 8) may improve the value of tests on CSF (Fig. 4.4).

SUMMARY

The symptoms of toxoplasmosis are insufficiently distinct to permit a reliable clinical diagnosis. Current laboratory techniques are useful in some clinical settings, but not in others. Serological methods provide the standard means of diagnosing acquired disease in healthy individuals. They are essential and perfectly adequate for the great majority of cases. In peripheral laboratories, one or two simple screening tests (HA, latex, ELISA, IF) provide a preliminary diagnosis. Which test is used will depend on individual laboratory circumstances and in some cases only IgG antibody will be measured. Preliminary tests can be used to exclude toxoplasmosis from the differential diagnosis.

Definitive diagnosis is best achieved in Reference Laboratories which have the facilities to carry out a series of tests for IgG and IgM antibody. Many still include the dye test (DT), a bioassay with a long pedigree which requires considerable expertise in performance and maintenance. The presence of specific IgM in conjunction with a significant DT titre is in most instances indicative of current infection. However, variable persistence of specific IgM tends to blur the distinction between current, recent, and persistent infection. This can pose problems in determining the precise time of infection, especially in pregnancy. Experienced interpretation of the results of all the IgM and IgG tests often helps to differentiate between these alternatives. Knowledge of the relative sensitivities of IgM tests (which now are usually ELISA tests) can also assist in defining the stage of infection. Determining past infection, by the absence of specific IgM, is usually not a problem. In diagnosing fetal or neonatal infection, maximum sensitivity is paramount, and the ISAGA or the newer, more complex ELIFA are probably the best available, but are still less than 90 per cent sensitive. Diagnosis then relies on the combined evidence from serological, isolation, scanning, and various non-specific methods.

Measurement of specific IgA or IgE, or IgG avidity, may supplement and perhaps improve on the evidence for current infection from IgM tests. Assays for detection of circulating antigen or immune complexes, again usually by ELISA tests, have proved disappointing. They often fail to confirm current infection, presumably because the antigenaemic phase is variable and frequently short. Detection of antigen in a priveleged site, such as amniotic fluid or CSF, is conclusive proof of infection. Antibody in CSF, or aqueous humour, provided it is proven to be locally synthesized, is also diagnostic of local, active infection. Again, however, detection rates are less than satisfactory, particularly in CSF. Better results are obtained from aqueous humour, but collection of this type of specimen is rarely attempted. Diagnosis of ocular toxoplasmosis then is virtually entirely based on clinical acumen, with the laboratory relegated to confirming past infection.

Identification or isolation of the parasite itself is of most use in the diagnosis of fetal infection or of cerebral toxoplasmosis in the immunocompromised. Toxoplasma tissue cysts are easier to see by staining techniques than tachyzoites, but as the latter may provide a better indication of active infection, the more appropriate stains are those which best illustrate them.

Peroxidase–anti-peroxidase appears to be the most sensitive. Culture of toxoplasma in animals is the most sensitive overall method of proving its presence. Its disadvantage is the length of time required before a result is available. Tissue culture on coverslips produces a more rapid result, especially if infected cells are identified by IF.

At present diagnosis of cerebral toxoplasmosis is often empirical, with supportive evidence from CT scans and serological evidence of past infection. Like ocular toxoplasmosis, the problem is diagnosis of reactivated infection for which standard serological tests are limited. Brain biopsy and demonstration of toxoplasma tachyzoites is often undertaken late, and reluctantly. Infection in the immunocompromised is the area in which improved diagnostic procedures are most urgently required. Preferably these should not be invasive and the results from studies using new, highly sensitive assays such as polymerase chain reaction, on accessible body fluids are awaited with anticipation.

REFERENCES

1 Remington, J. S. (1974). Toxoplasmosis in the adult. *Bull. N. Y. Acad. Med.*, **50**, 211–27.

2 Remington, J. S. and Cavanaugh, E. N. (1965). Isolation of the encysted form of *Toxoplasma gondii* from human skeletal muscle and brain. *N. Engl. J. Med.*, **273**, 1308–10.

3 Beverley, J. K. A. and Beattie, C. P. (1958). Glandular toxoplasmosis. A survey of 30 cases. *Lancet*, **ii**, 379–84.

4 Carey, R. M., Kimball, A. C., Armstrong, D., and Lieberman, P. H. (1973). Toxoplasmosis: clinical experiences in a cancer hospital. *Am. J. Med.*, **54**, 30–8.

5 Dutton, G. N. (1989). Toxoplasmic retinochoroiditis — a historical review and current concepts. *Ann. Acad. Med.*, **18**, 214–21.

6 Vischer, T. L., Bernheim, C., and Engelbrecht, E. (1967). Two cases of hepatitis due to *Toxoplasma gondii. Lancet*, **ii**, 919–21.

7 Cottrell, A. J. (1986). Acquired toxoplasma encephalitis. *Arch. Dis. Child.*, **61**, 84–5.

8 Balfour, A. H., Fleck, D. G., Hughes, H. P. A., and Sharp, D. (1982). Comparative study of three tests (dye test, indirect haemagglutination test, latex agglutination test) for the detection of antibodies to *Toxoplasma gondii* in human sera. *J. Clin. Pathol.*, **35**, 228–32.

9 Thorburn, H. and Williams, H. (1972). A stable haemagglutinating antigen for detecting toxoplasma antibodies. *J. Clin. Pathol.*, **25**, 762–7.

10 van Loon A. M. and van der Veen, J. (1980). Enzyme-linked immunosorbent assay for quantitation of toxoplasma antibodies in human sera. *J. Clin. Pathol.*, **33**, 635–9.

11 Walton, B. C., Benchoff, B. M., and Brooks, W. H. (1966). Comparison of indirect fluorescent antibody test and methylene blue dye test for detection of antibodies to *Toxoplasma gondii. Am. J. Trop. Med. Hyg.*, **15**, 149–52.

12 Sabin, A. B. and Feldman, H. A. (1948). Dyes as microchemical indicators of a new immunity phenomenon affecting a protozoon parasite (Toxoplasma). *Science*, **108**, 660–3.

13 Ritzmann, S. E. (1975). Immunoglobulin abnormalities. In *Serum protein abnormalities: diagnostic and clinical aspects.* (ed. S. E. Ritzman and J. C. Daniels), pp. 351–485. Little, Brown & Co, Boston.

14 Gray, J. J., Balfour, A. H., and Wreghitt, T. G. (1990). Evaluation of a commercial latex agglutination test for detecting antibodies to *Toxoplasma gondii. Serodiag. Immunotherap. Infect. Dis.*, **4**, 335–40.

15 Moyer, N. P., Hudson, J. D., and Hausler, W. J. Jr. (1987). Evaluation of MUREX SUDS Toxo Test. *J. Clin. Microbiol.*, **25**, 2049–53.

16 Balsari, A., Poli, G., Molina, V., Dovis, M., Petruzzelli, E., Boniolo, A. *et al.* (1980). ELISA for toxoplasma antibody detection: a comparison with other serodiagnostic tests. *J. Clin. Pathol.*, **33**, 640–3.

17 Joss, A. W. L., Skinner, L. J., Chatterton, J. M. W., Cubie, H. A., Pryde, J. F. D., and Campbell, J. D. (1989). Toxoplasmosis: effectiveness of enzyme immunoassay screening. *Med. Lab. Sci.*, **46**, 107–12.

18 Lin, T. M., Halbert, S. P., and O'Connor, G. R. (1980). Standardised quantitative enzyme-linked immunoassay for antibodies to *Toxoplasma gondii. J. Clin. Microbiol.*, **11**, 675–81.

19 Sutehall, G. M. and Wreghitt, T. G. (1989). False positive latex tests negative by ELISA for toxoplasma IgG. *J. Clin. Pathol.*, **42**, 204–5.

20 Balfour, A. H., Bridges, J. B., and Harford, J. P. (1980). An evaluation of the ToxHA test for the detection of antibodies to *Toxoplasma gondii* in human serum. *J. Clin. Pathol.*, **33**, 644–7.

21 Holliman, R. E., Johnson, J., Duffy, K., and New, L. (1989). Discrepant toxoplasma latex agglutination test results. *J. Clin. Pathol.*, **42**, 200–3.

22 Wilson, M., Ware, D. A., and Walls, K. W. (1987). Evaluation of commercial serodiagnostic kits for toxoplamosis. *J. Clin. Microbiol.*, **25**, 2262–5.

23 Holliman, R. E. (1988). Toxoplasmosis and the acquired immune deficiency syndrome. *J. Infect.*, **16**, 121–8.

24 Joss, A. W. L., Skinner, L. J., Chatterton, J. M. W., Chisholm, S. M., Williams, H. D., and Ho-Yen, D. O. (1988). Simultaneous serological screening for congenital cytomegalovirus and toxoplasma infection. *Public Health*, **102**, 409–17.

25 Lelong, M. and Desmonts, G. (1951). L'emploi du microscope à contraste de phase dans la réaction de Sabin–Feldman. *C. R. Séances Soc. Biol. Fil.*, **145**, 1660–1.

26 Feldman, H. A. and Lamb, G. A. (1966). A micromodification of the toxoplasma dye test. *J. Parasitol.*, **52**, 415.

27 Williams, K. A. B., Scott, J. M., MacFarlane, D. E., Williamson, J. M. W., Elias-Jones, T. F., and Williams, H. (1981). Congenital toxoplasmosis: a prospective survey in the west of Scotland. *J. Infect.*, **3**, 219–29.

28 Hughes, H. P. A., Hudson, L., and Fleck, D. G. (1986). *In vitro* culture of *Toxoplasma gondii* in primary and established cell lines. *Int. J. Parasitol.*, **16**, 317–22.

29 Payne, R. A. (1966). The effect of citrate radical on the dye test for toxoplasmosis. *Mon. Bull. Min. Health PHLS*, **25**, 161–4.

30 Desmonts, G. and Remington, J. S. (1980). Direct agglutination test for diagnosis of *toxoplasma* infection: method for increasing sensitivity and specificity. *J. Clin. Microbiol.*, **11**, 562–8.

31 Fleck, D. G. and Kwantes, W. (1980). The laboratory diagnosis of toxoplasmosis. *PHLS Monogr. Ser.*, No. 13. HMSO, London.

32 Remington, J. S. and Miller, M. J. (1966). 19S and 7S anti-toxoplasma antibodies in diagnosis of acute congenital and acquired toxoplasmosis. *Proc. Soc. Exp. Biol. Med.*, **121**, 357–63.

33 Pyndiah, N., Krech, U., Price, P., and Wilhelm, J. (1979). Simplified chromatographic separation of immunoglobulin M from G and its application to toxoplasma indirect immunofluorescence. *J. Clin. Microbiol.*, **9**, 170–4.

34 Filice, G., Meroni, V., Carnevale, G., Benzi-Cipelli, R., and Carosi, G. (1981). Reliability of IgM-IFA and IgM-IHA tests on pure IgM fractions obtained by a simple gel filtration method in acquired and congenital toxoplasmosis. *Biol. Res. Pregn.*, **2**, 114–6.

35 Joss, A. W. L., Skinner, L. J., Moir, I. L., Chatterton, J. M. W., Williams, H., and Ho-Yen, D. O. (1989). Biotin-labelled antigen screening test for toxoplasma IgM antibody. *J. Clin. Pathol.*, **42**, 206–9.

36 Payne, R. A., Joynson, D. H. M., Balfour, A. H., Harford, J. P., Fleck, D. G., Mythen, M. *et al.* (1987). Public Health Laboratory Service enzyme-linked immunosorbent assay for detecting toxoplasma specific IgM antibody. *J. Clin. Pathol.*, **40**, 276–81.

37 Herbrink, P., van Loon, A. M., Rotmans, J. P., van Knapen, F., and van Dijk, W. C. (1987). Interlaboratory evaluation of indirect enzyme-linked immunosorbent assay, antibody capture enzyme-linked immunosorbent assay, and immunoblotting for detection of immunoglobulin M antibodies to *Toxoplasma gondii*. *J. Clin. Microbiol.*, **25**, 100–5.

38 Remington, J. S., Miller, M. J., and Brownlee, I. (1968). IgM antibodies in acute toxoplasmosis. II. Prevalence and significance in acquired cases. *J. Lab. Clin. Med.*, **71**, 855–66.

39 Skinner, L. J., Chatterton, J. M. W., Joss, A. W. L., Moir, I. L., and Ho-Yen, D. O. (1989). The use of an IgM immunosorbent agglutination assay to diagnose congenital toxoplasmosis. *J. Med. Microbiol.*, **28**, 125–8.

40 Desmonts, G., Naot, Y., and Remington, J. S. (1981). Immunoglobulin M-immunosorbent agglutination assay for diagnosis of infectious diseases: diagnosis of acute congenital and acquired toxoplasma infections. *J. Clin. Microbiol.*, **14**, 486–91.

41 Weilaard, F., van Gruijthuijsen, H., Duermeyer, W., Joss, A. W. L., Skinner, L., Williams, H. *et al.* (1983). Diagnosis of acute toxoplasmosis by an enzyme immunoassay for specific immunoglobulin M antibodies. *J. Clin. Microbiol.*, **17**, 981–7.

42 van Loon, A. M., van der Logt, J. T. M., Heessen, F. W. A., and van der Veen, J. (1983). Enzyme-linked immunosorbent assay that uses labeled antigen for detection of immunoglobulin M and A antibodies in toxoplasmosis: comparison with indirect immunofluorescence and double-sandwich enzyme-linked immunosorbent assay. *J. Clin. Microbiol.*, **17**, 997–1004.

43 Verhofstede, C., van Renterghem, L., and Plum, J. (1989). Comparison of six commercial enzyme linked immunosorbent assays for dectecting IgM antibodies against *Toxoplasma gondii*. *J. Clin. Pathol.*, **42**, 1285–90.

44 Joynson, D. H. M., Payne, R. A., Balfour, A. H., Prestage, E. S., Fleck, D. G., and Chessum, B. S. (1989). Five commercial enzyme linked immunosorbent assay kits for toxoplasma specific IgM antibody. *J. Clin. Pathol.*, **42**, 653–7.

45 Araj, G. F. and Thorburn, H. (1986). Detection of toxoplasma specific IgM by a solid-phase haemadsorption assay. *Med. Lab. Sci.*, **43**, 356–9.

46 Camargo, M. E., Ferreira, A. W., Mineo, J. R., Takiguti, C. K., and Nakahara, O. S. (1978). Immunoglobulin G and immunoglobulin M enzyme-linked immunosorbent assays and defined toxoplasmosis serological patterns. *Infect. Immun.*, **21**, 55–8.

47 Araujo, F. G., Barnett, E. V., Gentry, L. O., and Remington, J. S. (1971). False-positive anti-toxoplasma fluorescent-antibody tests in patients with antinuclear antibodies. *Appl. Microbiol.*, **22**, 270–5.

48 Siegel, J. P. and Remington, J. S. (1983). Comparison of methods for quantitating antigen-specific immunoglobulin M antibody with a reverse enzyme-linked immunosorbent assay. *J. Clin. Microbiol.*, **18**, 63–70.

49 Wreghitt, T. G., Hakim, M., Gray, J. J., Balfour, A. H., Stovin, P. G. I., Stewart, S. *et al.* (1989). Toxoplasmosis in heart and heart and lung transplant recipients. *J. Clin. Pathol.*, **42**, 194–9.

50 Luft, B. J., Naot, Y., Araujo, F., Stinson, E. B., and Remington, J. S. (1983). Primary and reactivated toxoplasma infection in patients with cardiac transplants. Clinical spectrum and problems in diagnosis in a defined population. *Ann. Intern. Med.*, **99**, 27–31.

51 Luft, B. J. and Remington, J. S. (1985). Toxoplasmosis of the central nervous system. *Curr. Clin. Top. Infect. Dis.*, **6**, 315–58.

52 Hakes, T. B. and Armstrong, D. (1983). Toxoplasmosis: problems in diagnosis and treatment. *Cancer*, **52**, 1535–40.

53 van Renterghem, L. and van Nimmen, L. (1976). Indirect immunofluorescence in toxoplasmosis: frequency, nature and specificity of polar staining. *Zentralbl. Bakteriol., Mikrobiol, Hyg.*, **A 235**, 559–65.

54 Kelen, A. E., Ayllon-Leindl, L., and Labzoffsky, N. A. (1962). Indirect fluorescent antibody method in serodiagnosis of toxoplasmosis. *Can. J. Microbiol.*, **8**, 545–54.

55 Potasman, I., Araujo, F. G., and Remington, J. S. (1986). Toxoplasma antigens recognised by naturally occurring human antibodies. *J. Clin. Microbiol.*, **24**, 1050–4.

56 Suzuki, Y. and Remington, J. S. (1990). Importance of membrane-bound antigens of *Toxoplasma gondii* and their fixation for serodiagnosis of toxoplasmic encephalitis in patients with acquired immunodeficiency syndrome. *J. Clin. Microbiol.*, **28**, 2354–6.

57 Luft, B. J., Brooks, R. G., Conley, F. K., McCabe, R. E., and Remington, J. S. (1984). Toxoplasmic encephalitis in patients with acquired immune deficiency syndrome. *JAMA*, **252**, 913–7.

58 Thulliez, P., Remington, J. S., Santoro, F., Ovlaque, G., Sharma, S., and Desmonts, G. (1986). Une nouvelle réaction d'agglutination pour le diagnostic du stade évolutif de la toxoplasmose acquise. *Pathol. Biol.*, **34**, 173–7.

59 Suzuki, Y., Israelski, D. M., Dannemann, B. R., Stepick-Biek, P., Thulliez, P., and Remington, J. S. (1988). Diagnosis of toxoplasmic encephalitis in patients with acquired immunodeficiency syndrome by using a new serologic method. *J. Clin. Microbiol.*, **26**, 2541–3.

60 Holliman, R. E., Johnson, J., Burke, M., Adams, S., and Pepper, J. R. (1990). False-negative dye-test findings in a case of fatal toxoplasmosis associated with cardiac transplantation. *J. Infect.*, **21**, 185–9.

61 Naot, Y. and Remington, J. S. (1980). An enzyme-linked immunosorbent assay for detection of IgM antibodies to *Toxoplasma gondii*: use for diagnosis of acute acquired toxoplasmosis. *J. Infect. Dis.*, **142**, 757–66.

62 Naot, Y., Barnett, E. V., and Remington, J. S. (1981). Method for avoiding false-positive results occurring in immunoglobulin M enzyme-linked immunosorbent assays due to presence of both rheumatoid factor and antinuclear antibodies. *J. Clin. Microbiol.*, **14**, 73–8.

63 Naot, Y., Luft, B. J., and Remington, J. S. (1981). False-positive serological tests in heart transplant recipients. *Lancet*, **ii**, 590–1.

64 Naot, Y., Desmonts, G., and Remington, J. S. (1981). IgM enzyme-linked immunosorbent assay test for the diagnosis of congenital toxoplasma infection. *J. Pediatr.*, **98**, 32–6.

65 Duffy, K. T., Wharton, P. J., Johnson, J. D., New, L., and Holliman, R. E. (1989). Assessment of immunoglobulin-M immunosorbent agglutination assay (ISAGA) for detecting toxoplasma specific IgM. *J. Clin. Pathol.*, **42**, 1291–5.

66 Sulzer, A. J., Franco, E. L., Takafuji, E., Benenson, M., Walls, K. W., and Greenup, R. L. (1986). An oocyst-transmitted outbreak of toxoplasmosis: patterns of immunoglobulin G and M over one year. *Am. J. Trop. Med. Hyg.*, **35**, 290–6.

67 Hedman, K., Lappalainen, M., Seppaia, I., and Makela, O. (1989). Recent primary toxoplasma infection indicated by a low avidity of specific IgG. *J. Infect. Dis.*, **159**, 736–40.

68 Derouin, F., Sulcebe, G., and Ballet, J-J. (1987). Sequential determination of IgG subclasses and IgA specific antibodies in primary and reactivating toxoplasmosis. *Biomed. Pharmacother.*, **41**, 429–33.

69 Huskinson, J., Stepick-Biek, P. N., Araujo, F. G., Thulliez, P., Suzuki, Y., and Remington, J. S. (1989). Toxoplasma antigens recognized by immunoglobulin G subclasses during acute and chronic infection. *J. Clin. Microbiol.*, **27**, 2031–8.

70 Decoster, A., Darcy, F., Caron, A., and Capron, A. (1988). IgA antibodies against P30 as markers of congenital and acute toxoplasmosis. *Lancet*, **ii**, 1104–7.

71 Stepick-Biek, P., Thulliez, P., Araujo, F. G., and Remington, J. S. (1990). IgA antibodies for diagnosis of acute congenital and acquired toxoplasmosis. *J. Infect. Dis.*, **162**, 270–3.

72 Pinon, J. M., Toubas, D., Marx, C., Mougeot, G., Bonnin, A., Bonhomme, A. *et al.* (1990). Detection of specific immunoglobulin E in patients with toxoplasmosis. *J. Clin. Microbiol.*, **28**, 1739–43.

73 Pinon, J. M., Thoannes, H., and Gruson, N. (1985). An enzyme-linked immuno-filtration assay used to compare infant and maternal antibody profiles in toxoplasmosis. *J. Immunol. Methods*, **77**, 15–23.

74 van Knapen, F. and Panggabean, S. O. (1977). Detection of circulating antigen during acute infections with *Toxoplasma gondii* by enzyme-linked immunosorbent assay. *J. Clin. Microbiol.*, **6**, 545–7.

75 Araujo, F. G., and Remington, J. S. (1980). Antigemenia in recently acquired acute toxoplasmosis. *J. Infect. Dis.*, **141**, 144–50.

76 Brooks, R. G., Sharma, S. D., and Remington, J. S. (1985). Detection of *Toxoplasma gondii* antigens by a dot-immunobinding technique. *J. Clin. Microbiol.*, **21**, 113–6.

77 van Knapen, F., Panggabean, S. O., and van Leusden, J. (1985). Demonstration of toxoplasma antigen containing complexes in active toxoplasmosis. *J. Clin. Microbiol.*, **22**, 645–50.

78 Siegel, J. P. and Remington, J. S. (1983). Circulating immune complexes in toxoplasmosis: detection and clinical correlates. *Clin. Exp. Immunol.*, **52**, 157–63.

79 Hassl, A., Aspock, H., and Flamm, H. (1988). Circulating antigen of *Toxoplasma gondii* in patients with AIDS: significance of detection and structural properties. *Zentralbl. Bakteriol., Mikrobiol. Hyg., A.*, **270**, 302–9.

80 Candolfi, E., Derouin, F., and Kien, T. (1987). Detection of circulating antigens in immunocompromised patients during reactivation of chronic toxoplasmosis. *Eur. J. Clin. Microbiol.*, **6**, 44–8.

81 Huskinson, J., Stepick-Biek, P., and Remington, J. S. (1989). Detection of antigens in urine during acute toxoplasmosis. *J. Clin. Microbiol.*, **27**, 1099–101.

82 Hassl, A. and Aspock, H. (1990). Antigens of *Toxoplasma gondii* recognised by sera of AIDS patients before, during, and after clinically important infections. *Int. J. Med. Microbiol.*, **272**, 514–25.

83 Chang., G-N, Nemzek, J. A., Tjostem, J. L., and Gabrielson, D. A. (1985). Simple haemagglutination inhibition test for the diagnosis of toxoplasmosis. *J. Clin. Microbiol.*, **21**, 180–3.

84 Rougier, D. and Ambroise-Thomas, P. (1985). Detection of toxoplasmic immunity by multipuncture skin test with excretory–secretory antigen. *Lancet*, **ii**, 121–3.

85 Gehle, W. D., Smith, K. O., and Fuccillo, D. A. (1976). Radioimmunoassay for toxoplasmosis. *Infect. Immun.*, **14**, 1253–5.

86 Konishi, E. and Takahashi, J. (1983). Reproducible enzyme-linked immunosorbent assay with a magnetic processing system for diagnosis of toxoplasmosis. *J. Clin. Microbiol.*, **17**, 225–31.

87 Gordon, M. A., Duncan, R. A., and Kingsley, L. C. (1981). Automated immuno-fluorescence test for toxoplasmosis. *J. Clin. Microbiol.*, **13**, 283–5.

88 Safford, J. W., Abbott, G. G., Craine, M. C., and MacDonald, R. G. (1991). Automated microparticle enzyme immunoassays for IgG and IgM antibodies to *Toxoplasma gondii*. *J. Clin. Pathol.*, **44**, 238–42.

89 Pedersen, O. O. and Lorentzen-Styr, A-M. (1981). Antibodies against *Toxoplasma gondii* in the aqueous humour of patients with active retinochoroiditis. *Acta Ophthalmol.*, **59**, 719–26.

90 Saari., K. M., Turunen, H., Leinikki, P. O., Krause, U., Pohjanpelto, P. J., Parviainen, M. *et al.* (1981). Enzyme-linked immunosorbent assay (ELISA) for serological diagnosis of toxoplasmic chorioretinitis. *Acta Ophthalmol*, **59**, 800–9.

91 Kijlstra, A., Breebart, A. C., Baarsma, G. S., Bos, P. J. M., Rothova, A., Luyendijk, L. (1986). Aqueous chamber taps in toxoplasmic chorioretinitis. *Doc. Ophthalmol.*, **64**, 53–8.

92 Derouin, F., Thulliez, P., Candolfi, E., Daffos, F., and Forestier, F. (1988). Early prenatal diagnosis of congenital toxoplasmosis using amniotic fluid samples and tissue culture. *Eur. J. Clin. Microbiol. Infect. Dis.*, **7**, 423–5.

93 Potasman, I., Resnick, L., Luft, B. J., and Remington, J. S. (1988). Intrathecal production of antibodies against *Toxoplasma gondii* in patients with toxoplasmic encephalitis and the acquired immunodeficiency syndrome (AIDS). *Ann. Intern. Med.*, **108**, 49–51.

94 Kijlstra, A., Luyendijk, L., Baarsma, G. S., Rothova, A., Schweitzer, C. M. C., Timmerman, Z. *et al.* (1989). Aqueous humour analysis as a diagnostic tool in toxoplasma uveitis. *Int. Ophthalmol.*, **13**, 383–6.

95 van Loon, A. M. (1989). Laboratory diagnosis of toxoplasmosis. *Int. Ophthalmol.*, **13**, 377–81.

96 Derouin, F., Sarfati, C., Beauvais, B., Iliou, M-C., Dehen, L., and Lariviere, M. (1989). Laboratory diagnosis of pulmonary toxoplasmosis in patients with acquired immuno-deficiency syndrome. *J. Clin. Microbiol.*, **27**, 1661–3.

97 Conley, F. K., Jenkins, K. A., and Remington, J. S. (1981). *Toxoplasma gondii* infection of the central nervous system: use of the peroxidase–antiperoxidase method to demon-strate toxoplasma in formalin fixed, paraffin embedded tissue sections. *Human Pathol.*, **12**, 690–8.

98 Siim, J. C., and Nissen, N. I. (1958). Toxoplasmosis acquisita lymphonodosa in a 62-year-old woman: isolation of *Toxoplasma gondii* from lymph nodes and muscle biopsies. *Acta Pathol.*, **43**, 298–304.

99 Wanke, C., Tuazon, C. U., Kovacs, A., Dina, T., Davis, D. O., Barton, N. *et al.* (1987).

Toxoplasma encephalitis in patients with acquired immune deficiency syndrome: diagnosis and response to therapy. *Am. J. Trop. Med. Hyg.*, **36**, 509–16.

100 Desmonts, G., Couvreur, J., Ferrero, G., and Marcoux, M. (1974). L'isolement du parasite dans la toxoplasmose congenitale: interêt pratique et theorique. *Arch. Fr. Pédiatr.*, **31**, 157–66.

101 Hofflin, J. M. and Remington, J. S. (1985). Tissue culture isolation of toxoplasma from blood of a patient with AIDS. *Arch. Intern. Med.*, **145**, 925–6.

102 Dos Santos, Neto, J. G. (1975). Toxoplasmosis: a historical review, direct diagnostic microscopy, and report of a case. *Am. J. Clin. Pathol.*, **63**, 909–15.

103 Ruskin, J. and Remington, J. S. (1976). Toxoplasmosis in the compromised host. *Ann. Intern. Med.*, **84**, 193–9.

104 Behan, W. M. H., Behan, P. O., Draper, I. T., and Williams, H. (1983). Does toxoplasma cause polymyositis? *Acta Neuropathol.*, **61**, 246–52.

105 Daffos, F., Forestier, F., Capella-Pavlovsky, M., Thulliez, P., Aufrant, C., Valenti, D. *et al.* (1988). Prenatal management of 746 pregnancies at risk for congenital toxoplasmosis. *N. Engl. J. Med.*, **318**, 271–5.

106 Remington, J. S. and Desmonts, G. (1990). Toxoplasmosis. In *Infectious diseases of the foetus and newborn infant.* (3rd edn). (ed. J. S. Remington and J. O. Klein), pp. 89–195, W. B. Saunders, Philadelphia.

107 Wong, B., Gold, J. W. M., Brown, A. E., Lange, M., Fried, R., Grieco, M. *et al.* (1984). Central-nervous-system toxoplasmosis in homosexual men and parenteral drug abusers. *Ann. Intern. Med.*, **100**, 36–42.

108 Grose, C., Itani, O., and Weiner, C. P. (1989). Prenatal diagnosis of fetal infection: advances from amniocentesis to cordocentesis — congenital toxoplasmosis, rubella, cytomegalovirus, varicella virus, parvovirus and human immunodeficiency virus. *Pediatr. Infect. Dis. J.*, **8**, 459–68.

109 Frenkel, J. K. and Piekarski, G. (1978). The demonstration of toxoplasma and other organisms by immunofluorescence: a pitfall. *J. Infect. Dis.*, **138**, 265–6.

110 Cerezo, L., Alvarez, M., and Price, G. (1985). Electron microscopic diagnosis of cerebral toxoplasmosis. *J. Neurosurg.*, **63**, 470–2.

111 Andres, T. L., Dorman, S. A., Winn, Jr, W. C., Trainer, T. D., and Perl, D. P. (1981). Immunohistochemical demonstration of *Toxoplasma gondii. Am. J. Clin. Pathol.*, **75**, 431–4.

112 Saxen, L., Saxen, E., and Tenhunen, A. (1962). The significance of histological diagnosis in glandular toxoplasmosis. *Acta Pathol. Microbiol. Scand.*, **56**, 284–94.

113 Dorfman, R. F. and Remington, J. S. (1973). Value of lymph-node biopsy in the diagnosis of acute acquired toxoplasmosis. *New. Engl. J. Med.*, **289**, 878–81.

114 Desmonts, G., Daffos, F., Forestier, F., Capella-Pavlovsky, M., Thulliez, PH., and Chartier, M. (1985). Prenatal diagnosis of congenital toxoplasmosis. *Lancet*, **i**, 500–4.

115 Derouin, F., Mazeron, M. C., and Garin, Y. J. F. (1987). Comparative study of tissue culture and mouse inoculation methods for demonstration of *Toxoplasma gondii. J. Clin. Microbiol.*, **25**, 1597–1600.

116 Desmonts, G. and Couvreur, J. (1974). Toxoplasmosis in pregnancy and its transmission to the fetus. *Bull. N. Y. Acad. Med.*, **50**, 146–59.

117 Harrison, P. B., Clarke, S. D., and Silver, S. F. (1990). Focal brain lesions on computed tomography in patients with acquired immune deficiency syndrome. *J. Can. Assoc. Radiol.*, **41**, 83–6.

118 Mazer, S., Araujo, J. C., Ribeiro, R. C., and Kasting, G. (1983). Unusual computed tomographic presentation of cerebral toxoplasmosis. *Am. J. Neuroradiol.*, **4**, 458–60.

119 Zee, C-S., Segall, H. D., Rogers, C., Ahmadi, J., Apuzzo, M., and Rhodes, R. (1985). MR imaging of cerebral toxoplasmosis: correlation of computed tomography and pathology. *J. Comput. Assist. Tomogr.*, **9**, 797–9.

120   Luft, B. J. and Remington, J. S. (1988). Toxoplasmic encephalitis. *J. Infect. Dis.*, **157**, 1–6.
121   Collins, A. T. and Cromwell, L. D. (1980). Computed tomography in the evaluation of congenital cerebral toxoplasmosis. *J. Comput. Assist. Tomogr.*, **4**, 326–9.
122   Raymond, J., Poissonnier, M. H., Thulliez, P. H., Forestier, F., Daffos, F., and Lebon, P. (1990). Presence of gamma interferon in human acute and congenital toxoplasmosis. *J. Clin. Microbiol.*, **28**, 1434–7.
123   Cohn, J. A., McMeeking, A., Cohen, W., Jacobs, J., and Holzman, R. S. (1989). Evaluation of the policy of empiric treatment of suspected toxoplasma encephalitis in patients with the acquired immunodeficiency syndrome. *Am. J. Med.*, **86**, 521–7.

5

Treatment

ALEX W. L. JOSS

The list of suitable treatments for toxoplasmosis is not long but it is not always easy to decide which treatment is appropriate. The question is often whether treatment of any sort is required. As the majority of infections in non-pregnant, immune competent individuals are asymptomatic or not recognized (Chapter 3), the question of treatment is not considered. The majority of symptomatic cases in this group also resolve without treatment.[1] The dilemma arises in the minority in whom disease persists. Decisions become more urgent when infection is suspected during pregnancy, and the many questions on treatment during pregnancy are considered in Chapter 6. Different problems are faced when disease reactivates, as in ocular toxoplasmosis or infection in the immunocompromised. In cerebral or disseminated toxoplasmosis in the immunocompromised, there are several difficulties: the use of potentially toxic therapy in often very ill patients; the efficacy and length of treatment. There appears to be no effective way of eradicating infection once the parasite has encysted,[2] so therapy is aimed at containing the damage caused by any released tachyzoites.

The drugs used in treatment differ according to the clinical condition. Originally, therapy was only considered for active infection. Now the purpose of treatment can be either curative or prophylactic. Prophylaxis can be for a limited period, for example to prevent infection at the time of a heart transplant;[3] or, in other clinical settings, it can involve long term maintenance.[4,5] This chapter will describe the drugs which are regularly used in the treatment of toxoplasmosis, their modes of action, their distribution in the body and drug combinations. The various adverse effects will be considered and the measures to minimize such effects. Dosage, administration, and duration of therapy will be dealt with according to clinical condition. Treatment and management of infection in pregnancy and the neonate are described in Chapter 6.

THE DRUGS

There are only four drugs which are regularly used with any effect, and two of these, pyrimethamine and a sulphonamide (usually sulphadiazine) are almost invariably used in combination. These two drugs have the longest history in the treatment of toxoplasmosis[6,7] and are still the first choice for most conditions. The other two are spiramycin[8] and clindamycin.[9] Over the

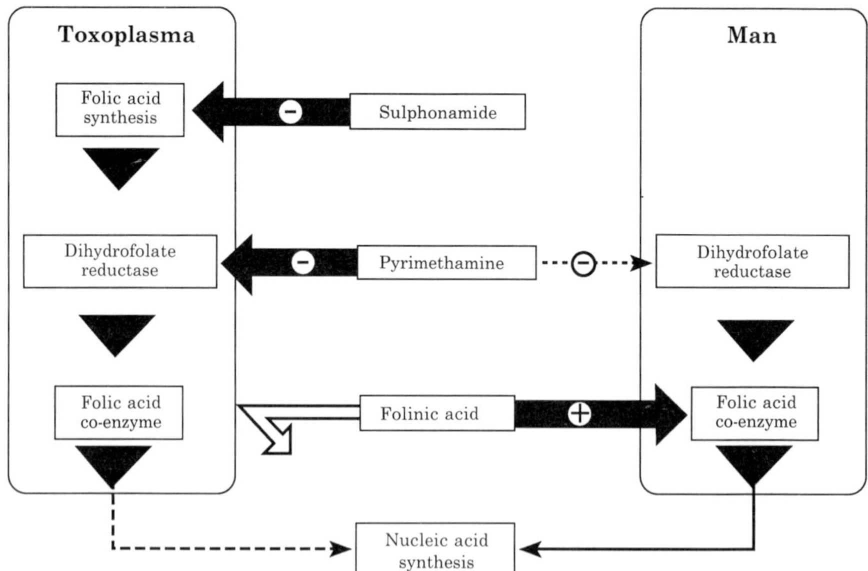

Fig. 5.1   Therapeutic intervention in folic acid metabolism.

years, alternatives such as co-trimoxazole[10] and tetracycline[11] have been tried and claimed to have been successful at the time. They have failed, however, to find general acceptance which is an indication that the results were mostly unconvincing. Now that longterm maintenance therapy is a serious proposition,[4] a wider choice of antibiotics is urgently needed to accommodate patients with drug reaction or resistance problems.[12]

Drugs act by interfering at some point in the metabolism of the parasite. At this chosen metabolic step, the more marked the dissimilarity between parasite and host, the more selectively toxic and effective the drug is likely to be. Unfortunately, protozoa such as toxoplasma have more in common with their hosts metabolically than bacteria, which limits the choice of therapy. The most advantageous identifiable evolutionary divergence that has arisen between toxoplasma and mammals is in the synthesis of the active form of the co-enzyme folic acid. It is at points on this pathway that sulphonamides and pyrimethamine act (Fig. 5.1). The other drugs, clindamycin and spiramycin, appear to rely more on their ability to reach and concentrate in the tissues which are most susceptible to damage. They presumably then achieve concentrations which have sufficient selective toxicity to check the proliferation of the parasite.

Sulphonamides

A sulphonamide is not an antibiotic, as none of the derivatives are made naturally. It is a chemically synthesized analogue of para-aminobenzoic acid which is an essential precursor for the synthesis of folic acid. Sulphonamide therefore competitively inhibits *de novo* synthesis in those micro-organisms which are obliged to make their own folic acid. The end effect is the inhibition of nucleic acid synthesis (Fig. 5.1). It limits multiplication rather than kills the parasite.[13] There are many derivatives of the simple para-aminobenzene sulphonamide molecule which have been successfully used as chemotherapeutic agents in a wide range of bacterial and parasitic infections. Sulphadiazine is the sulphonamide of choice. It is highly potent, less than 50 per cent protein bound and slowly excreted, thus maintaining adequate serum levels and permitting diffusion to important areas of infection such as cerebrospinal fluid (CSF) and tissues.[14] Now, it is usually used in combination with pyrimethamine, and achieves enhanced therapeutic effect by synergy.[13,14] It has also been used as one of the components of triple sulphonamide or sulphatriad therapy.[15] The others are sulphamerazine, sulphathiazole, and sulphamethazine. Various combinations containing sulphadiazine plus any two of the others, in roughly equal proportions, have been marketed. Sulphamerazine, despite being more highly protein bound than sulphadiazine, is excreted faster. It is only useful in combination with other sulphonamides. Sulphathiazole is highly potent, but liable to cause side effects. The other alternative, sulphamethazine, is much more soluble, with low risk of side effects. It is well absorbed and excreted fairly slowly, but unfortunately has a potency only one-eighth that of sulphadiazine.[14] Sulphadoxine and sulphametopyrazine are long acting derivatives, so that adequate serum levels can be maintained by once-weekly dosage.[14] Both have been used in combination with pyrimethamine (the former as Fansidar, an anti-malarial drug) for the treatment of toxoplasmosis[16,17] but are avoided now because of fear of adverse reactions.

Adverse effects

Although these drugs do not compete at a step which is necessary for human metabolism, they probably produce a greater range of side effects than any of the other anti-toxoplasma drugs. Some adverse effects (Table 5.1) are more severe than others, and their frequency varies according to which sulphonamide is used. Hypersensitivity is probably the most frequent problem which results from prolonged treatment. It arises because sulphonamides in general bind readily to proteins, thus acting as a hapten which may provoke an immune response. The commonest result is an urticarial or maculoerythematous rash with moderate fever, occurring about nine days after the start of the first treatment, but earlier in subsequent courses.[14] Longer persisting sulphonamides are more hazardous and are implicated in the extreme types of reaction, mucocutaneous erythema multiforme (Stevens–Johnson syndrome)[14] and toxic epidermal necrolysis (Lyell's syndrome).[14,16] These are fortunately rare but result in high mortality. Sulphamethoxazole, the sulphonamide component of co-trimoxazole, is a medium acting derivative

Table 5.1 Adverse effects of treatment

Affected organ	Frequency (severity) of adverse effect with each drug						
	PM	SDZ	TSP	SDX	Co-T	CL	SP
1. Central nervous system	±(+)	±	±	±	±		
2. Skin							
— hypersensitivity	+	++(++)	++(++)	+++(+++)	++(++)	+	±
3. Gastrointestinal	±	+	+	+	+	+++(++)	+(+)
— (pseudomembranous colitis)						++(+++)	
4. Hepatic		±	±	±	±	±	
5. Pulmonary eosinophilia		±(+)	±(+)	±(+)			
6. Haemopoietic	+++(+++)	+(+++)	+(+++)	+(+++)	±		
7. Renal blockage		+++(++)	+(++)		+(++)		
8. Fetal teratogenicity	±						

PM = pyrimethamine; SDZ = sulphadiazine; TSP = sulphatriad; SDX = sulphadoxine;
Co-T = co-trimoxazole; CL = clindamycin; SP = spiramycin.

which also occasionally gives rise to these reactions.[18] It has been reported that the problems of severe sulphadiazine reactions can be overcome, if considered essential to continue therapy, by desensitization, starting with a low dose and increasing it every second or third day until therapeutic levels are reached.[19] Other immunological complications associated with sulphonamides are systemic lupus erythematosus, polyarteritis nodosa, and myocarditis.[14] Renal damage is well recognized as a problem with some sulphonamides in certain circumstances. Sulphadiazine, sulphamerazine, and sulphathiazole all have poor solubility in urine, especially at lower pH, which can lead to crystal formation and renal blockage.[14,20] Therapy need not be altered if steps are taken to raise the urinary pH by alkaline rehydration.[20,21] Triple sulphonamide, or sulphatriad, preparations reduce the risk of crystalluria whilst maintaining the serum sulphonamide concentration at therapeutic levels.[20] Triple sulphonamides have been used in toxoplasmosis therapy but sulphadiazine (in combination with pyrimethamine) is the generally preferred option. Sulphamethoxazole has also been implicated in impaired renal function,[22] but long acting sulphonamides, because of their slow excretion rates, are not hazardous in this respect.

Other adverse effects attributed to sulphonamide therapy are quite rare.[14] They can suppress the bone marrow but use of sulphonamides on their own rarely results in agranulocytosis, anaemia, or thrombocytopenia. In contrast, this can be a serious problem when used in combination with pyrimethamine. Since both drugs act on the same metabolic pathway the effect is amplified but pyrimethamine exerts the principal effect. Neonatal haemolysis and methaemoglobulinaemia, as well as kernicterus in neonates are reasons why sulphonamides are avoided if possible in pregnancy and the newborn. Such reservations are overriden in proven congenital toxoplasmosis (Chapter 6). Nausea, diarrhoea, vomiting, headache, memory disturbance,[23] and hepatic injury[24] are rare. Administration with acid fluids, iron, or other heavy metal salts should be avoided. Concurrent hypoglycaemic agents, oral anticoagulants, and methotrexate[14] may lead to unwanted interactions.

Pyrimethamine

Pyrimethamine, like sulphonamide, is a synthetic anti-microbial. It selectively inhibits the enzyme, dihydrofolate reductase (Table 5.2) which carries out a key step in the generation of active folate co-enzyme by conversion of dihydrofolate to tetrahydrofolate (Fig. 5.1). Unlike the metabolic step at which sulphonamides are usually considered to act, this reaction occurs in both mammalian and protozoan cells. However, there has been marked evolutionary divergence so that the toxoplasma enzyme has little homology with its human counterpart.[25] Indeed, while the latter is a small monofunctional molecule, the protozoan enzyme is five times larger. In addition to its

Table 5.2   Points of action of anti-toxoplasma drugs

Toxoplasma growth process	Drugs	
	Current	Potential
1. Cell penetration	–	–
2. Intermediate synthesis (small molecules e.g. folic acid)	Pyrimethamine Sulphonamides Co-Trimoxazole	Methotrexate analogues Sulphones Trioxanes* Arprinocid*
3. Nucleic acid synthesis	–	–
4. Protein synthesis	Spiramycin Clindamycin	Clarithromycin Azithromycin Roxithromycin Doxycycline Minocycline
5. Parasite integrity	–	Electron carrier dyes* γ-interferon Interleukin-2
6. Parasite respiration	–	Hydroxynaphthoquinones*

* Mode of action not yet fully understood.

dihydrofolate reductase activity, it has subunits which catalyse one of the other steps in the parasite's folate metabolism.[26]

The toxicity of pyrimethamine for the parasite enzyme is 2000 × that for the human enzyme.[14] This is enhanced by the simultaneous administration of sulphonamide. The resulting synergy produces a therapeutic effect which is up to ten times that expected from mere addition of the individual drug effects.[13,14,27] It is usually thought to be achieved by two drugs acting at different steps on the same metabolic pathway (Fig. 5.1), but there is some suggestion that sulphonamides also potentiate the pyrimethamine activity by binding directly to the dihydrofolate reductase.[28] Another difference between host and parasite folate metabolism which is used to therapeutic advantage is the ability to use preformed folate. The parasite synthesizes folate but cannot absorb it preformed, while the reverse is true of humans and animals. Therefore the adverse effects of therapy can be substantially reduced by feeding preformed co-enzyme (as folinic acid, not dried Brewer's yeast as sometimes used before) without diminishing the drug effect on the parasite.[29] Parasite growth is stopped when its endogenous supply of folinic acid is diluted by cell division. Pyrimethamine is thought to be parasiticidal because treatment results in toxoplasma cells which are visibly damaged[13,27] and therefore presumably non-viable. Although the folic acid pathway is the specific target for both sulphonamides and pyrimethamine, nucleic acid synthesis, through the role of folic acid co-enzyme in it, is where the crucial effect of these drugs is felt[25] (Fig. 5.1).

Pyrimethamine is well absorbed but has a variable half-life, from one to four days.[30] Daily dosages which are too low may result in inadequate therapeutic levels in some patients, but higher doses may lead to accumulation and adverse effects in others, especially children.[31] As with sulphonamides, its penetration into CSF is acceptable,[32] achieving concentrations approximately one-fifth that of serum.[33] There is still, however, inadequate information on the optimum concentration required in clinical situations to kill the parasite.[30]

Adverse effects

Depression of the bone marrow is the principal problem[32] and is exacerbated by co-administration of sulphonamides (Table 5.1). Leukopenia, megaloblastic anaemia, and thrombocytopenia are the result.[34,35] The effect is worse when bone marrow function is already compromised by haematological disease, malignancy, or cytotoxic therapy.[32] Blood cell and platelet counts should be monitored during therapy[36] and pyrimethamine and/or sulphonamide may have to be discontinued or modified if the white cell count becomes unacceptably low.[37,38] The standard antidote is folinic acid, preferably given before side effects manifest themselves. Leukopenia may still develop, however, in spite of adequate folinic acid treatment.[33,38] The frequency of side effects in patients with cerebral toxoplasmosis, usually AIDS patients, is remarkably high (50–60%), 40–55 per cent having to discontinue treatment.[5,34,38] Pyrimethamine therapy has also been claimed to be responsible for bringing normal resting bone marrow stem cells into a reproductive cycle, thus making them as vulnerable to killing by cytotoxics as tumour cells.[39]

Other reactions occasionally caused by pyrimethamine are a transient rash, vomiting, tachycardia, respiratory distress, cyanosis, or convulsions.[23] Caution is required in patients with renal or hepatic impairment, or receiving benzodiazepines. Threats to fetal haemopoiesis and the largely unproven fear of teratogenicity, usually restrict pyrimethamine and sulphadiazine to late pregnancy (Chapter 6).

Clindamycin

Clindamycin is the most active of the lincosamide group of antibiotics. Like spiramycin, its action is on protein synthesis (Table 5.2). It binds to ribosomes and interferes with one or more of the synthetic steps which occur there.[14] Why it should be selectively toxic to parasitic cells is not fully understood. The effect may be direct, arising from differences between parasite and host ribosomes; or it may be indirect, reflecting greater access to parasite ribosomes as a result of differences in cell permeability. Clindamycin is highly lipid-soluble, and penetrates the eye[40] and dense tissue like bone. It is 94 per cent protein bound[14] which has a significant delaying effect on its excretion. It is perhaps these attributes, allowing it to concentrate in tissues where it can encounter parasites released from cysts, which confer the most therapeutic value. Its principal use is treatment of ocular toxoplasmosis.[41] Interestingly, it is increasingly used as an alternative to one or other of the pyrimethamine and sulphonamide combination in the treatment of cerebral toxoplasmosis[30,34,38,42] yet its penetration of CSF is poor.[14] On its own, it is a poor substitute, merely suppressing rather than fully inhibiting the spread of infection in tissues.[43]

Adverse effects

Diarrhoea is the commonest complication encountered with clindamycin therapy[14] (Table 5.1). It can occur following parenteral and topical as well as oral administration of clindamycin. The symptoms occur in 10–30 per cent of patients and in their extreme form (1–2%) are accompanied by severe abdominal pain, profound dehydration, and mucosal ulceration (pseudo-membranous colitis). They usually begin early after initiation of treatment and are the indirect result of suppression of normal gut flora by the drug, which allows toxin-producing *Clostridium difficile* to proliferate. As mortality can be high and symptoms are likely to persist even on discontinuing clindamycin, anti-clostridial therapy (preferably vancomycin administrated orally) is frequently necessary. Other side effects are rare: skin rash, nausea, vomiting, jaundice, leukaemia, and hypersensitivity. Periocular injection for treatment of ophthalmic toxoplasmosis carries a high risk of side effects (papillitis, optic atrophy, diplopia) especially with larger doses.[9] Care is required in treating patients with asthma, those receiving neuromuscular blocking agents, and in prolonged treatment of patients with impaired liver or renal function.[23] Although there is no evidence of abnormalities following treatment in pregnancy,[14] avoidance is usually advised. Apart from the very real one of colitis, the problems are usually less than with pyrimethamine and sulphonamide.

Spiramycin

Spiramycin is a macrolide antibiotic with a completely different molecular structure to clindamycin, yet apparently active at the same stage of protein synthesis (Table 5.2). Its *in vitro* activity is deceptively poor in comparison to the more well-known member of its group, erythromycin. Its strength lies in its exceptional persistence in tissues, so that *in vivo* it achieves therapeutic concentrations which are probably unattainable by alternative antibiotics.[14] Unfortunately, its penetration of the CSF is very poor,[18] so it has no logical role in the treatment of cerebral toxoplasmosis. Because it does accumulate in the placenta, its principal application is in the treatment of infected pregnant women,[16] where it can prevent spread of infection to the fetus (Chapter 6). Its limited clinical usefulness suggests it is only parasitostatic, containing the proliferation of parasites released from cysts[2] and thus preventing their penetration of the fetal circulation.

Adverse effects

The persistence of spiramycin in tissues is tolerable because it produces few side effects.[14] Gastrointestinal upset with nausea, vomiting, dry mouth, and abdominal pain can occur, more likely with high doses. Skin rashes are rare[18] (Table 5.1). There are no significant interactions with other drugs.[18] Most important, given that its main application is treating toxoplasmosis in pregnancy, it does not appear to have any teratogenic activity. Although cost and lack of availability are not adverse effects on the patient, they probably exert an inhibitory effect on the ready use of drugs. Such a problem exists with spiramycin in the UK, where it is only available on a named patient basis and the expense of a full course of therapy during pregnancy is significant.[44]

OTHER DRUGS

Trimethoprim (in co-trimoxazole)

Trimethoprim is, like pyrimethamine, a diaminopyrimidine (Table 5.2). It has very marked selective toxicity for bacterial dihydrofolate reductase.[14,45] Although its selective toxicity for the protozoan enzyme is considerably lower[14] and less than pyrimethamine,[13] it is of potential value.[13,14] It has no apparent therapeutic value on its own against toxoplasma.[46,47] However, it has occasionally been used in combination with sulphamethoxazole, as co-trimoxazole.[10,48] A ratio of 5:1 sulphamethoxazole: trimethoprim is designed to harmonize absorption and excretion of the two drugs.[18] Their ability to distribute rapidly to many tissues, including the CSF, and the lower toxicity of trimethoprim compared with pyrimethamine, makes co-trimoxazole an alternative in exceptional circumstances but not a first choice.[2,49] Recent studies show some success in AIDS patients with cerebral toxoplasmosis[34,50] and as toxicity problems frequently arise with pyrimethamine and sulphonamide, co-trimoxazole may have an increasing role in these patients.

Adverse effects

Many of the side effects of co-trimoxazole (renal and hepatic damage, hypersensitivity) are due to the sulphamethoxazole component (Table 5.1). As an inhibitor of folic acid metabolism, the haematological problems associated with pyrimethamine are uncommon[18] with trimethoprim. Rashes with co-trimoxazole are usually erythematous, but Stevens–Johnson and

Lyell's syndromes have been reported.[18] Co-trimoxazole has been estimated to cause nausea, vomiting and anorexia in 3 per cent of patients, diarrhoea in 0.5 per cent, and glossitis in 0.4 per cent. Pseudomembranous colitis is very rare.[18] As with pyrimethamine and sulphadiazine, teratogenic effects have not been reported in treatment early in pregnancy but because of the theoretical possibility, these drugs are usually avoided.[18] Although co-trimoxazole is safer than pyrimethamine and sulphadiazine, there are several cautionary notes. When switching to co-trimoxazole, haemopoiesis might already be depressed because of pyrimethamine treatment; side effects caused by sulphadiazine may not be improved by changing to sulphamethoxazole; and, finally, there is a general observation that side effects with co-trimoxazole are more frequent in AIDS patients.[18]

Tetracyclines

In bacteria, tetracyclines inhibit protein synthesis at the stage where amino acids attach to the ribosome[18] (Table 5.2). It is not certain whether there is similar toxicity for parasite protein synthesis, but oxytetracycline has been claimed to relieve symptoms of human infection.[11] More recently, apparently useful inhibitory activity has been recorded under experimental conditions with minocycline[51] and doxycycline.[52] In mice, better results were obtained when drug combinations were used. A combination of minocycline and sulphadiazine was comparable to that of pyrimethamine and sulphadiazine;[51] plasma doxycycline levels were only just within the therapeutic range, but in combination with pyrimethamine cure rates of infected mice were impressive.[52] The fact that promising results were obtained with these particular tetracyclines is significant. All tetracyclines are well absorbed and distribute to most tissues fairly efficiently. Minocycline and doxycycline, however, are the most lipid soluble and best able to penetrate the brain[18] especially if the meninges are inflamed. Therefore, these drugs may have a role as alternatives in the treatment of cerebral toxoplasmosis.

Adverse effects

There is a daunting list of adverse effects associated with tetracycline therapy:[18] gastrointestinal upset, deposition in bones, bacterial superinfections, photosensitivity, hepatotoxicity, nephrotoxicity, and neurotoxicity. Hypersensitivity and haematological side effects have also been recorded. Nevertheless, tetracyclines are widely used and the likelihood of serious consequences is probably significantly less than would be expected with pyrimethamine and sulphonamide. As a result of deposition in bone, tetracyclines are not recommended in children under twelve years of age. Teratogenicity is not considered a problem, but the increased likelihood of 'fatty liver' during pregnancy contraindicates its use then, particularly in the third trimester.[18]

NEWER DRUGS

During an infection, there are many points in the multiplication process of a micro-organism at which therapeutic intervention is possible (Table 5.2). The drugs described so far for the treatment of toxoplasmosis largely act at only two of these points, synthesis of co-enzymes and of proteins, although the former has knock-on effects on nucleic acid synthesis. Perhaps there are other stages in toxoplasma growth yet to be exploited. However, many of the newly developed drugs are variations of the present anti-toxoplasma themes.

Antifolates

Two new dihydrofolate reductase inhibitors, the methotrexate analogues, piritrexim[53] and trimetrexate,[54] have considerable potential. In *in vitro* experiments their inhibitory activity against toxoplasma were, respectively, 10–45 times and 250–750 times that of pyrimethamine.[13,53] The concentrations at which inhibition was achieved were 100-fold and 1000-fold, respectively, less than can be maintained in human sera.[53,55] Furthermore, both drugs showed synergy with sulphonamides and their inhibitory activity was unaffected by the presence of folinic acid.[53,54] Obviously if similar levels of effectiveness can be confirmed in the treatment of toxoplasma infections in humans, either drug could be used at concentrations low enough to reduce the risks of bone marrow toxicity. A possible alternative to sulphonamides is the sulphone, dapsone, or one of its derivatives.[56] This drug has a similar inhibitory effect to sulphadiazine on folate synthesis and has been used for over 40 years in long term therapy for leprosy which suggests that it may be a substitute for sulphadiazine when toxicity problems arise.

Macrolides

Three new macrolides are currently being assessed as potential anti-toxoplasma drugs (Table 5.2). Roxithromycin has a similar anti-microbial spectrum to erythromycin but has a longer half-life and achieves higher serum concentrations.[57] *In vitro*, it inhibits and has some killing effect on toxoplasma within mouse macrophages[58] and its *in vivo* efficacy is enhanced by synergy with γ-interferon.[59] Azithromycin has high tissue penetration,[57] and concentrates within fibroblasts and phagocytic cells. This could prove useful in dealing with intracellular parasites and could allow concentration of the drug at sites of inflammation due to infection.[60] It has been shown to be effective in treating mice with intracerebral toxoplasmosis.[61] The third, clarithromycin or A-56268 (TE-031), has also been shown to be protective in animals,[62] where it achieves high tissue levels.[57] Azithromycin and clarithromycin appear to be parasitostatic rather than cidal, and they have no synergistic effect with pyrimethamine.[63] It remains to be seen how effective any of these drugs will be in treating human toxoplasmosis.

Trioxanes

Trioxanes[64] such as artemisinin (qinghaosu), or other derivatives of artemisinin[65] successfully eliminate toxoplasma from tissue cultures of macrophages[64] or fibroblast cells.[64,65] Their effect is on intracellular growth, with activity comparable to that of pyrimethamine plus sulphadiazine. They do not affect the viability of extracellular parasites, nor do they confer protection on pretreated cells. They appear to act as oxidizing agents, somehow blocking synthesis of nucleotides, the raw material for nucleic acid synthesis in intracellular parasites[64] (Table 5.2). Artemisinin has low toxicity and is currently the subject of international clinical trials for use against malaria.[64] If these are successful, it is likely that its application in toxoplasmosis therapy will be further investigated.

Miscellaneous

Arprinocid, a purine analogue, has been shown to be toxoplasmacidal *in vitro* and *in vivo*.[66] Cationic electron carriers such as methylene blue or crystal violet dyes in very low concentrations are also parasitocidal against intracellular toxoplasma *in vitro*.[67] The mechanisms of action are unknown and much research has yet to be done before it can be decided if they have a role in treating human toxoplasmosis.

Immune mediators

The release and proliferation of toxoplasma parasites from latent tissue cysts is presumably kept in check by immune mechanisms in healthy individuals (Chapter 7). In the immunocompromised, this protection is eroded by deficiency or malfunction of one or more components of the immune system. Adding them back exogenously may restore the capacity to contain disease proliferation. Administration of recombinant interleukin-2[68] or γ-interferon[69] enhances survival of infected mice. The latter substantially decreased inflammatory foci in the brains of mice with encephalitis, but the effect was only transient.[69] This would therefore appear to be an expensive form of maintenance therapy, but may have a role if used in conjunction with drug therapy.[59]

DOSE, DURATION, AND EFFICACY

In this section, the way in which the principal drugs are used in different clinical situations will be described. Pyrimethamine plus sulphadiazine is usually the first choice, but the dose and duration of therapy may have to be adjusted according to the severity of illness. For some situations alternative therapies may be more appropriate, for instance in early pregnancy or perhaps for ocular disease. In severe illness in the immunocompromised where prolonged treatment may be essential, toxicity problems may require alteration of therapy. The efficacy of treatment under all these circumstances has to be considered.

Acquired infection in the immunocompetent

Therapy is generally accepted to be unnecessary for acquired toxoplasmosis in the immunocompetent unless it affects a vital organ, e.g. brain, lung, liver, heart. When symptomatic individuals with lymphadenopathy have been treated with pyrimethamine and sulphadiazine, clinical and serological improvement has been noted.[70] The clinicians who reported this observation now accept that uncomplicated lymphadenopathy resolves just as well spontaneously,[33] but still see advantages in therapy to reduce fever, if present. Whether treatment would help those patients whose lymphadenopathy is prolonged, and perhaps accompanied by malaise and fatigue, is unknown. There is little evidence in the literature for or against therapy in these cases.

Reports on treatment of the more serious forms of acquired disease tend to describe only small numbers of cases. Often the patient groups appear to include cases of reactivated infection, which is not the subject of this section. In those cases of toxoplasmosis associated with polymyositis in which there was evidence of recent infection, the response to therapy was mixed.[71,72] Most improved, but sometimes slowly. There is evidence that polymyositis may develop after acute toxoplasmosis despite adequate therapy.[73] Toxoplasmic myocarditis and pericarditis can be cured with therapy, although some patients recover without it.[74] In other cases, with pericardial effusions due to toxoplasma, treatment with pyrimethamine and spiramycin failed, but surgery was successful.[75] Likewise hepatitis, a rare usually late complication of toxoplasmosis, responds to therapy[76,77] but it may not always be necessary.[78] The most serious form of the disease, toxoplasmic encephalitis, is usually associated with reactivation in the immunocompromised, but sporadic cases do occur following acquired disease in the otherwise healthy.[79–81] Treatment is essential in these cases. A variety of skin disorders associated with toxoplasmosis can respond to therapy.[82] Ophthalmic disease occasionally results from acquired toxoplasmosis,[83] but it is usually due to reactivated congenital infection and will be discussed separately in the next section. Finally, acquired infection in pregnancy has potentially serious consequences and must be treated to prevent congenital infection (Chapter 6). Treatments at the various stages following infection are summarized in Table 5.3, as is prophylaxis for accidental infection (Appendix 2).

For acquired infection in the immunocompetent, pyrimethamine and

Table 5.3 Treatment of the immunocompetent

Clinical	Drugs and choice	Dosage, route, frequency
Acquired		
Disseminated	1. PM + SDZ + FA	High, or high then standard, oral, daily
(vital organ)	2. PM + TSP + FA	High, or high then standard, oral, daily
Persistent	1. PM + SDZ + FA	Standard, oral, daily
	2. PM + TSP + FA	Standard, oral, daily
	3. Co-T	Standard, oral, daily
Prophylaxis	1. PM + FA	Standard, oral, daily
(self inoculation)	2. Co-T	Standard, oral, daily
Ocular	1. PM + SDZ + FA	High, or high then standard, oral, daily
	2. PM + TSP + FA	High, or high then standard, oral, daily
	3. CL ± SDZ	Standard, oral, daily
Pregnancy		
Infected mother	1. SP	High, oral, daily (till decision on fetal infection, or term)
Infected fetus	1. PM + SDZ + FA	Standard, oral, daily
	2. SP (1 and 2 alternating 3-weekly)	High, oral, daily (till term)
Infected neonate	1. PM + SDZ + FA	Standard, oral, daily
	2. SP (1 for 6 wk or 6 mo* then alternating with 2, 4-weekly)	High, oral, daily (for a total of 1 year)
Possibly infected	1. PM + SDZ + FA	Standard, oral, daily
neonate	2. SP (1 or 2#)	High, oral, daily (for at least 1 month)

* Depending on severity of infection.   # Depending on likelihood of infection.
Drugs as Table 5.1; FA = folinic acid.

sulphonamide is the treatment of choice for most situations. Pyrimethamine is usually given at 25 mg per day orally. It may be increased to 50 mg per day and the higher dose would appear to be more appropriate for serious illness, when it is desirable to achieve and maintain therapeutic levels as early as possible. For one encephalitis case,[79] a loading dose of 3 × 50 mg per day for two days was followed by daily doses of 3 × 25 mg (Table 5.3). Daily sulphonamide dosages range from 2–6 g, with at least 4 g being the general recommendation. Sulphadiazine is the preferred form, but sulphatriad or sulphadimidine have also been used. Folinic acid was not always used, but would be advisable. Doses used range from 5 to 15 mg per day. The duration of treatment is also variable depending on the response, but at least one month appears to be normal. Steroids (prednisone or prednisolone) are frequently administered in conjunction, to control inflammation, but doses are quickly tapered down as soon as clinical improvement is evident.[71,73,74,76,81,83]

For less serious clinical cases, drugs of lower toxicity have had success. One hepatitis case was treated with spiramycin (2 g/day) and sulphadiazine (4 g/day) for 10 days, followed by spiramycin and pyrimethamine (50 mg /day), again for 10 days.[76] Recommended follow-up treatments of skin infections (after an initial four weeks of pyrimethamine and sulphadoxine) are spiramycin (2–3 g/day) or clindamycin (2 g/day) or co-trimoxazole (2 × 960 mg/day).[82] It is encouraging to note that co-trimoxazole is reported to have been successful in treating toxoplasma lymphadenopathy[10,84] which was sometimes of prolonged duration.[84] Because it is less hazardous than pyrimethamine and sulphonamide, it may provide the answer to the problem of how to deal with patients who suffer prolonged malaise and fatigue following acquired infection (Table 5.3). Treatment may have to be maintained for several months.[48]

Ocular toxoplasmosis

In one controlled trial the benefits of treating toxoplasma retinochoroiditis were far from spectacular.[85] None of the treatment regimes reduced the duration of inflammatory activity in comparison with untreated patients. Only pyrimethamine and sulphadiazine (given with folinic acid and corticosteroids) significantly reduced the size of the retinal lesion (by 52 per cent compared with 25 per cent in controls). Clindamycin, sulphadiazine, and corticosteroids showed a marginal improvement (32% reduction) and co-trimoxazole plus corticosteroids were ineffectual. Anti-toxoplasma therapy for all cases of ocular toxoplasmosis is not usually advocated. Its necessity is generally accepted when central visual function is threatened either by focal or general inflammation, but peripheral retinal pathology is usually allowed to resolve spontaneously under observation.[86] Also it has been claimed that it is possible to distinguish active proliferation of toxoplasma in the eye, which occurs in 10 per cent of patients and should be treated, from a self-limited lesion caused by cyst rupture, which occurs in 90 per cent of patients and may not require therapy.[87] However, since the diagnosis depends on antibody tests on aqueous humour, which are infrequently available (Chapter 4), the distinction may be difficult to make. The decision to treat is therefore likely to be based on clinical information.

Therapy appears to involve a choice between pyrimethamine plus sulphadiazine or clindamycin plus sulphadiazine (Table 5.3). The controlled trial[85] and earlier literature favour the former while a review of the more recent literature favours the latter.[41] In one report all three have been used on the same patient.[88] In the controlled trial,[85] 100 mg pyrimethamine was given on the first day, followed by 2×25 mg per day, sulphadiazine at 4×1 g per day and clindamycin at 4×300 mg per day. Clindamycin, like the other drugs should be given orally, as peri-ocular injection to avoid gastrointestinal side effects is too hazardous.[41,88] Folinic acid is advisable when using pyrimethamine, and oral corticosteroids are generally added to the therapy to decrease tissue damage. The duration of therapy is generally at least three weeks, but one author recommends continuation until a reduction in inflammation is demonstrable and then continuation for a further two weeks at half dose.[86] If it is decided that anti-parasitic therapy is unnecessary, management of inflammation with corticosteroids is possible[87] but runs the risk of worsening the condition.[88,89]

AIDS

Acute infection

AIDS is the most important single factor in bringing about a resurgence of interest in finding appropriate treatments for toxoplasmosis. Disease almost always arises from reactivation of latent infection. It can become disseminated leading to rare forms of toxoplasmosis, such as pulmonary infection,[36] but the principal concern by far is toxoplasmic encephalitis. This must be treated as a matter of urgency, often before an absolute diagnosis is available (Chapter 4). The efficacy of treatment in the short term varies between 68 and 90 per cent[5,32,38,42,90–92] provided appropriate therapy is instituted early enough. The principal reason for failure is delay in starting therapy, which in turn can stem from failure to diagnose.[36,92] Comatose patients are less likely to respond than alert patients.[34,38,50,90] Clinical improvement is often rapid, within 24 to 48 hours,[36,50,92] and the majority of improvements are evident in 10 days to 4 weeks.[5,36,50,92] However it is often transient[42] or partial[91] and the longterm prognosis is poor.[34,50] AIDS patients face death from many causes other than toxoplasmosis, but without longterm,[38] even lifelong[30,42] anti-toxoplasma maintenance therapy, toxoplasmic encephalitis is very likely to recur.

There is no standard treatment regime which can be administered with confidence to all AIDS patients with toxoplasmic encephalitis. Patient vari-

Table 5.4   Treatment of the immunocompromised

Clinical	Drugs and choice	Dosage, route, frequency
AIDS		
treatment	1. PM + SDZ + FA	V. high, then high, oral, daily
	2. PM + TSP + FA	V. high, then high, oral, daily
	3. PM + CL	V. high, then high, IV or oral, daily
	4. PM + Co-T	V. high, then high, IV or oral, daily
	5. Co-T	High IV, then standard oral, daily
	6. Other	
Maintenance	1. PM + SDZ + FA	Standard, oral, 2–7 d/wk
	2. PM + TSP + FA	Standard, oral, 2–7 d/wk
	3. PM + CL	Standard (-high), oral, daily
	4. Co-T	Standard (-high), oral, daily
Transplant		
Treatment	1. PM + SDZ + FA	Standard (-high), oral, daily
	2. PM + TSP + FA	Standard (-high), oral, daily
	3. SP + PM + TSP + FA	Standard (-high), oral, daily
Prophylaxis	1. PM + FA	Standard, oral, daily
Other		
Treatment	1. PM + SDZ + FA	Standard (-high), oral, daily
	2. PM + TSP + FA	Standard (-high), oral, daily

Drugs as Table 5.3.

ability in drug half-life and the gravity of the disease have led clinicians to use higher drug doses, however toxicity problems frequently force alterations to therapy. Pyrimethamine and sulphadiazine are generally the initial choice. In the earlier reports, standard doses of pyrimethamine (25 mg, sometimes increased to 50 mg/day) were most often used.[5,12,90] Occasionally, up to 100 mg pyrimethamine per day was used, especially early in treatment,[90,91] and the most recent preference is for a loading dose of 100–200 mg per day, reduced to 50–100 mg per day for at least 6 weeks[30,42] (Table 5.4). Part of the explanation for a higher dose requirement may be interaction with azidothymidine (AZT), given to combat HIV infection. *In vitro* and *in vivo* experiments prove that the latter inhibits pyrimethamine at low doses and nullifies its synergy with sulphadiazine.[93] Doses of sulphadiazine are also high (4–8 g/day). Consequently folinic acid has tended to be administered at higher concentrations than previously described, with maxima of between 10 and 50 mg per day.[5,38,91] In one report, the authors adjusted the folinic acid dose according to blood count, which was frequently monitored.[91] Steroids, usually dexamethasone, but also prednisone or hydrocortisone, are frequently given to reduce oedema.[4,34,38,42,92]

Not surprisingly, given the high doses used, adverse effects are common, often affecting 50–60 per cent of patients.[5,34,38] Substitution of clindamycin (at 0.6–2.4 g/day intravenously or orally) for sulphadiazine, whilst continuing with pyrimethamine, is an acceptable solution,[4,38,94] (Table 5.4) although not always successful.[5] It is an improvement on pyrimethamine alone.[92,95] In one patient, a response was only achieved after clindamycin was added to pyrimethamine and sulphadiazine.[95] It has also been used successfully as initial therapy for encephalitis, usually in conjunction with pyrimethamine.[96] However, there is little evidence to support regular first use of clindamycin in preference to sulfadiazine-pyrimethamine. Co-trimoxazole has also on occasion been successfully used as an alternative to sulphadiazine[90] or sulphadiazine and pyrimethamine.[34] One group has used it as first choice at high dose (96 mg/kg/day intravenously for 3 weeks, then 30 mg/kg/day orally) with good response in the short term.[50] However its regular use is questionable as encephalitis has developed in patients receiving co-trimoxazole for *Pneumocystis carinii* infection.[34] There is no evidence that spiramycin has a role in this condition.[5,91] A preliminary trial of one of the newer drugs, trimetrexate, as an alternative therapy for nine sulphonamide-intolerant AIDS patients proved disappointing.[97] Transiently, the response was good but eventually all deteriorated despite sustained therapy, so if trimetrexate is to have a role it will have to be in combination with other drugs. In conclusion, therefore, pyrimethamine, sulphadiazine plus folinic acid at higher than normal dosages would appear to be the recommended first approach, followed by substitution of sulphadiazine by clindamycin or co-trimoxazole in the event of toxicity (Table 5.4).

Maintenance therapy

None of the drugs regularly used in treating toxoplasmosis are considered to eliminate the tissue cyst,[30] although very recent experiments with a hydroxynaphthoquinone appear to have eradicated cysts in mice.[98] If similarly effective in humans, this drug would be of enormous benefit to the treatment of AIDS (and other immunocompromised or congenitally infected) patients. Additionally, its promise for the AIDS patient is enhanced by its apparent activity against *P. carinii*.[99] However, until such a drug becomes available, or alternatively until AIDS itself is curable, it is generally agreed that these patients will require lifelong maintenance therapy. Without it, successfully treated patients face a high probability of relapse.[4,38] Many clinicians favour continuing therapy with the same drug combinations in the same daily doses as used at the end of treatment.[5,38,92] Others believe the doses and/or frequency can be reduced. As an example, intermittent therapy with pyrimethamine (25 mg) and sulphadiazine (75 mg/kg) twice a week, plus folinic acid (10 mg) daily was fully effective.[4] In contrast, pyrimethamine (25 mg) plus clindamycin (2.4 g) twice a week was only 60 per cent effective.[4] Pyrimethamine alone has been tried occasionally[38,100] even on alternate days at 25 mg.[100] Relapses are likely, but may be corrected by adding clindamycin for a period.[95,100] A future alternative might be to substitute dapsone for sulphonamide.[101] It has good synergy with pyrimethamine, a longer half-life which has advantages in maintenance therapy and, if current trials of its efficacy for *P. carinii* pneumonia are favourable, it might offer a suitable dual maintenance therapy. Maintenance therapy therefore is advocated for treated AIDS patients, preferably with low dose pyrimethamine and sulphadiazine, perhaps given less frequently (Table 5.4). If an alteration to a less toxic combination or monotherapy is used, recrudescence of disease should be promptly treated by restarting appropriate combination therapy.

Other immunocompromised patients

Toxoplasma reactivation occurs in patients with other forms of immunocompromise such as neoplasia, organ transplants, and collagen-vascular disorders.[32] Neurological sequelae predominate but a variety of other manifestations are possible, including dissemination[1] (Chapter 7). Pyrimethamine and sulphadiazine, with folinic acid, are again the treatments of choice[32,33,70,102] (Table 5.4) with 80 per cent success rates recorded.[32,70] Interestingly, spiramycin (2 g/day) in conjunction with pyrimethamine (50 mg/day) and sulphatriad (2 g/day) was successfully used to treat two patients with toxoplasmosis following a heart transplant.[3] Therapy was continued for 4 and 10 months. The efficacy of spiramycin in this situation probably derived from its ability to concentrate in dense tissue. Because the source of immunocompromise is often the therapy (steroids or cytotoxic drugs) for the underlying condition, they can sometimes be temporarily stopped[102] to help overcome the toxoplasma infection. Also as they are not intended to be given continuously throughout the patient's life, longterm anti-toxoplasma maintenance is not essential. Short-term prophylaxis with pyrimethamine (25 mg/day for 6 weeks) has been used for patients receiving heart and heart–lung transplants, when a seronegative recipient was in receipt of an organ from a seropositive donor[3] (Table 5.4). Infections were reduced by 75 per cent. Low toxicity anti-toxoplasma therapy, for example co-trimoxazole, has been suggested for other patients on longterm immunosuppression.[103]

SUMMARY

The principal drugs for treatment of toxoplasmosis have been in use for 40 years or more, despite the concern that prolonged courses of therapy frequently result in quite serious side effects. Pyrimethamine and sulphonamides both exploit the best recognized difference in the growth requirements of toxoplasma and its host. They selectively inhibit the synthesis of folic acid co-enzyme which eventually disrupts nucleic acid synthesis. Because they act synergistically, they are almost invariably used in combination. Sulphadiazine is the most popular sulphonamide. Adverse effects are produced by both drugs: renal damage arising from drug insolubility, hypersensitivity reactions in the case of sulphadiazine, and depression of bone marrow by pyrimethamine. The latter can result in severe leukopenia, thrombocytopenia, and anaemia which is counteracted, not always successfully, by co-administration of folinic acid. Sulphonamide solubility can be improved by substituting with triple sulphonamide or by alkaline rehydration. Adverse effects from pyrimethamine or sulphonamides however may still be sufficiently serious to require discontinuation or interruption of therapy. These drugs are the first choice therapy and the dose is generally increased according to the gravity of the consequences of uncontrolled infection.

Spiramycin is at present the only recommended drug for infection in the first two trimesters of pregnancy. Its preeminence derives from its low toxicity and absence of teratogenic activity. Its efficacy stems from its exceptional ability to concentrate in tissues such as the placenta where it somehow interferes with parasite protein synthesis. Its only accepted role is prevention of transmission to the fetus. Clindamycin has a similar action to spiramycin on the protein synthesis of toxoplasma, an action also potentiated by an ability to concentrate in dense tissue. Until recently it has mostly been used in the treatment of ocular toxoplasmosis, but is now the first alternative when toxicity problems arise in treating cerebral toxoplasmosis. Pseudomembranous colitis is a major side effect.

Acquired infections in otherwise healthy individuals generally do not require any treatment. Even ophthalmic disease is usually left to resolve spontaneously, unless central vision is threatened. Only in pregnancy is treatment strongly recommended to prevent congenital damage. In immunocompromised patients, however, toxoplasmosis rapidly becomes life threatening and the limited choice of suitable drugs is exposed. Short-term prophylaxis with pyrimethamine and folinic acid appears to be a viable solution for susceptible transplant patients. However when immunocompromise persists indefinitely, as in AIDS patients, standard anti-toxoplasma therapy is more likely to encounter problems of toxicity. Currently available alternatives, such as co-trimoxazole, have been tried as substitutes with limited success. There is now a considerable body of research devoted to finding new less toxic drugs. Some look promising experimentally, and results of clinical trials may soon determine whether they will have a routine application. The ultimate goal is a form of therapy which can achieve what none of the existing drugs is capable of doing, namely the eradication of tissue cysts and so prevent recrudescence of disease.

REFERENCES

1   Remington, J. S. (1974). Toxoplasmosis in the adult. *Bull. N. Y. Acad. Med.*, **50**, 211–27.

2   Nguyen, B. T. and Stadtsbaeder, S. (1983). Comparative effects of cotrimoxazole (trimethoprim-sulphamethoxazole), pyrimethamine–sulphadiazine and spiramycin during avirulent infection with *Toxoplasma gondii* (Beverley strain) in mice. *Br. J. Pharmacol.*, **79**, 923–8.

3   Wreghitt, T. G., Hakim, M., Gray, J. J., Balfour, A. H., Stovin, P. G. I., Stewart, S. *et al.* (1984). Toxoplasmosis in heart and heart and lung transplant recipients. *J. Clin. Pathol.*, **42**, 194–9.

4   Pedrol, E., Gonzalez-Clemente, J. M., Gatell, J. M., Mallolas, J., Miro, J. M., Graus, F. *et al.* (1990). Central nervous system toxoplasmosis in AIDS patients: efficacy of an intermittent maintenance therapy. *AIDS*, **4**, 511–17.

5   Wanke, C., Tuazon, C. U., Kovacs, A., Dina, T., Davis, D. O., Barton, N. *et al.* (1987). Toxoplasma encephalitis in patients with acquired immune deficiency syndrome: diagnosis and response to therapy. *Am. J. Trop. Med. Hyg.*, **36**, 509–16.

6   Eyles, D. E. and Coleman, N. (1952). Tests of 2,4-diaminopyrimidines on toxoplasmosis. *Public Health Rep.*, **67**, 249–52.

7   Sabin, A. B. and Warren, J. (1942). Therapeutic effectiveness of certain sulfonamides on infection by an intracellular protozoan (Toxoplasma). *Proc. Soc. Exp. Biol. Med.*, **51**, 19–23.

8   Garin, J-P. and Eyles, D. E. (1958). Le traitement de la toxoplasmose experimentale de la souris par la spiramycine. *Presse Med.*, **66**, 957–8.

9   Tate, G. W. and Martin, R. G. (1977). Clindamycin in the treatment of human ocular toxoplasmosis. *Can. J. Ophthalmol.*, **12**, 188–95.

10 Norrby, R., Eilard, T., Svedhem, A., and Lycke, E. (1975). Treatment of toxoplasmosis with trimethoprim-sulphamethoxazole. *Scand. J. Infect. Dis.*, **7**, 72–5.

11 Fertig, A., Selwyn, S., and Tibble, M. J. K. (1977). Tetracycline treatment in a food-borne outbreak of toxoplasmosis. *Br. Med. J.*, **ii**, 1064.

12 Luft, B. J., Brooks, R. G., Conley, F. K., McCabe, R. E., and Remington, J. S. (1984). Toxoplasmic encephalitis in patients with acquired immune deficiency syndrome. *JAMA*, **252**, 913–7.

13 Derouin, F. and Chastang, C. (1989). In vitro effects of folate inhibitors on *Toxoplasma gondii*. *Antimicrob. Agents Chemother.*, **33**, 1753–9.

14 Garrod, L. P., Lambert, M. R., and O'Grady, F. (ed.) (1981). *Antibiotic and chemotherapy*, (5th edn). Churchill Livingstone, Edinburgh.

15 Kayhoe, D. E., Jacobs, L., Beye, H. K., and McCullough, N. B. (1957). Acquired toxoplasmosis. Observations on two parasitologically proved cases treated with pyrimethamine and triple sulfonamides. *N. Engl. J. Med.*, **257**, 1247–54.

16 Daffos, F., Forestier, F., Cappella-Pavlovsky, M., Thulliez, P., Aufrant, C., Valenti, D. *et al.* (1988). Prenatal management of 746 pregnancies at risk for congenital toxoplasmosis. *N. Engl. J. Med.*, **318**, 271–5.

17 Tolentino, P. (1971). Treatment of toxoplasmosis with long-acting sulphonamides. In *Toxoplasmosis* (ed. D. Hentsch), pp. 251–5. Hans Huber, Bern.

18 Kucers, A. and Bennett, N. McK. (1987). *The use of antibiotics*, (4th edn). William Heinemann, London.

19 Bell, E. T., Tapper, M. L., and Pollock, A. A. (1985). Sulphadiazine desensitisation in AIDS patients. *Lancet*, **i**, 163.

20 Oster, S., Hutchison, F., and McCabe, R. (1990). Resolution of acute renal failure in toxoplasmic encephalitis despite continuance of sulfadiazine. *Rev. Infect. Dis.*, **12**, 618–20.

21 Ventura, M. G., Wybran, J., and Farber, C. M. (1989). Sulfadiazine revisited. *J. Infect. Dis.*, **160**, 556–7.

22 Kalowski, S., Nanra, R. S., Mathew, T. H., and Kincaid-Smith, P. (1973). Deterioration in renal function in association with co-trimoxazole therapy. *Lancet*, **i**, 394–7.

23 ABPI *Data Sheet Compendium*. (1989–90). Datapharm, London.

24 Dujovne, C. A., Chan, C. H., and Zimmerman, H. J. (1967). Sulfonamide hepatic injury. Review of the literature and report of a case due to sulfamethoxazole. *N. Engl. J. Med.*, **277**, 785–8.

25 Hightower, R. C. and Santi, D. V. (1988). Drug action and drug resistance: antifolate resistance in parasitic protozoa. In *Parasitic infections*. (ed., J. H. Leech, M. A. Sande, and R. K. Root), pp. 81–94. Churchill Livingstone, New York.

26 Kovacs, J. A., Allegra, C. J., and Masur, H. (1990). Characterization of dihydrofolate reductase of *Pneumocystis carinii* and *Toxoplasma gondii*. *Exp. Parasitol.*, **71**, 60–8.

27 Sheffield, H. G. and Melton, M. L. (1975). Effect of pyrimethamine and sulphadiazine on the fine structure and multiplication of *Toxoplasma gondii* in cell cultures. *J. Parasitol.*, **61**, 704–12.

28 Poe, M. (1976). Antibacterial synergism: a proposal for chemotherapeutic potentiation between trimethroprim and sulfamethoxazole. *Science*, **194**, 533–5.

29 Frenkel, J. K. and Hitchings, G. H. (1957). Relative reversal by vitamins (*p*-aminobenzoic, folic, and folinic acids) of the effects of sulfadiazine and pyrimethamine on toxoplasma, mouse and man. *Antibiot. Chemother.*, **7**, 630–8.

30 Luft, B. J. and Hafner, R. (1990). Toxoplasmic encephalitis. *AIDS*, **4**, 593–5.

31 Nichol, C. A. (1973). Long plasma half-life of pyrimethamine. *Lancet*, **ii**, 161.

32 Ruskin, S. and Remington, J. S. (1976). Toxoplasmosis in the compromised host. *Ann. Intern. Med.*, **84**, 193–9.

33 Hakes, T. B. and Armstrong, D. (1983). Toxoplasmosis: problems in diagnosis and treatment. *Cancer*, **52**, 1535–40.

34 Haverkos, H. W. (1987). Assessment of therapy for toxoplasma encephalitis. *Am. J. Med.*, **82**, 907–14.

35 Rose, M. S., Black, P. J., and Barkhan, P. (1973). Fatal outcome after combined therapy for myeloblastic leukaemia and toxoplasmosis. *Lancet*, **i**, 600.

36 Holliman, R. E. (1988). Toxoplasmosis and the acquired immune deficiency syndrome. *J. Infect.*, **16**, 121–8.

37 Nye, F. J. (1979). Treating toxoplasmosis. *J. Antimicrob. Chemother.*, **5**, 244–6.

38 Cohn, J. A., McMeeking, A., Cohen, W., Jacobs, J., and Holzman, R. S. (1989). Evaluation of the policy of empiric treatment of suspected toxoplasma encephalitis in patients with the acquired immunodeficiency syndrome. *Am. J. Med.*, **86**, 521–7.

39 Price, L. A. and Bondy, P. K. (1973). Fatal outcome after combined therapy for myeloblastic leukaemia and toxoplasmosis. *Lancet*, **i**, 727.

40 Tabbara, K. F. and O'Connor, G. R. (1980). Treatment of ocular toxoplasmosis with clindamycin and sulfadiazine. *Ophthalmology*, **87**, 129–34.

41 Dutton, G. N. (1989). Toxoplasmic retinochoroiditis — a historical review and current concepts. *Ann. Acad. Med.*, **18**, 214–21.

42 Rossitch, E. Jr., Carrazana, E. J., and Samuels, M. A. (1990). Cerebral toxoplasmosis in patients with AIDS. *Am. Fam. Physician*, **41**, 867–73.

43 Piketty, C., Derouin, F., Rouveix, B., and Pocidalo, J-J. (1990). In vivo assessment of antimicrobial agents against *Toxoplasma gondii* by quantification of parasites in the blood, lungs and brain of infected mice. *Antimicrob. Agents Chemother.*, **34**, 1467–72.

44 Joss, A. W. L., Chatterton, J. M. W., and Ho-Yen, D. O. (1990). Congenital toxoplasmosis: to screen or not to screen? *Public Health*, **104**, 9–20.

45 Hitchings, G. H. (1969). Species differences among dihydrofolate reductases as a basis for chemotherapy. *Postgrad. Med. J.*, **45**, (suppl. Nov.), 7–10.

46 Feldman, H. A. (1973). Effects of trimethoprim and sulfisoxazole alone and in combination on murine toxoplasmosis. *J. Infect. Dis.*, **128**, (suppl. Nov.), S774–6.

47 Remington, J. S. (1973). *Experimental data published in an addendum to Ref 46*, p. S776.

48 Norrby, R. and Eilard, T. (1976). Recurrent toxoplasmosis: case report. *Scand. J. Infect. Dis.*, **8**, 275–6.

49 Grossman, P. L. and Remington, J. S. (1979). The effect of trimethoprim and sulfamethoxazole on *Toxoplasma gondii* in vitro and in vivo. *Am. J. Trop. Med. Hyg.*, **28**, 445–55.

50 Solbreux, P., Sonnet, J., and Zech, F. (1990). A retrospective study about the use of cotrimoxazole as diagnostic support and treatment of suspected cerebral toxoplasmosis in AIDS. *Acta Clin. Belg.*, **45**, 85–96.

51 Tabbara, K. F., Sakuragi, S., and O'Connor, G. R. (1982). Minocycline in the chemotherapy of murine toxoplasmosis. *Parasitology*, **84**, 297–302.

52 Chang, H. R., Comte, R., and Pechere, J-C. (1990). In vitro and in vivo effects of doxycycline on *Toxoplasma gondii*. *Antimicrob. Agents Chemother.*, **34**, 775–80.

53 Kovacs, J. A., Allegra, C. J., Swan, J. C., Drake, J. C., Parrillo, J. E., Chabner, B. A. *et al.* (1988). Potent antipneumocystis and antitoxoplasma activities of piritrexim, a lipid-soluble antifolate. *Antimicrob. Agents Chemother.*, **32**, 430–3.

54 Allegra, C. J., Kovacs, J. A., Drake, J. C., Swan, J. C., Chabner, B. A., and Masur, H. (1987). Potent in vitro and in vivo antitoxoplasma activity of the lipid-soluble antifolate trimetrexate. *J. Clin. Invest.*, **79**, 478–82.

55 Rogers, P., Allegra, C. J., Murphy, R. F., Drake, J. C., Masur, H., Poplack, D. J. *et al.* (1988). Bioavailability of oral trimetrexate in patients with acquired immunodeficiency syndrome. *Antimicrob. Agents Chemother.*, **32**, 324–6.

56 Allegra, C. J., Boarman, D., Kovacs, J. A., Morrison, P., Beaver, J., Chabner, B. A. *et al.* (1990). Interaction of sulfonamide and sulfone compounds with *Toxoplasma gondii* dihydropteroate synthase. *J. Clin. Invest.*, **85**, 371–9.

57 Kirst, H. E. and Sides, G. D. (1989). New directions for macrolide antibiotics: pharmacokinetics and clinical efficacy. *Antimicrob. Agents Chemother.*, **33**, 1419–22.

58 Chang, H. R. and Pechere, J-C. F. (1988). In vitro effects of four macrolides (roxithromycin, spiramycin, azithromycin (CP-62, 993), and (A-56268) on *Toxoplasma gondii*. *Antimicrob. Agents Chemother.*, **32**, 524–9.

59 Hofflin, J. M. and Remington, J. S. (1987). In vivo synergism of roxithromycin (RU 965) and interferon against *Toxoplasma gondii*. *Antimicrob. Agents Chemother.*, **31**, 346–8.

60 Gladue, R. P. and Snider, M. E. (1990). Intracellular accumulation of azithromycin by cultured human fibroblasts. *Antimicrob. Agents Chemother.*, **34**, 1056–60.

61 Araujo, F. G., Guptill, D. R. and Remington, J. S. (1988). Azithromycin, a macrolide antibiotic with potent activity against *Toxoplasma gondii*. *Antimicrob. Agents Chemother.*, **32**, 755–7.

62 Chang, H. R., Rudareanu, F. C., and Pechere, J-C. (1988). Activity of A-56268 (TE-031), a new macrolide, against *Toxoplasma gondii* in mice. *J. Antimicrob. Chemother.*, **22**, 359–61.

63 Derouin, F. and Chastang, C. (1990). Activity *in vitro* against *Toxoplasma gondii* of azithromycin and clarithromycin alone and with pyrimethamine. *J. Antimicrob. Chemother.*, **25**, 708–11.

64 Chang, H. R., Jefford, C. W., and Pechere, J-C. (1989). In vitro effects of three new 1,2,4-trioxanes (pentratroxane, thiahexatroxane, and hexatroxanone) on *Toxoplasma gondii*. *Antimicrob. Agents Chemother.*, **33**, 1748–52.

65 Ou-Yang, K., Krug, E. C., Marr, J. J., and Berens, R. L. (1990). Inhibition of growth of *Toxoplasma gondii* by qinghaosu and derivatives. *Antimicrob. Agents Chemother.*, **34**, 1961–5.

66 Pfefferkorn, E. R., Eckel, M. E., and McAdams, E. (1988). *Toxoplasma gondii*: in vivo and in vitro studies of a mutant resistant to arprinocid-N-oxide. *Exp. Parasitol.*, **65**, 282–9.

67 Chang, H. R. and Pechere, J-C. (1989). In-vitro toxoplasmacidal activity of cationic electron carriers. *J. Antimicrob. Chemother.*, **23**, 229–35.

68   Sharma, S. D., Hofflin, J. M., and Remington, J. S. (1985). In vivo recombinant interleukin-2 administration enhances survival against a lethal challenge with *Toxoplasma gondii*, *J. Immunol.*, **135**, 4160–3.

69   Suzuki, Y., Conley, F. K., and Remington, J. S. (1990). Treatment of toxoplasmic encephalitis in mice with recombinant gamma interferon. *Infect. Immun.*, **58**, 3050–5.

70   Carey, R. M., Kimball, A. C., Armstong, D., and Lieberman, P. H. (1973). Toxoplasmosis: clinical experiences in a cancer hospital. *Am. J. Med.*, **54**, 30–8.

71   Behan, W. M. H., Behan, P. O., Draper, I. T., and Williams, H. (1983). Does toxoplasma cause polymyositis? *Acta Neuropathol.*, **61**, 246–52.

72   Kagen, L. J., Kimball, A. C., and Christian, C. L. (1974). Serological evidence of toxoplasmosis among patients with polymyositis. *Am. J. Med.*, **56**, 186–91.

73   Adams, E. M., Hafez, G. R., Carnes, M., Wiesner, J. K., and Graziano, F. M. (1984). The development of polymyositis in a patient with toxoplasmosis: clinical and pathological findings and review of the literature. *Clin. Exp. Rheumatol.*, **2**, 205–8.

74   Theologides, A. and Kennedy, B. J. (1969). Toxoplasmic myocarditis and pericarditis. *Am. J. Med.*, **47**, 169–74.

75   Sagrista-Sauleda, J., Permanyer-Miralda, G., Juste-Sanchez, C., de Buen-Sanchez, M. L., Pujadas-Capmany, R., Arcalis-Arce, L. *et al.* (1982). Huge chronic pericardial effusion caused by *Toxoplasma gondii. Circulation*, **66**, 895–7.

76   Visher, T. L., Bernheim, C., and Engelbrecht, E. (1967). Two cases of hepatitis due to *Toxoplasma gondii. Lancet*, **ii**, 919–21.

77   Vethanyagam, A. and Bryceson, A. D. M. (1976). Acquired toxoplasmosis presenting as hepatitis. *Trans. R. Soc. Trop. Med. Hyg.*, **70**, 524–5.

78   Weitberg, A. B., Alper, J. C., Diamond, I., and Fligiel, Z. (1979). Acute granulomatous hepatitis in the course of acquired toxoplasmosis. *N. Engl. J. Med.*, **300**, 1093–6.

79   Bach, M. C. and Armstrong, R. M. (1983). Acute toxoplasmic encephalitis in a normal adult. *Arch. Neurol.*, **40**, 596–7.

80   Cottrell, A. J. (1986). Acquired toxoplasma encephalitis. *Arch. Dis. Child.*, **61**, 84–5.

81   Grant, S. C. and Klein, C. (1987). *Toxoplasma gondii* encephalitis in an immunocompetent adult. *S. Afr. Med. J.*, **71**, 585–7.

82   Binazzi, M. (1986). Profile of cutaneous toxoplasmosis. *Int. J. Dermatol.*, **25**, 357–63.

83   Saari, M., Vuorre, I., Neiminen, H., and Raisanen, S. (1976). Acquired toxoplasmic chorioretinitis. *Arch. Ophthalmol.*, **94**, 1485–8.

84   Thomaidis, T., Anastassea-Vlachou, K., Mandalenaki-Lambrou, C., Theodoridis, C., and Vrahnou, E. (1977). Chronic lymphoglandular enlargement and toxoplasmosis in children. *Arch. Dis. Child.*, **52**, 403–7.

85   Rothova, A., Buitenhuis, H. J., Meenken, C., Baarsma, G. S., Boen-Tan, T. N., de Jong, P. T. V. M. *et al.* (1989). Therapy of ocular toxoplasmosis. *Int. Ophthalmol.*, **13**, 415–19.

86   Dutton, G. N. (1989). Recent developments in the prevention and treatment of congenital toxoplasmosis. *Int. Ophthalmol.*, **13**, 407–13.

87   Desmonts, G. (1966). Definitive serological diagnosis of ocular toxoplasmosis. *Arch. Ophthalmol.*, **76**, 839–51.

88   Lackhanpal, V., Schocket, S. S., and Nirankari, V. S. (1983). Clindamycin in the treatment of toxoplasmic retinochoroiditis. *Am. J. Opthalmol.*, **95**, 605–13.

89   O'Connor, G. R. and Frenkel, J. K. (1976). Dangers of steroid treatment in toxoplasmosis. Periocular injections and systemic therapy. *Arch. Ophthalmol.*, **94**, 213.

90   Wong, B., Gold, J. W. M., Brown, A. E., Lange, M. Fried, R., Grieco, M., *et al.* (1984). Central-nervous-system toxoplasmosis in homosexual men and parenteral drug abusers. *Ann. Intern. Med.*, **100**, 36–42.

91   Leport, C., Vilde, J. L., Katlama, C., Regnier, B., Matheron, S., and Saimot, A. G. (1987). Toxoplasmose cérébrale de l'immunodeprimé: diagnostic et traitement. *Ann. Méd. Interne*, **138**, 30–3.

92   Navia, B. A., Petito, C. K., Gold, J. W. M., Cho, E-S., Jordan, B. D., and Price, R. W.

(1986). Cerebral toxoplasmosis complicating the acquired immunodeficiency syndrome: clinical and neuropathological findings in 27 patients. *Ann. Neurol.*, **19**, 224–38.

93 Israelski, D. M., Tom, C., and Remington, J. S. (1989). Zidovudine antagonises the action of pyrimethamine in experimental infection with *Toxoplasma gondii. Antimicrob. Agents Chemother.*, **33**, 30–4.

94 Leport, C., Bastuji-Garin, S., Perronne, C., Salmon, D., Marche, C., Bricaire, F. *et al.* (1989). An open study of pyrimethamine-clindamycin combination in AIDS patients with brain toxoplasmosis. *J. Infect. Dis.*, **160**, 557–8.

95 Rolston, K. V. I. and Hoy, J. (1987). Role of clindamycin in the treatment of central nervous system toxoplasmosis. *Am. J. Med.*, **83**, 551–4.

96 Dannemann, B. R., Israelski, D. M., and Remington, J. S. (1988). Treatment of toxoplasmic encephalitis with intravenous clindamycin. *Arch. Intern. Med.*, **148**, 2477–82.

97 Polis, M. A., Masur, H., Tuazon, C., Kovacs, J., Katz, D., Hilts, D. *et al.* (1989). Salvage trial of trimetrexate-leucovorin for treatment of cerebral toxoplasmosis in AIDS patients. *Clin. Res.*, **37**, 437A.

98 Araujo, F. G., Huskinson, J., and Remington, J. S. (1991). Remarkable in vitro and in vivo activities of the hydroxynaphthoquinone 566C80 against tachyzoites and tissue cysts of *Toxoplasma gondii. Antimicrob. Agents Chemother.*, **35**, 293–9.

99 Hughes, W. T., Kennedy, W., Shenep, J. L., Flynn, P. M., Hetherington, S. V., Fullen, G. *et al.* (1991). Safety and pharmacokinetics of 566C80, a hydroxynaphthoquinone with anti-*Pneumocystis carinii* activity; a phase I study in human immunodeficiency virus (HIV) infected men. *J. Infect. Dis.*, 163, 843–8.

100 Bhatti, N. and Larson, E. (1990). Low-dose alternate-day pyrimethamine for maintenance therapy in cerebral toxoplasmosis complicating AIDS. *J. Infect.*, **21**, 119–20.

101 Derouin, F., Piketty, C., Chastang, C., Chau, F., Rouveix, B., and Pocidalo, J. J. (1991). Anti-toxoplasma effects of dapsone alone and combined with pyrimethamine. *Antimicrob. Agents Chemother.*, **35**, 252–5.

102 McGregor, C. G. A., Fleck, D. G., Nagington, J., Stovin, P. G. I., Cory-Pearce, R., and English, T. A. H. (1984). Disseminated toxoplasmosis in cardiac transplantation. *J. Clin. Pathol.*, **37**, 74–7.

103 Smith, G. M., Leyland, M. J., Crocker, J., and Geddes, A. M. (1987). Cerebral toxoplasmosis in a patient with Hodgkin's disease. *J. Infect.*, **14**, 243–5.

6

Pregnancy

JEAN M. W. CHATTERTON

During pregnancy *Toxoplasma gondii* parasites have two targets: the mother and the developing baby. Infection in 90 per cent of healthy women will pass unnoticed as symptoms are usually mild or absent. This apparently 'innocent' infection can have grave consequences. The fetus is vulnerable and if the parasite crosses the placenta and infects the baby a spectrum of clinical features results.[1,2] Since toxoplasma was first identified as a cause of congenital encephalomyelitis in 1939[3] cases of congenital toxoplasmosis have been reported from all parts of the world. Vast literature on infection in pregnancy and its effects has accumulated. There is great potential to underestimate the problem. Severe clinical disease occurs in about 10 per cent of babies infected *in utero*.[4] If maternal infection passes undetected the remaining infected infants are not investigated nor treated. Most will not remain symptom free but may develop ocular and neurological sequelae in later years.[5,6]

Attitudes to toxoplasmosis in pregnancy vary in different countries. Some like France and Austria consider it is a significant public health problem; others like the UK and USA are unconvinced. This is reflected in different approaches to identifying, managing, and treating pregnant women with primary toxoplasmosis. For many years France and Austria have advocated and operated antenatal screening programmes. The UK and USA as yet do not. Is the problem so different in different countries? Are France and Austria over reacting or are the UK and USA too complacent? Recently in the UK media attention has highlighted the association of toxoplasma infection in pregnancy with the birth of damaged babies.[7,8] Increased public awareness has caused anxiety and led to demands for information and for testing.[7] Such recognition and awareness is important but hysteria is just as dangerous as complacency. It is important to consider the problem factually and not emotionally. The aim of this chapter is to examine present knowledge of toxoplasmosis in pregnancy, and consider what can and should be done about it.

RISK OF INFECTION

If a pregnant woman becomes infected with toxoplasma for the first time (a primary infection) there is a risk that the infection will be passed on to her baby. The mother's immune system may not be mobilized quickly enough to

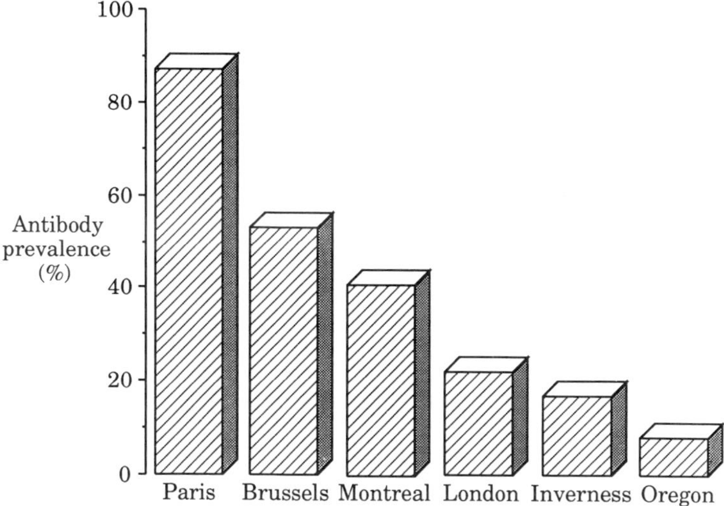

Fig. 6.1   Prevalence of toxoplasma antibody in different populations of pregnant women.[9-14]

control the infection and prevent the parasite crossing the placenta. Overall, approximately 60 per cent of untreated women with primary toxoplasmosis in pregnancy will transmit infection to their fetuses (congenital infection).[4] Pregnant women who have been infected before conception are immune because they have developed specific toxoplasma antibodies in their blood (seropositive). Those who have not been infected lack these antibodies (seronegative) and are susceptible to toxoplasma infection. The number of seronegative women becoming seropositive (seroconversion) represents the rate of infection in the population. The risk of infection in pregnancy is the percentage of women who are susceptible in the population multiplied by the infection rate in their age group.

Seroepidemiological studies

These studies measure antibodies to toxoplasma in large groups of people. From this information the overall immunity in pregnant women at any one time (seroprevalence) can be found. Striking differences have been reported in the proportion of women who are immune in different populations[9-14] (Fig. 6.1). Variability in sample size, the different types of serological tests and the level of antibody which constitutes immunity may contribute to these differences but do not explain them. Immunity by childbearing age reflects past toxoplasma infection and is influenced by a wide range of factors such as climate, behaviour, and place of residence (Chapter 1).

In France toxoplasma infection is common and by childbearing age up to 87 per cent of women are already immune. In the UK however, only 20 per cent of women are immune at the start of pregnancy (Fig. 6.1). Unfortunately results from large population studies may hide regional variations. Thus, in the Paris area more than 80 per cent of pregnant women are immune, while in other areas of France the figure can be as low as 40 per cent.[15] This difference in seroprevalence has been attributed to the fondness of Parisians for undercooked meat. Ethnic groups within a population may also have different rates of infection. In the Paris area Desmonts and Couvreur[9] reported 87 per cent of pregnant French women had toxoplasma antibody compared to 70 per cent of immigrant women (Spaniards, North African Moslems, and Portuguese). However, some 20 years later, there is now good evidence that immunity has fallen markedly in Paris to 71 per cent in women of French origin and 51 per cent in immigrant women.[16] Changes in age, geographical origin, or the specificity or sensitivity of tests used could not explain this fall in immunity. Nevertheless, a fall in the domestic cat population, a reduction in consumption of undercooked meat, changes in types of meat consumed and the increasing use of ready cooked food may all have contributed to the drop in the rate of infection.

The proportion of women who are immune increases with age[9–14] (Fig. 6.2). From the prevalence of antibodies in different age groups the risk of infection to pregnant women can be estimated by calculating the average annual increase in immunity corrected for the nine months of a pregnancy. The frequency of infection is usually expressed as the number of primary infections per 1000 pregnant women in the population[10–14,17] (Table 6.1). Antibody acquisition however, shows high and low rates in different age groups. This means that the risk of infection is not the same for women of all ages. A study in the Netherlands found the incidence of primary infection was higher in younger women and fell from 1.62 per cent per 9 months at 17.5–20 years to 0.37 per cent between the ages of 35–45.[18] This is not true of all populations. In Scotland, exactly the opposite was found in the pregnant women studied. Antibody prevalence remained constant at 15 per cent until the late 20s and then rose in the late 30s by 1.1 per cent per 9 months to 27.7 per cent in the 36–40 age group.[13] This suggests older women are more at risk of acquiring infection in pregnancy. Another study in London similarly suggested women aged 36–40 years had a higher risk of infection than women aged 15–35 years, possibly associated with keeping cats as pets.[19] Thus not only is overall immunity by childbearing age different in different populations but behavioural patterns may lead to differences in the age when infection is likely to be acquired.

In countries where toxoplasma infection is common, such as France and Austria, it is believed to be worthwhile to test for infection during pregnancy while in countries where infection rates are low, toxoplasmosis is considered to be rare and testing is not undertaken. In fact the incidence of primary toxoplasmosis in women of childbearing age can be similar despite striking differences in infection rates in populations. Comparison of the immunity

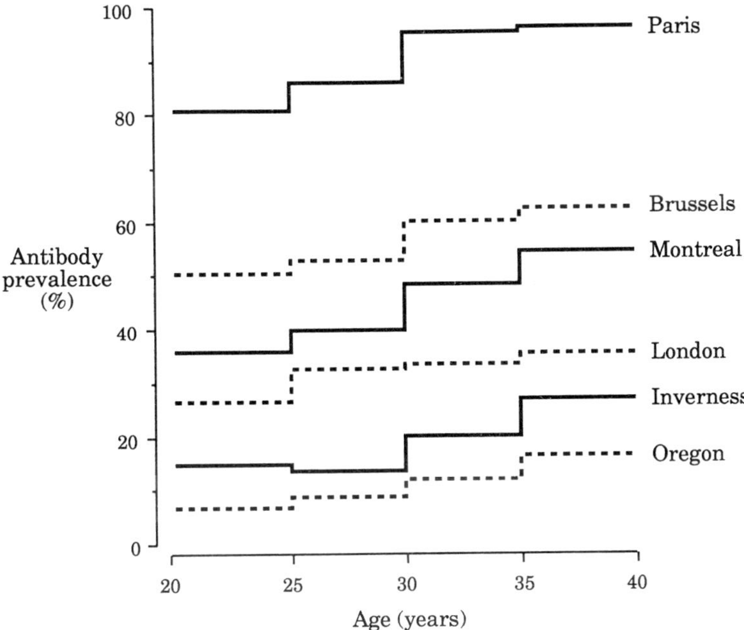

Fig. 6.2   Variations in toxoplasma antibody in pregnant women with age and geography. [9-14] (From Ho-Yen, D. O. and Chatterton, J. M. W. (1990). Congenital toxoplasmosis — why and how to screen. *Reviews in Medical Microbiology*, **1**, 229–35)

Table 6.1   Frequency of toxoplasma infection per 1000 pregnancies[10-14,17]

Location	Sample size	Primary toxoplasmosis		Frequency			
		*Confirmed	#Possible	Observed			Estimated
				Minimum		Maximum	
Brussels	2986	20	17	7	–	12	6
Montreal	4136	2	50	<1	–	13	7
London	3187	7	2	2	–	3	4
Inverness	4748	4	6	1	–	2	3
Oregon	10 298	22	–		2		5
Glasgow	10 677	21	65	2	–	8	2

From Ho-Yen, D. O. and Chatterton, J. M. W. (1990). Congenital toxoplasmosis — why and how to screen. *Reviews in Medical Microbiology*, **1**, 229–35.
* Seroconversion.
# High titre, low level seroconversion or IgM in first specimen.
Minimum = no. of confirmed cases.
Maximum = no. of confirmed cases plus possible cases.

patterns in pregnant women in London and Paris shows that at the start of the childbearing years (18 years) there is a marked difference in immunity: 80 per cent in Paris and 17 per cent in London but the increase in immunity over the childbearing years (18–36 years) is the same in the two populations. Thus at 36 years the final immunity is 93 per cent in Paris and 30 per cent in London.[1] Where toxoplasmosis is common a high risk of infection applies but only to the small number of women who are still susceptible at the start of pregnancy. Conversely in populations with low rates of infection a small risk applies but there are large numbers of susceptible women. Overall the incidence of primary infection in pregnancy can be the same for both groups. This balance of the infection rate in the population and the proportion of women who are susceptible can be upset if women move between areas for holidays or immigration. Susceptible women from areas of low prevalence of toxoplasma can be at increased risk of infection if local habits like eating undercooked meat are adopted. There is some evidence for this in immigrant women in France.[15]

Prospective studies

Prospective studies test women for toxoplasma antibody as they progress through pregnancy. As several samples are taken from each patient they are more difficult to carry out than seroprevalence studies but have the advantage of providing observed rather than estimated figures for infection in pregnancy. Demonstration of seroconversion during pregnancy provides unequivocal evidence of primary infection. These represent the minimum number of confirmed cases of toxoplasmosis[10–14,17] (Table 6.1). Other women may have serological evidence of infection in pregnancy: high or increasing amounts of antibody (rising titre) or the presence of antibody of the type known to be produced early in infection (IgM antibody). Inclusion of these women gives the total possible number of observed primary cases in pregnant women (Table 6.1).

The effectiveness of prospective studies depends on how and when during pregnancy women are tested. This differs from one study to another. Most test as early as possible, i.e. at the first antenatal visit but the frequency of follow-up testing of seronegative women varies. Some test at 6 week intervals throughout pregnancy[10] while others test again only in the second and third trimester.[12,17] In some cases women are only tested again at delivery.[13] Early and frequent testing optimizes detection of infection in pregnancy. Prospective studies have problems with sample size and patient follow-up. To be representative of pregnant women in the population the study group must be large enough to cover all age ranges and social subgroups. This can be difficult to achieve. Similarly the accuracy of the figure for primary infection depends on complete testing of all women from the beginning to the end of pregnancy. The loss of any who may have been infected underestimates infection in pregnancy. Unfortunately most studies have found women are lost to follow-up testing. Prospective studies by local laboratories have severe limitations. The only way to accurately assess the risk of toxoplasma infection in pregnancy is properly resourced, pilot, national testing of pregnant women.

EFFECTS OF INFECTION

When a healthy pregnant woman is infected with toxoplasma for the first time there may be a generalized infection with involvement of any of the body's organs.[1,2] Infected white cells effectively transport and distribute the parasite throughout the body.[20] Toxoplasma actively penetrates body cells and multiplies within them (Chapter 2). This results in cell destruction and the release of more parasites to invade and destroy other cells. The ability to control the multiplication and spread of toxoplasma depends on the immune system.[2,20] Both antibodies (humoral immunity) and cells (cellular immunity) are involved (Chapter 7). Cell mediated immunity as measured by lymphocyte transformation may not be demonstrable for weeks or even months after infection with toxoplasma.[21] In a pregnant woman the immune system may be suppressed by pregnancy itself.[2] Thus pregnant women may be less able to deal effectively with toxoplasma infection than non-pregnant women. The fetus is only capable of a limited immunological response to infection. The cells of the immune system are developing and up until a gestation age of about 6 months the fetus relies mainly on specific antibody from the mother which is passively transferred across the placenta. Humoral immunity is the first to develop and the fetus starts to manufacture antibody during the second half of intrauterine life.[20] Cell mediated immunity does not develop until much later and full immune competence is not achieved until 6–12 months after birth.[20] In pregnancy both the mother and her baby are targets for toxoplasma and the effect of infection on both must be considered.

The pregnant woman

Most pregnant women acquiring toxoplasmosis have no significant symptoms (asymptomatic).[1,2,9] The evidence for this comes largely from prospective testing of pregnant women where primary infection is diagnosed serologically irrespective of the presence or absence of symptoms. Some studies report significant symptoms were absent[12,22] or rare[11,23] in the infected women they identified. In others the clinical picture is not clear: it is impossible to determine if there were no symptoms or just no information.[17,24] In one study details were provided on only 12/122 women and six reported some symptoms.[17] Certainly non-specific symptoms such as fever and malaise may often be present without being recognized. Support for this comes from retrospective questioning of apparently asymptomatic women who have given birth to damaged babies with congenital toxoplasmosis. Such women may recall mild symptoms during their pregnancy[25] which resolved without specific treatment. The symptoms were simply forgotten and the diagnosis of toxoplasmosis would never have been made had the baby not been clinically affected and investigated.

It is easy to dismiss toxoplasma infection in pregnancy as asymptomatic. One recent study considered 150 clinical and laboratory observations on pregnant women and correlated these with toxoplasma serology.[23] They concluded that maternal infection was more frequently associated with illness than previously thought. Thrombophlebitis and asthma were common findings which have not been previously reported.[23] Lymphadenopathy, the most common clinical manifestation of toxoplasmosis in healthy individuals[26,27] (Chapter 3) occurs in pregnant women.[9,28,29] Infectious mononucleosis-like illness however, seems to be the exception not the rule. In prospective studies, only one of 122,[17] one of 52,[11] and 1 of 15[23] women reported infectious mononucleosis-like illness. However direct questioning in one study revealed mild symptoms in 40 per cent of cases of maternal infection, usually persistent occipital or submaxillary adenopathy.[29] Other recognized clinical features of toxoplasmosis, such as eye complaints or involvement of liver or spleen probably do occur but are not frequently recognized in pregnant women. There is no correlation between symptoms in the mother and fetal infection.

The placenta

There is good evidence that the placenta can be infected when toxoplasmosis is acquired in pregnancy. One investigation showed that the parasite could be grown from placenta in 26 per cent (25/97) of such women but from none known to have been infected before conception.[9] The placenta is more frequently infected when toxoplasmosis is acquired later in pregnancy; with positive isolates from 25 per cent (4/16), 54 per cent (7/13) and 65 per cent (15/23) of untreated women infected in the first, second and third trimesters, respectively.[4] This is probably due to the increased blood flow through the growing placenta as pregnancy progresses. Although infection is not always transmitted to the baby when the placenta is infected there is a very close correlation between congenital toxoplasmosis and isolation of the parasite from the placenta: 88 per cent of 43 cases in one study[30] and of 36 cases in another.[31] Most evidence suggests isolation studies are always negative in uninfected infants.[32] We have experience of one case where toxoplasma was isolated from placenta yet the baby seemed to be healthy and uninfected.[25] Longer follow-up may have revealed a subclinical infection in this infant.

In cases of proven congenital toxoplasma infection the placenta may appear grossly normal.[33] However, one study found 10 of 13 placentas showed signs of prolonged fetal distress and or haematogenous infection.[34] The parasite may be seen microscopically, usually as cysts,[2,28] although tachyzoite forms (Chapter 2) have also been described.[34,35] In one report 40 toxoplasma cysts were counted in a single section of placenta[33] but microscopic examination can often be negative when isolation is positive.[28] The precise mechanism of transmission from the placenta to the fetus is not clear.[36] Rupture of infected cells or cysts probably results in parasites being discharged into the fetal circulation.

Twins account for about one per cent of births. In 75 per cent of twins each fetus is independent and has a separate placenta (dizygotic twins). Both may escape infection but, either or both placentas can become infected so it is possible to find one infected and one uninfected twin.[29,33,37] Equally one twin may be clinically damaged while in the other infection may be sub-clinical.[33,37] Thus, investigation and diagnosis of congenital toxoplasmosis in one twin should prompt thorough investigation of the other. Some 25 per cent of twins are produced from a single fertilized egg (monozygotic) and may share a common placenta. Consequently the risk of fetal infection is similar in both twins. As might be expected the clinical pattern in congenitally infected monozygotic twins is usually very similar.[33,37]

The fetus

Toxoplasma infection acquired in pregnancy represents a serious threat to the fetus. If the parasite crosses the placenta and infects the fetus there is the potential for severe damage.[1,2] Congenital toxoplasmosis however, has a wide range of manifestations and indeed it was soon recognized that infected babies may appear well at birth.[38] For this reason most of the information on the fate of the fetus comes from prospective surveys of pregnant women.[4,9–14,16,17,22,24,28,29,31,39,40] These studies set out to identify women with primary infection, monitor their pregnancies, determine whether or not the fetus becomes infected, and assess the clinical effects. This may seem logical and possible in theory but in practice it is far from straightforward.

Results of maternal infection

There is great variability in how successfully pregnant women are monitored. Few studies have been able to monitor all of the infected women they identify and in some less than half[10,17] were followed completely enough to determine whether or not the fetus was infected (Table 6.2). It is understandable that women are sometimes reluctant to submit babies to the necessary examinations and blood tests, particularly if they appear to be healthy and the initial examination results were favourable. It is difficult to take account of cases 'lost to follow-up' or diagnosed as 'probably uninfected' or 'possibly infected' when trying to estimate the frequency of fetal infection. The maximum transmission rate is the number of cases of fetal infection

Table 6.2   Consequences of primary infection in pregnant women[10–12,16,17,24]

Location	Seroconversion in pregnant women		Fetal infections	Maternal fetal transmission, %	
	Detected	Follow up completed (%)		Minimum	Maximum
Brussels	20	8 (40)	2	10	25
Montreal	2	2 (100)	1	50	50
London	7	7 (100)	0	0	0
Paris	33	17 (52)	11	33	65
Glasgow	21	9 (43)	3	14	33
New York	6	6 (100)	2	33	33

Minimum — no. of fetal infections as a percentage of women with seroconversion.
Maximum — no. of fetal infections as a percentage of women with seroconversion and complete follow-up.

detected in those women followed completely enough to allow a final diagnosis. Alternatively, if it may be assumed that when cases were lost or incompletely followed up the fetus was healthy and uninfected, this gives the minimum rate of fetal infection (Table 6.2). Interpretation of the results in some studies is complicated by treatment given to pregnant women. This must affect the results but it is difficult to assess to what extent.

Fetal infection

All studies agree that women with primary infection in pregnancy do not always pass infection to the fetus. The observed incidence of fetal infection has varied widely from 0 to more than 50 per cent[11,12] (Table 6.2). The most complete information on maternal–fetal transmission of toxoplasma comes from French studies carried out in Paris over a 20 year period and published in 1979.[4] The comparitively large numbers of women studied and effective follow-up of the pregnancies makes the data more significant. The results showed that fetal infection was confirmed in 178 (33%) of 542 cases of acquired maternal toxoplasmosis and excluded in 357 (66%). This data had the advantage of being complete enough to analyse factors which appeared to be important to the fate of the fetus; treatment of pregnant women and when in pregnancy infection was acquired.[4] Treatment appeared to reduce transmission of toxoplasma.[4,9,28] There were 23 per cent (89/388) infected offspring in treated women compared to 58 per cent (89/154) in untreated women.[2,4] It was possible to divide 500 women according to the probable date of infection. The rate of fetal infection increased with gestational age from 14 per cent (17/126) when infection was acquired in the first trimester to 29 per cent (73/246) in the second and 59 per cent (76/128) in the third[4] (Fig. 6.3). These figures include both treated and untreated women.

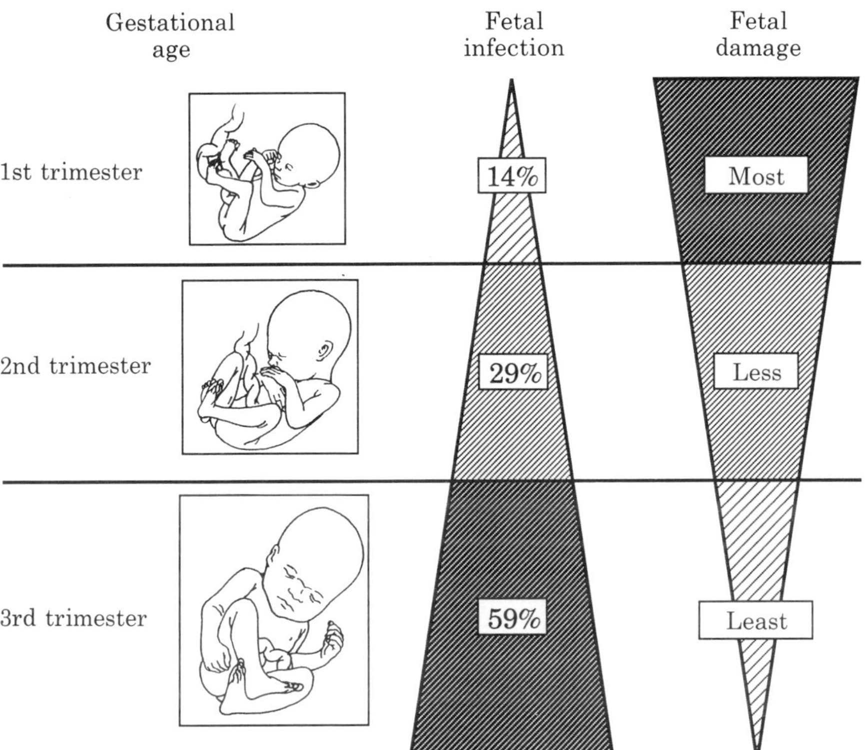

Fig. 6.3 Gestational age, fetal infection, and fetal damage associated with toxoplasma infection in pregnancy.

Fetal damage

The clinical effects of toxoplasmosis on the fetus can also be related to when in pregnancy maternal infection was acquired. Infection early in pregnancy usually resulted in severe disease while fetal infection following third trimester infection was usually subclinical[4,9,28] (Fig. 6.3). Infection can lead to spontaneous abortion but the evidence from prospective studies monitoring women with current infection suggests this is uncommon. In one study it only occurred in 1.3 per cent (5/378) of pregnancies with evidence of recent toxoplasma infection.[28] In others no spontaneous abortions were reported.[10,12,24] Similarly, attempts to isolate the parasite from abortion material have failed to prove toxoplasma infection.[12,41] It is believed that the earlier fetal infection occurs, the greater the risk of abortion. Studies can be criticized for excluding women who acquire infection very early in pregnancy. First specimens from pregnant women are usually taken at the end of the second month or later in pregnancy. Women who miscarry before this will not be

Table 6.3 Relative frequency of clinical features of congenital toxoplasmosis[38,42]

Clinical features	Babies studied (no.)		Total (%)	
	152	73	225	
Retinochoroiditis	121	52	173	(77)
Fever	78	0	78	(35)
Jaundice	82	20	102	(45)
Splenomegaly	85	13	–	–
Hepatomegaly	71		–	–
Anaemia	97	0	97	(43)
Intracranial calcification	41	28	69	(31)
Hydrocephalus or microcephaly	32	40	72	(32)
Convulsions	52	0	52	(23)

examined for evidence of toxoplasmosis. The ideal study would require specimens to be collected at or shortly before conception.

The clinical manifestations of congenital toxoplasmosis vary from asymptomatic to severe. The first description of congenital toxoplasmosis was in a child with severe neurological damage: eye lesions (retinochoroiditis), large head (hydrocephalus), and widespread infection of the brain (encephalitis followed by intracerebral calcification).[3] This description is recognized as the classic triad of congenital toxoplasmosis. Such a severe clinical picture is not easily missed and has resulted in a large percentage of reported cases.[42] However, in prospective studies, severely affected babies are rare. Sadly these studies are more difficult to interpret because of therapeutic intervention, such as abortion and specific treatment. In an attempt to determine the relative, but probably not actual, frequency of the clinical findings, the results of two early studies[38,42] are considered (Table 6.3). Early severe brain damage, manifest by hydrocephalus/microcephaly and intracerebral calcification occurred in about a third of cases. Convulsions were not recorded in one study but probably manifest in as many cases. Although a clinical picture of meningoencephalitis can be found in 41 per cent of cases diagnosed in the first 6 months of life, it is not seen after the second year.[42] The picture of jaundice and hepatosplenomegaly is common in many congenital infections and is not diagnostic. The most common finding is retinochoroiditis (77%, Table 6.3) and this is very easily overlooked in newborn babies.[43] Other non-specific manifestations include anaemia, fever, lymphadenopathy, vomiting, diarrhoea, pneumonia, rash, and hypothermia.[38] Premature delivery is common both in congenital infants with clinically apparent disease[2] and those subclinically infected. One study found prematurity was a feature in 50 per cent (5/10) of subclinical cases.[43]

Although any of the clinical manifestations of acquired infection (Chapter 3) may occur in congenital toxoplasmosis infection may well be overlooked in an apparently normal baby. Abnormalities, especially minor retino-choroiditis, may be missed;[43] and many quite severe clinical features (such as neurological disorders and intracranial calcification) may not be diagnosed until adulthood.[42] Deafness has been associated with congenital toxo-plasmosis usually as a feature of severe disease.[2] However, recently bilateral deafness was identified as one of the most significant findings in children examined at 7 years of age.[23] Only greatly increased awareness of the pos-sibility of infection, probably as a result of antenatal screening, will ensure that clinically affected babies are not wrongly categorized as normal.

One great difficulty for clinicians is that congenital toxoplasmosis is an uncommon disease. Even among cases with retinochoroiditis alone or associated with neurological disturbances, the diagnosis of toxoplasmosis is only correct in 70 per cent.[42] Similarly toxoplasmosis is not an important cause of microcephaly, hydrocephaly, or cerebral palsy.[44] Overall, the best indication of the frequency of congenital toxoplasmosis is that infection is transmitted to the fetus in about 60 per cent of untreated infected women;[2,4] and of these, the majority (67%) will have subclinical toxoplasmosis, 23 per cent clinical toxoplasmosis (36% severe, 64% mild), and 10 per cent will result in neonatal death.[4] It used to be felt that those with subclinical infec-tion or mild disease were not at further risk.[40] However, it is now realized that asymptomatic babies may develop eye problems, including blindness, up to 18 years later.[6]

INFECTION BEFORE PREGNANCY

It is important to distinguish between toxoplasmosis acquired during a preg-nancy and infection acquired before the pregnancy. Primary toxoplasmosis in pregnancy can certainly result in abortion or the birth of an infected baby.[1,2] The effect of infection acquired before pregnancy is less clear. Late parasitaemia has been reported 14 months after acute toxoplasmosis.[45] Furthermore, recovery from toxoplasmosis does not mean the parasite has been eradicated from the body. Toxoplasma persists in many tissues as quiescent tissue cysts (latent infection). Persistence in the uterus has been found in 12 per cent (4/32) of women with stable low antibody titres and no evidence of active infection.[46] There has been much speculation that these uterine cysts represent a threat to the fetus in future pregnancies.

Periconceptional

The demonstration of seroconversion during pregnancy proves that infection was acquired after conception. A high antibody titre or specific IgM in the first specimen examined during a pregnancy indicates recent infection.

Table 6.4 Consequences of proven or possible primary maternal infection[10–13,16,17,24,28,31,39]

Location	[a]Proven maternal infection	Fetal infections (%)	[b]Possible maternal infection	Fetal infections (%)
Brussels	8	2 (25)	9	2 (22)
Montreal	2	1 (50)	50	1 (2)
London	7	0 (0)	2	0 (0)
Inverness	4	1 (25)	6	0 (0)
Paris	33	11 (33)	0	0 (0)
Glasgow	15	3 (20)	107	1 (1)
New York	6	2 (33)	52	1 (2)
Paris	176	55 (31)	191	4 (2)
Rouen	109	36 (33)	64	0 (0)
Oslo	8	3 (38)	21	0 (0)

[a] Seroconversion.
[b] High titre, IgM positive, rising titre.

Depending on when the specimen was taken this could have been before or after conception. When the pregnancies of these two groups of women are compared in prospective studies the difference is significant (Table 6.4) with an average of 31 per cent (114/368) fetal infection in women who seroconvert during pregnancy compared to 2 per cent (9/502) in women with evidence of recent infection. On critical examination eight of these nine cases fail to provide convincing evidence of transmitted periconceptional infection. Three of the women with apparently infected babies were initially tested at 5[11] and 7 months[10,24] of pregnancy and it is quite possible that infection was acquired after conception. Proof of congenital infection was lacking in five other babies. In the Scottish survey[17] the baby was lost to follow-up after 6 weeks and congenital infection could not be confirmed. In the French study[28] the four infants were first examined between the ages of 14–45 months by which time it was impossible to exclude acquired infection. In only one case was there good evidence of maternal infection before but near conception (high antibody titres, IgM negative at 8 weeks gestation) and proven infection in the baby with toxoplasma isolated from neonatal blood and CSF.[10] Taking these factors into consideration, the rate of fetal infection is reduced from 2 per cent (9/502) to 0.2 per cent (1/502).

In another study of 21 women with evidence of recent toxoplasmosis, 18 were considered to have been infected shortly before conception.[39] Seven had high antibody titres and no detectable IgM in the first trimester and 11 had moderate antibody titres and positive IgM results in the first half of pregnancy. There were no cases of fetal infection. Among 195 French women with recent infection 14 had toxoplasma lymphadenopathy diagnosed around the time of conception.[28] There was only one spontaneous abortion in the symptomatic group and no proof that it was due to toxoplasma. It seems there is little risk to the fetus if maternal infection is acquired shortly before or around the time of conception. Nevertheless as infection may take some time to settle, women are usually advised to wait 6 months after the acute infection before becoming pregnant.[47] However, if there are persistent symptoms or signs, further advice should be sought from the Reference Laboratories. These laboratories might advise further tests such as the polymerase chain reaction (Chapter 8).

Before conception

An alarming German study[48] provided evidence that during pregnancy dormant toxoplasma cysts in the uterus of chronically infected women could rupture causing abortion or repeated congenital infection. They reported one proven and one likely case of repeated congenital infection and the isolation of toxoplasma from 36 specimens (placenta, fetal tissue, menstrual blood, milk) from 23/70 women with habitual or repeated miscarriages, premature births, or stillbirths. Nineteen of these women had serology consistent with long past toxoplasmosis; inexplicably four had no serological evidence of toxoplasmosis.[48] This controversial work challenged the belief that only primary maternal infection in pregnancy threatens the fetus. The evidence was called into question when other workers failed to repeat the results.[12,41] Indeed the author himself cast doubt on the findings by admitting to some problems with pollen grain contamination which meant it was impossible in retrospect to determine in which cases toxoplasma cysts or pollen grains had been seen.[2] Furthermore he was reportedly unable to repeat his initial findings when he studied cases similar to those he had studied earlier.[49] Nevertheless there are enormous physical changes in the uterus during pregnancy and it is plausible that rarely these changes could rupture uterine cysts and produce abortion.

Abortion

Most studies on abortion which have attempted to isolate toxoplasma from placenta or products of conception have been unsuccessful.[12,41] These failures may have been due to sample preparation or isolation technique but they suggest past maternal infection is an uncommon cause of abortion. Toxoplasma tachyzoites were observed by immunofluorescence in endometrial biopsies and menstrual blood from 14.6 per cent (6/41) of patients with habitual abortion, 0 per cent (0/20) with sporadic abortion and 1.7 per cent (1/59) of obstetric controls.[50] On this basis an association between toxoplasma infection of the uterus and habitual abortion might be suggested. However, the fact that the parasite was not isolated from any of the biopsies and 5/7 biopsy positive women had no antibody to toxoplasma must cast considerable doubt on these results.

Table 6.5 Toxoplasma antibody in women with abortion and obstetric controls[41,50,51,53–56]

Location	No. of abortions (% positive)		No. of obstetric controls (% positive)
	[a]Sporadic	[b]Habitual	
New York	868 (38)	73 (33)	3891 (31)
Oslo	61 (25)	96 (25)	59 (19)
California	113 (29)	0 (0)	442 (27)
India	51 (24)	124 (18)	100 (9)
India	69 (20)	171 (14)	150 (7)
Jordan	0 (0)	55 (58)	46 (26)
Dallas	0 (0)	25 (20)	47 (21)

[a] 1, 2, or more non-consecutive abortions.
[b] 3 or more consecutive abortions.

Good serological evidence of past toxoplasma infection is only present in a few reports of toxoplasma abortion in women.[51,52] Spontaneous abortion occurred at 5 weeks gestation in a woman with positive toxoplasma antibody titres.[51] The parasite was not isolated from abortion tissues but a positive isolate was obtained one year later when she aborted again. Identical antibody titres one year apart is unequivocal evidence of past infection.[51] Three cases of habitual abortion and two of recurrent abortion were attributed to chronic toxoplasmosis.[52] Toxoplasma was isolated from only one of these. This woman, after two normal pregnancies, suffered three spontaneous abortions. At this time she had antibody to toxoplasma. Some 20 months later toxoplasma was isolated from placenta when she aborted for the fourth time.[52] It should be emphasized that such cases with isolation of the parasite and well documented maternal serology are rare.

Less convincing evidence of an aetioloical role for toxoplasma is found in the many studies which link past toxoplasma infection and abortion by comparing the prevalence of toxoplasma antibodies in women with a history of abortion and obstetric controls[41,50,51,53–56] (Table 6.5). Some report significantly higher seroprevalence rates in abortion groups[41,53,54,55] and this seemed to be associated with sporadic rather than habitual abortion. Others find little difference between those women who abort and the controls.[50,51,56] Serological surveys are relatively easy to carry out but their value is questionable. Detection of toxoplasma antibody provides evidence of infection but does not prove a causative role in abortion.

Treatment of latent maternal toxoplasmosis has been advocated for women with recurrent or habitual abortion.[50,52,57] The success of such treatment with subsequent normal deliveries of healthy babies is indirect evidence of a role for toxoplasma. There may indeed be a case for treating women in whom all other causes of abortion have been excluded and who show high or rising antibody titres to toxoplasma at or shortly after abortion.[52] However, such treatment should be approached with caution.[2,49] If latent maternal toxoplasmosis were a significant cause of abortion then it would be expected to be most evident in countries like France where 80 per cent of women have been infected before childbearing years. Such a problem has not been identified. Present experience suggests latent toxoplasmosis is an uncommon cause of abortion.

Congenital infection

If latent maternal toxoplasmosis were responsible for congenital infection, cases of toxoplasmosis would be expected in future siblings. Careful follow-up of more than 400 babies born to mothers who had previously given birth to an infected baby revealed no congenital cases.[49] Examination of more than 800 cases of congenital toxoplasmosis in Paris revealed 14 pairs of twins but no other affected siblings.[9] It is possible that subclinical cases could have been missed but given the expertise and experience of these workers this seems unlikely. There are sporadic reports of siblings with eye lesions believed to be characteristic of congenital toxoplasmosis.[58,59] The affected siblings and their mothers had serological evidence of previous toxoplasma infection but since the children were initially examined between 6 and 19 years of age, it is not possible to differentiate acquired from congenital infection.

Toxoplasma was isolated from the blood of a woman 14 months after the birth of her congenitally infected baby and a few weeks before conception of her second child. Despite this parasitaemia the baby was apparently uninfected.[45] Extensive studies in France have failed to identify a single case of congenital toxoplasmosis when mothers were immunocompetent and infected before pregnancy.[4,9,28] It is generally accepted that past maternal toxoplasmosis does not represent a threat to the fetus. Indeed one of the major advantages in diagnosing congenital toxoplasmosis or past maternal infection is to reassure women that subsequent pregnancies are unlikely to be affected.

Immunocompromised state

There is no doubt that an immunocompromised state as a result of disease, treatment, or infection can cause reactivation of latent toxoplasmosis (Chapter 7). Congenital toxoplasma infection has been attributed to reactivation of maternal toxoplasmosis in a few cases.[2] It should be emphasized that these cases are unusual but illustrate a potential danger in pregnancy. A woman who had had a stable toxoplasma titre for 5 years gave birth to an infected baby after steroid treatment for systemic lupus erythematosus.[60] Similarly, a subclinically infected baby was born to a woman suffering from Hodgkin's disease who had had latent toxoplasma infection for more than 4 years.[2]

Reactivation in this case was accompanied by a rise in toxoplasma antibody titres. A subsequent pregnancy 2 years later produced an uninfected baby.

The role of human immunodeficiency virus (HIV) in reactivating toxoplasma infection in acquired immunodeficiency patients is well established (Chapter 7). While the dangers to the fetus from reactivated toxoplasmosis in pregnant HIV patients have yet to be evaluated, cases of congenital infection are being recognized.[61–63] In one mother toxoplasma reactivation was believed to be localized to the central nervous system and the baby was not thought to be at risk. In another, toxoplasma encephalitis was never diagnosed yet her third and fifth babies were congenitally infected.[62] Co-infection with HIV means the immunological response in such infants and their mothers may be impaired and this may explain the severity of the lesions seen. One infant with proven toxoplasma encephalitis when 4 months old was seronegative for toxoplasma at 5, 6, and 8 months of age.[62] As HIV spreads in the heterosexual community more women will become infected. Dual infection is likely to become more common particularly in populations where toxoplasmosis is prevalent. How frequently HIV infection results in reactivated toxoplasmosis and fetal infection has yet to be established but indications are that this will be a significant problem in the future.

MANAGEMENT

When acute toxoplasmosis is suspected in a pregnant woman what can and should be done? Diagnosis usually depends on serological tests.[47,64,65] which detect specific antibodies in the blood. A wide variety of tests have been developed (Chapter 4) and treatment of toxoplasma infections is established (Chapter 5). This section will only consider the diagnosis and treatment of toxoplasma infection in the pregnant woman, the fetus, and the neonate.

The pregnant woman

A choice of commercial assays is available to laboratories for toxoplasma testing. Positive results in a pregnant woman must be interpreted with care. The critical question is whether or not infection was acquired before or after conception. High titres of toxoplasma antibody can persist for months after acute infection and depending on which test is used, specific IgM may be found several months[66,67] or more than a year[68,69] after documented primary toxoplasmosis, i.e. before conception. The possibility of false positive results must also be considered (Chapter 4). The implications of toxoplasma infection in pregnancy are grave and accurate diagnosis is of paramount importance. Sera showing evidence of toxoplasma infection by whatever test is employed should be sent for confirmation to a Toxoplasma Reference

Laboratory. These centres provide a complete range of specialized tests, the most important of which is the toxoplasma dye test[70] (Chapter 4). If the results suggest acute toxoplasmosis (positive dye test; specific IgM detected) then a second sample, and any earlier samples available, would be requested. Demonstration of seroconversion in a pregnant women is proof of current primary infection. Often interpretation of the results is far less straightforward and in these circumstances careful questioning about symptoms may be useful in determining when infection was acquired. Only on the basis of all of these results should the diagnosis of infection in pregnancy be made.

The parents must be counselled and advised on the risks to the pregnancy. It is important to recognize that toxoplasmosis in pregnancy is not a hopeless situation and at least 40 per cent of untreated infected women will have uninfected babies.[4] The stage of pregnancy when maternal infection is acquired is important to the fate of the fetus. In early maternal toxoplasmosis there is a low risk of fetal infection but when it occurs damage is usually severe. Conversely the later in pregnancy maternal infection is acquired the greater the risk of fetal infection but damage is less[4,9,28] (Fig. 6.3). Acquiring infection in the second trimester represents the greatest hazard as it is associated with the highest risk of fetal disease.[2] The more accurately the time of maternal infection can be established the more accurately the risk to the fetus can be assessed. Each case must be considered individually but dating maternal infection is usually difficult because specimens are taken late in the course of infection. It is important to recognize that leaving the outcome of the pregnancy to chance is only one of the available choices. A scheme for the management of such patients (Fig. 6.4) and the implications of each stage should be explained and considered before deciding on how the pregnancy will be managed.

Termination

Therapeutic termination is one course of action. As there are legal limits applied to this procedure it is not an option available to all.[2,71,72] In France abortion may be carried out at any time providing two doctors consent to the procedure.[71] In the UK and USA the upper limit is usually 24 weeks but can be at any time if there is evidence of severe fetal abnormality. The indications for termination should be carefully and compassionately discussed. If termination is carried out on the basis of maternal infection alone then it is certain that healthy babies will be lost. When placentas from seven of ten therapeutic abortions carried out because of maternal seroconversion were cultured for toxoplasma all were negative.[10] Of 215 women with evidence of maternal toxoplasmosis 205 chose to continue their pregnancy under treatment; 198 delivered healthy uninfected babies.[71] If these women had been offered and chosen abortion 198 babies would have been unnecessarily sacrificed.[71]

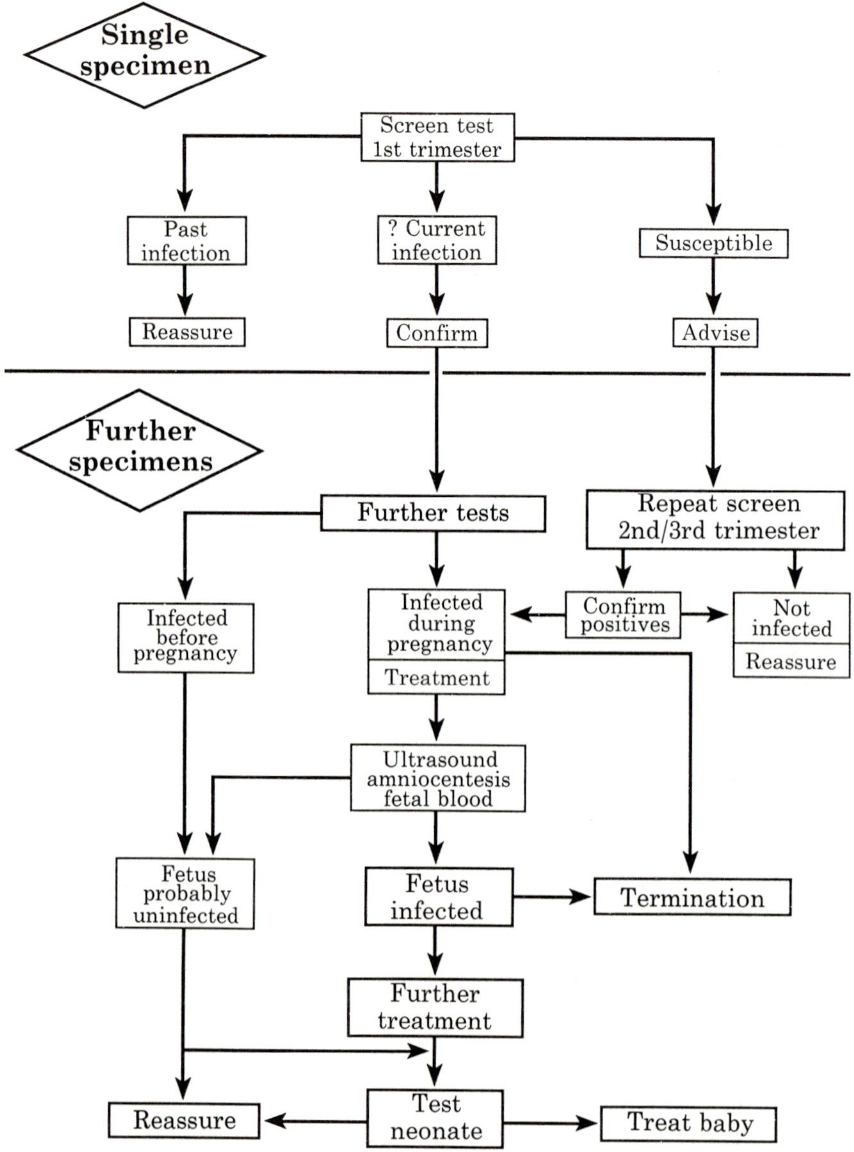

Fig. 6.4 A scheme for diagnosing and treating toxoplasma infection in pregnancy. (From Ho-Yen, D. O. and Chatterton, J. M. W. (1990). Congenital toxoplasmosis — why and how to screen. *Reviews in Medical Microbiology*, **1**, 229–35).

Clinicians may feel termination is only justified in cases where they consider there is a risk of severe fetal damage. Parents may find this unacceptable. They want a healthy baby and are unwilling to continue the pregnancy if there is any risk at all.[71] Termination is a drastic solution but parents may welcome an end to the uncertainty and favour this course of action. When a pregnancy is terminated for fear of toxoplasmosis, placenta and fetal tissue should be examined by a Toxoplasma Reference Laboratory for evidence of infection. Parents having made the decision to terminate may not wish to know the results but the information is useful. For some, termination will not be an option, either for religious or ethical reasons or perhaps because it is a 'special baby'. Others may be prepared to consider termination but wish to explore any other options available to them before making a final decision.

Treatment

The aim of treatment is firstly to prevent the parasite crossing the placenta and infecting the fetus and secondly to reduce the damage caused if infection has already taken place. Initial therapy is with the macrolide antibiotic spiramycin (Table 6.6). The value of spiramycin therapy in pregnant women was reported in French prospective studies in the 1960s and 1970s[4,9,28] which showed that on average 58 per cent of untreated women had infected babies compared with 23 per cent of treated mothers.[2,4] Furthermore attempts to isolate toxoplasma from placenta proved positive in 50 per cent (26/52) of untreated women compared with 19 per cent (51/269) of treated women.[4] The proportion of babies with clinical disease, however, was the same in both groups. Spiramycin appeared to reduce the incidence of fetal infection but had little effect on the already infected fetus.[4,9,28] These studies can be criticized as these untreated mothers do not represent a good control group. More women were treated when infection occurred early in pregnancy and it is well established that maternal infection later in pregnancy is more often transmitted to the fetus.[4,9,28] Nevertheless fetal infections and positive isolates from placenta were less frequent among treated women in each trimester of pregnancy so, even acknowledging the bias, spiramycin treatment seemed to have a positive effect.[4,9,28]

A recent study[29] showed that early and continuous treatment with 3 g spiramycin/day throughout the pregnancy was most useful. The percentage of infected fetuses was about 70 per cent less than in earlier comparable trials where treatment consisted of 2–3 g spiramycin/day for 3 weeks either arbitrarily repeated at 2 week intervals until delivery or stopped after one course.[28] Toxoplasma was isolated from placenta in 8 per cent (7/89) of treated cases of maternal infection in the first trimester of pregnancy compared with 19 per cent (28/144) in the second and 44 per cent (16/36) in the third trimester.[4,32] The best results were achieved treating women infected between 6 and 16 weeks of pregnancy when 3.7 per cent (18/487) transmitted infection to their fetus.[29] In women infected between 17 and 25

Table 6.6 Treatment of primary toxoplasmosis in pregnant women and neonates[2,29]

Pregnant women	Prenatal tests in the fetus		
	Not done	Negative	Positive

(1) Spiramycin
    3 g/day
(2) Pyrimethamine**
    1 mg/kg/day (25 mg)*
    Sulphadiazine
    50–100 mg/kg/day (3 g)*
    Folinic acid
    3–10 mg/day
	(1) given immediately until delivery		(1) and (2) alternate every 3 weeks until delivery

Neonates	Infected		?Infected as maternal infection	
	Clinical damage	Subclinical	Proven	Suspected

(1) Pyrimethamine**
    1 mg/kg/day
    Sulphadiazine
    85 mg/kg/day
    Folinic acid
    5 mg/3 days
(2) Spiramycin
    100 mg/kg/day
(3) Corticosteroids
    1.5 mg/kg/day
	(1) for 6 months
	   then
	(1) and (2) alternate
	   monthly for 6 months
	   add
	(3) if evidence of
	   inflammatory
	   process
		(1) for 6 weeks
		   then
		(2) for 6 weeks
		   then
		(1) and (2) alternate
		   4 weeks and 6 weeks
		   respectively for 1 year
			(1) for 4 weeks
			monitor infants to determine if
			further treatment is necessary
				(2) for 4 weeks

* Recommended maximum daily drug dose.

** Haematology checked twice a week.

weeks of pregnancy fetal infection still occurred in 20 per cent (20/100) of treated cases.[29] In cases of proven congenital toxoplasmosis the parasite was isolated from placenta in 89 per cent (41/46) of untreated or poorly treated women compared with 75 per cent (89/118) of treated women.[32] The available evidence indicates that spiramycin does have a preventative role in maternal–fetal transmission of toxoplasma and is most useful in cases of early maternal infection. Accordingly the recommended treatment of confirmed acute maternal toxoplasmosis is 3 g of spiramycin per day,[2,29,32,73,74] given without delay and continued for the remainder of the pregnancy (Table 6.6) or, until further decisions are taken (Fig. 6.4).

Spiramycin concentrates in the placenta (×4 times serum concentration) and acts directly on the parasite reducing the severity and duration of toxoplasma placentitis.[32] It does not however, eradicate cysts (Chapter 5) therefore continuous treatment is necessary to maintain the high levels required to act on any parasites which are released. Treatment is of little help once the fetus has been infected since, although the antibiotic crosses the placenta, only low fetal concentrations (47 per cent maternal plasma levels) are achieved. Consequently although spiramycin is considered to be safe for use in pregnant women it is not the best therapy. The most effective regime is the combination of pyrimethamine plus sulphadiazine.[2,29,73,74] The addition of pyrimethamine and sulphadiazine reduced positive isolates from placenta from 75 per cent to 50 per cent in cases of congenital toxoplasmosis[32] and apparently improved the clinical outcome of the infected fetus.[29] Pyrimethamine unfortunately has side effects which limit its use in treating pregnant women. It is known to cause bone marrow suppression and is possibly teratogenic, so it is wise to be cautious and pyrimethamine should not be used in the first trimester. In Austria pyrimethamine and sulphonamide treatment is routinely and successfully used after 20 weeks gestation while in France because of the potential hazard it is usually limited to the last 2 months of pregnancy,[2] although treatment is carried out earlier in cases of proven or strongly suspected fetal infection (Table 6.6). Before pyrimethamine therapy is recommended further investigations should be undertaken to determine whether or not the fetus has been infected.[2,29] Spiramycin therapy must continue until pyrimethamine is started (Table 6.6) or till delivery when fetal infection is not proven.

Investigations

The most effective and direct way of diagnosing fetal infection is to obtain samples directly from the fetus and examine them for evidence of toxoplasmosis (prenatal testing).[75] Under ultrasound guidance a sample of amniotic fluid can be removed (amniocentesis) and 1–3 ml of fetal blood can be taken from the umbilical cord (cordocentesis) (Fig. 6.4). Cordocentesis is usually performed after a gestational age of 17 weeks, as an outpatient under local anaesthetic.[75,76] As with any invasive procedure fetal sampling is not without risk. Rates of fetal loss of 1–2 per cent have been reported[75,76] but loss cannot always be attributed to the procedure. In a series of 606 consecutive fetal blood samplings from 562 pregnancies there were seven cases of fetal loss in 359 completely documented pregnancies (1.9%). Only two of these (0.6%) seemed to be related to the procedure.[76] Congenital infection does not necessarily mean a damaged baby. Fortnightly ultrasound examination should be performed from the time of fetal sampling to the end of pregnancy,[29] for signs of abnormality e.g. hydrocephalus, hepatomegaly, or intracranial calcification.[29,71,74] If fetal sampling is performed then fetal blood and amniotic fluid should be examined in a Toxoplasma Reference Laboratory for evidence of infection.[71] Prenatal testing offers two advantages. Identification of infected fetuses allows selective use of abortion or if pregnancy is continued, more aggressive therapy with pyrimethamine and sulphadiazine is suggested (Table 6.6).

The fetus

Diagnosis of fetal infection at present depends on isolating toxoplasma from amniotic fluid and/or fetal blood by animal culture.[29,71] Additionally, fetal blood[29,71] is examined for antibodies to toxoplasma. Detection of specific IgM is diagnostic unless there is maternal contamination. Several non-specific tests have also proved useful indicators of infection: white cell, platelet, eosinophil counts, total IgM, γ-glutamyltransferase, and lactic dehydrogenase assays should be performed.[29,71] The results of French studies using these tests have been very encouraging. They obtained samples at 20–24 weeks and, of 218 pregnant women at risk of giving birth to an infected child fetal infection was diagnosed prenatally in nine cases. Of the remaining 209 there was only one case of congenital toxoplasmosis.[71] In a study of 746 pregnant women prenatal diagnosis was negative in three of 42 cases of congenital toxoplasmosis (7%).[29] In a subsequent report this increased to nine of 89 cases (10%).[74] A false negative rate of 7–10 per cent is disappointing but this does not necessarily show poor sensitivity of the procedure. Failure may be equally due to transmission of the parasite to the fetus after the time of prenatal testing.[29,74]

The results of IgM assays are available immediately but are limited by poor sensitivity (21–44%)[29,71,77] probably due to the immaturity of the fetal immune system. Isolation of the parasite from fetal blood and amniotic fluid is much more sensitive (70–90%).[29,71,77] However, animal culture is not ideal because it takes at least 3 weeks or often longer to isolate the parasite. Furthermore if non-specific signs of infection are present but parasite isolation proves negative, repeat sampling is indicated. Arguably this limits the usefulness of the technique as such delays may make legal abortion impossible. Tissue culture offers the advantage of results in 2–3 days but is less sensitive with positive results reduced from 7/9 to 4/9[78] and 7/10 to 4/10.[77]

The recent application of the polymerase chain reaction technique (PCR) (Chapter 8) shows considerable promise and may prove to be as rapid as IgM testing and as sensitive as animal culture.[77]

In prenatal testing there has been much emphasis placed on the value of fetal blood sampling. Cordocentesis however, is a relatively new technique and is only available in specialized centres. Amniocentesis on the other hand was first performed in 1952[75] and is a widely available, well established and accepted technique. It seems that its potential value in prenatal diagnosis of congenital toxoplasmosis may have been overlooked. Combining the results of three studies[29,77,78] of 51 positive isolates in animals, 26 were obtained from both fetal blood and amniotic fluid, 15 only from blood and 10 only from amniotic fluid. Clearly the best prenatal results are obtained when both fetal blood and amniotic fluid are examined. If cordocentesis is unavailable or is felt to be too hazardous then amniocentesis alone may be a valuable aid in prenatal diagnosis.

The French now offer termination only if there is ultrasound evidence of severe lesions or proof of maternal infection early in pregnancy.[74] If the pregnancy is terminated, fetal brain and placenta should be biopsied to confirm infection. If the pregnancy is continued, aggressive chemotherapy with pyrimethamine and sulphadiazine is indicated.[2,29,32,74] Pyrimethamine and sulphadiazine alternating with spiramycin is recommended (Table 6.6). The standard dose of pyrimethamine is 25 mg daily[2] but 50 mg has been successfully used.[29,74] Folinic acid must be included to reduce the risk of bone marrow suppression and women should be monitored haematologically. Treatment was well tolerated by both mothers and fetuses[29,74] and the clinical outcome was better than expected for maternal infection acquired before 25 weeks gestation with 11/15 congenital infants asymptomatic and four others only moderately affected.[29] It appears that the addition of pyrimethamine and sulphadiazine allows the fetus to be treated earlier[29] and further studies will hopefully confirm this advantage.

The neonate

After delivery the priority is to determine if the baby is infected. Only 10 per cent of babies of infected mothers will have obvious clinical symptoms at birth; 90 per cent appear well.[4] Detailed clinical examination including skull radiography, cranial ultrasound and ophthalmoscopy[29,74] may reveal cerebral calcification or eye lesions which indicate infection. Positive prenatal tests for toxoplasma would also strongly indicate an infected baby. In most cases diagnosis of congenital infection depends to a large extent on serological tests. At birth placenta, amniotic fluid, neonatal blood, and maternal serum should be examined. There is no single definitive diagnostic test and a combination of procedures is necessary. Detection of specific neonatal IgM is diagnostic but tests will be negative in 25 per cent of infected babies.[79,80]

Toxoplasma IgG antibody in the baby's blood may not be produced by the baby but may be from the mother. Only demonstration of persisting IgG antibody after the disappearance of maternal antibody is proof of congenital infection. This requires follow-up testing over a period of several months. Positive isolation of toxoplasma from placenta, CSF or neonatal blood is good evidence of congenital toxoplasmosis but negative results do not exclude infection.[29] The future application of PCR techniques will hopefully provide more rapid and reliable diagnosis.[81]

The key question is whether or not treatment is necessary. The preferred treatment is pyrimethamine and sulphadiazine alternating with spiramycin (Table 6.6). The regimen given is that developed by Dr Jacques Couvreur in Paris based on extensive clinical experience.[2] Folinic acid must always be given with pyrimethamine and sulphadiazine to prevent bone marrow toxicity. The infant's blood count should be regularly monitored and if toxicity becomes severe pyrimethamine treatment should be discontinued until the abnormality is corrected. Treatment can be given immediately in cases with good evidence of toxoplasmosis, e.g. positive IgM, clinical signs, or positive prenatal diagnosis in the fetus. In cases with overt clinical damage pyrimethamine and sulphadiazine is given for the first 6 months then alternated monthly with spiramycin for a further 6 months.[2] Other supportive measures for the management of any infants with severe congenital disease, such as drainage of the hydrocephalus (Fig. 6.5), may also be required. Corticosteroids (prednisone or methylprednisolone) should be given if there is evidence of an inflammatory process (retinochoroiditis, high level of protein in the cerebrospinal fluid, generalized infection or jaundice).

In subclinical cases the indications for treatment are less clear. Understandably there may be reluctance to treat apparently healthy babies with potentially toxic drugs. However, such babies may not escape the aftermath of infection.[5,6] Up to 80 per cent may develop retinochoroiditis in childhood.[6] Indications are that early treatment protects against development of late ocular lesions.[82,83] Pyrimethamine and sulphadiazine has proved more effective than spiramycin alone[82] but because of its toxicity should only be used in cases of proven congenital toxoplasmosis. The recommendation is that treatment should be given for the first year of life[2] (Table 6.6). This may seem excessive but evidence suggests that brief treatment with pyrimethamine and sulphadiazine was less successful in preventing the development of sequelae.[2,83] After one 2 month course of therapy the risk of recurrence of retinochoroiditis was 26 per cent (5 of 19) compared with 5.8 per cent of 153 children given two to five courses.[83]

Clinical judgement may dictate therapy. With mild disease that responds well, a minimum of 4–6 months of therapy may be prudent; longer courses are indicated in more severe disease.[84] Because of the difficulty in diagnosing subclinical congenital infection and to avoid unnecessary therapy a modified treatment regime has been suggested.[2] Healthy babies born to mothers

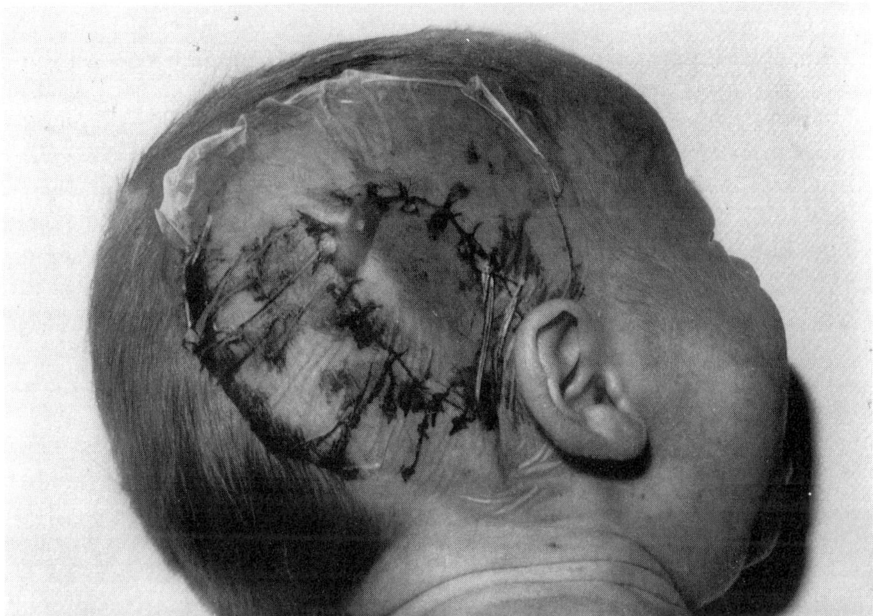

Fig. 6.5   Severe congenital toxoplasmosis with drainage of hydrocephalus.

definitely infected in pregnancy are given one course of pyrimethamine and sulphadiazine; those born to mothers possibly infected in pregnancy receive one course of spiramycin (Table 6.6). These babies should be monitored clinically and serologically and if congenital infection is confirmed further treatment can be given.

SCREENING

Diagnosis depends on the recognition of possible toxoplasmosis usually because of symptoms. The fact that toxoplasmosis is often a 'silent' infection means that diagnosis in a pregnant woman or congenitally infected baby is unlikely[25] and effective treatment[71,29,85] is not given. Only prospective serological testing of pregnant women can effectively identify those who acquire infection. Accordingly, for more than 10 years maternity care in France[29,71] and Austria[85] has included toxoplasma screening, while in the UK and USA demands for a government funded pilot screening programme[7,72] have been countered with claims that more information is needed before routine testing of pregnant women is introduced.[8,86]

Why screen?

The first and most important reason for screening is to prevent congenital toxoplasmosis and the second is to prevent *ad hoc* testing of pregnant women. Concerned women are increasingly asking to be tested. Doctors should not refuse to test anxious women and indeed may be open to litigation if they do refuse.[7] Such haphazard testing is of limited value. For effective serological diagnosis of acute primary infection in pregnant women the appropriate tests must be performed in a timely manner and the results correctly interpreted. It is doubtful that *ad hoc* testing of pregnant women would fulfill these requirements. There is a danger that wrong decisions may be made on the basis of lack of knowledge, unreliable results or wrong information.[87] Screening is a service offered routinely to all women and the advantages must not be outweighed by the disadvantages. It is therefore important to consider whether congenital toxoplasmosis fulfills the recognized criteria for screening programmes: a suitable disease; acceptable tests; effective treatment; and cost-benefit.[88,89]

Suitable disease

Two of the most important factors are that the natural history of the condition should be adequately understood and there should be a recognizable early or symptomatic stage[88,89] to allow intervention. The third factor is establishing whether or not the condition can be considered to be an important health problem.[88,89] Assessing the magnitude of the problem of congenital toxoplasmosis is difficult. Accurate data on fetal infection and clinical damage is not easily obtained from recent prospective studies. The outcome is biased by treatment since it would now be considered unethical to withhold therapy. Perhaps the most useful data comes from French studies in the 1960s and 1970s[4,9,28] and applied to the rates of maternal infection in any population. The best estimates suggest an incidence of congenital infection of perhaps 1–2 per 1000 in the UK and 3–6 per 1000 in Europe.[7] Overall it seems that only 10 per cent of infected babies will be severely damaged clinically at birth.[4] A French study monitored 210 infants with congenital toxoplasmosis until the age of 11 months and found abnormalities in 39 formerly considered subclinical.[90] Longterm follow-up of subclinically infected babies show most develop sequelae, usually eye lesions.[5,6] These studies have been criticized because the number of cases was small and biased by treatment.[8] Nevertheless in one study, 23 per cent (3/13) of the asymptomatic cases identified prospectively and 80 per cent (8/10) of children diagnosed retrospectively, when they presented with abnormal eye findings, became blind in one or both eyes.[5] This must surely be considered significant.

Another measure of the problem of congenital toxoplasmosis is the number of cases diagnosed each year. Surveys of laboratory reports in England, Wales, Northern Ireland[91] and Scotland[25] showed that the number of cases reported was very much less than expected when calculated from maternal infection rates in antenatal surveys and from French data on fetal infection and clinical damage. This may indicate that the data in the 1960s and 1970s do not apply in the 1990s or that French data do not apply to other countries. However the limitations of such passive surveillance are significant and make it an unsuitable method of assessing this disease. Variations in reporting criteria and clinical information make data difficult to interpret[91] and many cases are probably never recorded because sequelae develop later in life when diagnosis is difficult.[25] In an attempt to improve surveillance, in 1989 congenital toxoplasmosis was included in the list of diseases reported to the British Paediatric Surveillance Unit.[92] Through monthly reports from some 800 paediatricians it is hoped to actively quantify the number of cases of congenital toxoplasmosis, how they are diagnosed, and the extent of the damage and disability caused. However patients presenting to opticians or ophthalmologists are unlikely to be included. We believe the numbers affected, the seriousness of the disease, and the availability of treatment make congenital toxoplasmosis a suitable disease for a screening programme.[7] It should be emphasized that this view is not shared by all specialists who feel that more information on the disease is still required before an informed decision can be made.[8]

Acceptable tests

A variety of serological tests are available for detection of toxoplasma antibodies (Chapter 4). Availability does not mean these tests can be used to diagnose primary toxoplasmosis in a pregnant woman. It is vital to appreciate the limitations of the different tests and there is a danger that to exploit the demand for screening pregnant women commercial kits of doubtful specificity and sensitivity will flood the market. The latex agglutination test has been recommended for screening pregnant women[93] and enzyme-linked immunosorbent assays (ELISAs) have been evaluated as screening tests.[13,68,94,95] ELISA tests can be automated and easily applied to testing large numbers of sera. Concern has been expressed that even highly specific tests will inevitably generate false positive results when the condition is uncommon.[8,96,97] False positive results are only catastrophic if the screening test is considered diagnostic and the result acted upon. This is not advocated. The specificity of the overall surveillance scheme depends on the quality of confirmatory and follow-up testing. The best results will be obtained if screen positive results are confirmed at Toxoplasma Reference Laboratories. It has been estimated that the overall false positive rate would be 0.02 per cent or less.[98] It is equally important to ensure that screening tests do not generate false negative results and fail to identify women who should be investigated. There is no doubt suitable screening tests are available providing they are correctly used and interpreted. Blood samples are already taken from pregnant women for other purposes and there would seem to be no reason why the addition of toxoplasma testing should not be welcomed.

Effective treatment

The treatment of acutely infected pregnant women is well established in France[29,71,73,99] and Austria[85,100] and combined with a screening programme has proved to be effective in reducing the incidence and effect of congenital toxoplasmosis. Congenital toxoplasmosis was the most frequent cause of fetal infections in France in the 1950s; severe cases are now reported to be rare.[99] In Austria the incidence of congenital toxoplasmosis has dropped from five to seven cases per 1000 births before 1975 to about one case per 10 000 births.[100] The few severe cases which have been seen have been in mothers not examined in pregnancy or not adequately treated.[99,100] It is arguable that this is not the result of treatment but reflects a general decrease in the rate of toxoplasma infection in pregnant women due to behaviour changes or health education.

The efficacy of spiramycin treatment has been questioned, emphasizing the lack of controlled trials,[8,101,102] and the wisdom of a long course of treatment with potentially harmful effects. The French with extensive experience of using the drug remain convinced of its safety and therapeutic value.[29,73,99] Its use on thousands of women has shown it is well tolerated.[32] Prenatal sampling can be hazardous and there is a risk of fetal loss[8] but it overcomes the major problem of unnecessary terminations associated with screening. The value of pyrimethamine and sulphadiazine is open to question,[86,103] but it is acknowledged that results are encouraging.[86,103] It is doubtful that controlled trials on the efficacy of treatment can now proceed as it would seem unethical not to offer the best available treatment to acutely infected women. The Toxoplasmosis Trust,[104] founded in 1989, publicizes the problem of congenital toxoplasmosis and campaigns for a screening programme in the UK. Their views may therefore be considered biased but it is their experience that most infected women wish to continue their pregnancy under treatment.[105]

Cost-benefit

Resources are limited and any screening programme should show cost-benefit. Studies have attempted to quantify and compare the preventable costs of congenital toxoplasmosis with screening and treatment costs.[98,106–108] These studies cannot be directly compared because the parameters considered were not identical. The preventable costs depend on the incidence of congenital infection and the morbidity of disease caused. Without active serological screening of pregnant women and follow-up of infected babies this can only be estimated from prospective studies. These costs should include medical costs, special education, residential care requirements, and lost income due to disability or death. It is more difficult to put a price on intangible costs like the stress and difficulty of caring for a handicapped child. Nevertheless these costs are important. The percentage of cases which might be prevented by screening and treatment is an important factor and has been variously estimated at 50[106]–80 per cent.[98] The estimated costs of caring for these babies was: 127–196 million francs in France (1973 prices);[106] $221.9 million dollars (1975 prices)[108] to between $369 million and $8.7

billion dollars (1989 prices)[109] in the USA; £1.6–16 million in the UK (1980 prices);[107] £674 000 in Scotland (1989 prices).[98]

The cost of testing for toxoplasmosis in pregnancy depends on the number of pregnancies, the tests employed, immunity in the population, and the number of specimens examined from susceptible women during pregnancy. A minimum of two sera, one at the beginning and one near or at the end of pregnancy must be tested to detect seroconversion. Monthly testing[71] allows earlier detection and treatment but is more expensive. Three specimens taken in the first, second, and third trimesters is a favoured option.[7,98,100] In the 1990s costs must also take account of confirmatory testing and treatment costs, the cost of confirming fetal infection, and the resultant therapy. The levels of immunity in France (80%) and Austria (60%) mean that few women require further testing. In the UK and USA where 80 per cent of women are susceptible, most require further testing, hence the cost of prevention of a case of congenital toxoplasmosis is much higher. Based on 80 per cent immunity, a 1974 analysis of the screening service in France concluded it was of cost-benefit.[106] In 1990, when immunity in some areas of France has fallen to 45 per cent, many more seronegative women will need to be followed throughout pregnancy and this could have serious cost implications.[101]

One American analysis concluded that a screening programme would save one-half of the cost of caring for children born each year with congenital toxoplasmosis.[108] The British analysis of 1980–81 concluded screening would not be of cost-benefit.[107] Yet in 1989 another British analysis proposed schemes which were of cost-benefit.[98] One scheme would cost 66 per cent of the estimated preventable costs on the assumption that the overall effectiveness of surveillance and intervention would be 80 per cent.[7] While the cost of care has increased over the years, improved technology has meant the cost of testing has fallen and will hopefully continue to fall. Ready availability of drugs such as spiramycin and bulk ordering of commercial tests would lead to further reductions in costs. There is little point in further analysis of historical data. The only certain way to provide accurate answers on cost-benefit in the screening for toxoplasmosis in pregnancy is properly resourced and funded pilot screening programmes.[7,98,108] Such programmes will provide the bonus of information on the disease, the acceptability of tests and treatments, and answer once and for all the imponderable questions on toxoplasma screening.

How to screen?

The scheme of testing and its administration requires careful thought and planning to ensure the programme is acceptable and successful. It must also be recognized that screening programmes can have psychological costs which limit their benefits.[110] Women expect pregnancy to be a time of well-being

Table 6.7   Advice for pregnant women on avoiding toxoplasmosis

Sources of infection	Prevention measures
Oocysts in cat faeces	Do not clean cat litter trays, but if you must, clean them daily and wear gloves and wash hands Wash hands after gardening or contact with children's sandpits Wash vegetables which have contact with soil
Tissue cysts in meat	Do not eat undercooked meat (especially on foreign holidays) and wash hands after handling raw meat
Tachyzoites in milk	Do not drink unpasteurized milk

From Ho-Yen, D. O. and Chatterton J. M. W. (1990). Congenital toxoplasmosis — why and how to screen. *Reviews in Medical Microbiology*, **1**, 229–35.

and the possibility of illness may have a profound effect. Screening may result in knowledge of an illness which was not perceived and the anxiety caused in the minds of pregnant women must be considered. Health promotion has an important role to play.[7] Individuals then understand what is happening and are better able to deal with bad news. Pretest and posttest counselling is important.[8] Women should have knowledge of the illness and be given detailed information on the limitations and implications of screening test results.

Health promotion

Health promotion has long been advocated as a method of preventing toxoplasma infection in pregnancy and hence congenital toxoplasmosis.[111–113] The sources of infection are well defined (Chapter 2) and can be avoided by taking simple precautions (Table 6.7). Providing information on how to avoid becoming infected is simple and inexpensive. It has sometimes been concluded that health promotion would be the most cost-effective way of dealing with congenital toxoplasmosis.[97,107,112] Undoubtedly preventative measures could be effective if they were followed by all women but how successfully can health promotion persuade women to avoid high risk behaviour? It has been suggested that a decrease in primary infection of 50 per cent or more might be expected[112] but there is little documented evidence of a positive effect of health education.

The effectiveness of this type of programme has not been adequately assessed. The rate of seroconversion among susceptible pregnant women was not significantly reduced when patients were given a written list of precautions at the first prenatal visit or, where possible, before a pregnancy was planned.[114] A ten minute teaching package was given to women in early pregnancy and their behaviour changes were evaluated later in the same pregnancy.[115] The results were encouraging with positive behavioural changes in cat hygiene and to a more limited extent in food and personal hygiene.[115] Because of its low cost and risks health promotion should be available to all women either before or early in pregnancy. Efforts should be made to determine the best time and method of encouraging women to avoid toxoplasmosis in pregnancy.[115] It has recently been suggested that improved health promotion and not screening may be more appropriate for prevention of toxoplasmosis in France in the 1990s.[101] The strongest argument for this is cost, not proven effectiveness. The value of health promotion cannot be assessed without serological surveillance. It would seem wise to combine health education with a screening programme to assess the benefits of both. It should be remembered that health education is of no benefit to women who become infected in pregnancy.

Scheme of testing

Surveillance schemes should maximize early detection of primary toxoplasmosis in pregnancy and allow optimum intervention. Any scheme must reliably diagnose current infection. A single test for toxoplasma antibodies at the start of pregnancy is of limited value (Fig. 6.4). Women who are immune can be reassured while susceptible women can be advised on how to avoid becoming infected (Table 6.7). Results which suggest recent or current toxoplasmosis should be confirmed. Susceptible women must not be falsely reassured by a negative result. Further tests and action is required to identify those who acquire infection in pregnancy (Fig. 6.4). All necessary tests, treatment, and investigations should be available within the scheme (Fig. 6.4).

When and how frequently women are tested determines how early infection will be detected. The French national programme recommends screening at prenuptial and first antenatal visits with repeat testing every 4–8 weeks until delivery.[99] In Austria women are tested in the first trimester and again in the second and third trimesters of pregnancy.[100] The French scheme maximizes detection and intervention; the Austrian scheme, although less thorough, is less expensive and easier to administer. The simplest and cheapest option is to test one specimen at the beginning and another at the end of pregnancy. This could be done using blood already routinely taken for syphilis and rubella testing and cord blood.[13] If women acquire infection after initial testing such a scheme would not allow intervention; however infected babies could be identified and treated. Testing three specimens during pregnancy maximizes the use of resources.[7,98] The first determines immunity, the second allows the option of abortion or further investigations and treatment if the pregnancy is continued, and the third ensures that any infected neonate is treated.

Administration of the scheme is of critical importance. Toxoplasma screening will involve general practitioners, obstetricians, paediatricians, and laboratory personnel working together. Laboratories can set up systems for screening[98] but success depends on information flow. The best results will probably be achieved if one clinician acts as the principal co-ordinator for individual patients.[7] Health promotion and facilities for specialized individual counselling should be provided. In France laboratories have been advised to include a letter describing hygiene measures with all negative serological reports.[99] France and Austria have included legal requirements in their screening schemes. In France prenuptial testing is obligatory and thereafter toxoplasma screening in pregnancy is voluntary.[99] In Austria the tests are virtually compulsory[100] as they are one of the criteria for full maternity benefit.[116] There is no reason to believe that mandatory testing makes screening more successful. Women are highly motivated to protect their pregnancy and it seems likely that they will take advantage of a screening service if it is offered.

SUMMARY

The association of toxoplasma infection in pregnancy with the birth of damaged babies makes it an important and emotive problem. Fetal infection is a consequence of primary infection in the mother. Maternal infection prior to conception poses no threat except when latent toxoplasmosis is reactivated due to immunocompromise. The magnitude of the problem of congenital toxoplasmosis depends on the number of pregnant women who become infected and how frequently the parasite crosses the placenta. Estimates from seroprevalence and prospective studies vary but suggest an infection rate of two per 1000 pregnant women in the UK. Infection in pregnancy does not mean an infected baby. Overall about 60 per cent of untreated women might transmit toxoplasma to their fetus.

There is some evidence that toxoplasma infection may be becoming less common. Seroprevalence in pregnant women in Paris dropped from 87 per cent in 1960–70 to 71 per cent in 1981–83. This does not necessarily mean fewer pregnant women will become infected. A low risk of infection applied to large numbers of susceptible women appears to equate with a high risk applied to small numbers. Fetal infection and clinical outcome is related to when in pregnancy maternal infection was acquired. The risk of transmission increases from 14 per cent in the first trimester to 29 per cent in the second and 59 per cent in the third. Conversely, clinical damage decreases from about 80 per cent in the first to 10 per cent in the third trimester.

Generalized disease, mental retardation, and/or visual handicap occurs in some 25 per cent of infected neonates; 75 per cent appear to be healthy. However, most subclinical cases will develop ocular and neurological sequelae at some time with new lesions appearing up to to 18 years later. Early diagnosis and treatment of these babies can reduce sequelae but is unlikely unless maternal infection is detected. There have been considerable advances in the management of infected women. Despite reservations on its efficacy, treatment to reduce the risk of fetal infection and damage is well established. Prenatal testing to identify infected fetuses is gaining acceptance and allows women to make a rational choice on termination or treatment. These options are only available if infection in pregnancy is diagnosed. Suitable serological tests are available but, as 90 per cent of infected women may be asymptomatic, diagnosis is unlikely unless prospective testing is undertaken.

The media and parental groups have created a public demand for toxoplasma screening to prevent congenital toxoplasmosis. *Ad hoc* screening is unlikely to be successful and may be harmful. Wrong decisions may be made on the basis of wrong information and results. Properly funded national screening and treatment programmes can be successful as has been demonstrated in France and Austria. Screening is expensive but with improved technology schemes costing less than the costs of caring for infected babies can be devised. There is concern that screening could lead to unnecessary treatment and terminations. Rigorous confirmatory testing and prenatal investigations should minimize this. Health promotion has an important role in any scheme. If women understand what is happening there is better acceptance and compliance. Educating women to avoid toxoplasma infection in pregnancy could eliminate congenital toxoplasmosis. If health promotion could be shown to be effective it would certainly be the most cost effective method of prevention. Combining health promotion with serological testing maximizes prevention opportunity. The success of different health promotion approaches can be evaluated and infected women and their babies diagnosed and treated. Although the effects of congenital toxoplasmosis can be devastating, severe cases are fortunately rare. Nevertheless all reasonable measures to prevent infection must be taken. Nothing less will be acceptable to those who might have to live with the consequences of congenital toxoplasmosis.

REFERENCES

1 Remington, J. S. and Desmonts, G. (1976). Toxoplasmosis. In *Infectious diseases of the fetus and newborn infant.* (ed. J. S. Remington and J. O. Klein), pp. 191–332. W. B. Saunders, Philadelphia.

2 Remington, J. S. and Desmonts, G. (1990). Toxoplasmosis. In *Infectious diseases of the fetus and newborn infant.* (3rd edn). (ed. J. S. Remington and J. O. Klein), pp. 89–195. W. B. Saunders, Philadelphia.

3 Wolf, A., Cowen, D., and Paige, B. H. (1939). Toxoplasmic encephalomyelitis III. A new case of granulomatous encephalomyelitis due to a protozoon. *Am. J. Pathol.*, **15**, 657–95.

4 Desmonts, G., and Couvreur, J. (1979). Congenital toxoplasmosis. A prospective study of the offspring of 542 women who acquired toxoplasmosis during pregnancy. Pathophysiology of congenital disease. In *Perinatal medicine*, Sixth European Congress. (ed. O. Thalhammer, K. Baumgarten, and A. Pollak), pp. 51–60. Georg Thieme, Stuttgart.

5 Wilson, C. B., Remington, J. S., Stagno, S., and Reynolds D. W. (1980). Development of adverse sequelae in children born with subclinical congenital toxoplasma infection. *Pediatrics*, **66**, 767–74.

6 Koppe, J. G., Loewer-Sieger, D. H., and de Roever-Bonnet, H. (1986). Results of 20-year follow-up of congenital toxoplasmosis. *Lancet*, **i**, 254–6.

7 Ho-Yen, D. O. and Chatterton, J. M. W. (1990). Congenital toxoplasmosis — why and how to screen. *Rev. Med. Microbiol.*, **1**, 229–35.

8 Lancet. (1990). Antenatal screening for toxoplasmosis in the UK. **336**, 346–8.

9 Desmonts, G. and Couvreur J. (1974). Toxoplasmosis in pregnancy and its transmission to the fetus. *Bull. N. Y. Acad. Med.*, **50**, 146–59.

10 Foulon, W., Naessens, A., Volckaert, M., Lauwers, S., and Amy, J. J. (1984). Congenital toxoplasmosis: a prospective survey in Brussels. *Br. J. Obstet. Gynaecol.*, **91**, 419–23.

11 Viens, P., Auger, P., Villeneuve, R., and Stefanescu-Soare, I. (1977). Serological survey for congenital toxoplasmosis among 4,136 pregnant women. *Trans. R. Soc. Trop. Med. Hyg.*, **71**, 136–9.

12 Ruoss, C. F. and Bourne, G. L. (1972). Toxoplasmosis in pregnancy. *J. Obstet. Gynaecol. Br. Commonwealth*, **79**, 1115–18.

13 Joss, A. W. L., Skinner, L. J., Chatterton, J. M. W., Chisholm, S. M., Williams, H. D., and Ho-Yen, D. O. (1988). Simultaneous serological screening for congenital cytomegalovirus and toxoplasma infection. *Public Health*, **102**, 409–17.

14 Beach, P. G. (1979). Prevalence of antibodies to *Toxoplasma gondii* in pregnant women in Oregon. *J. Infect. Dis.*, **140**, 780–3.

15 Papoz, L., Simondon, F., Saurin, W., and Sarmini, H. (1986). A simple model relevant to toxoplasmosis applied to epidemiologic results in France. *Am. J. Epidemiol.*, **123**, 154–61.

16 Jeannel, D., Niel, G., Costagliola, D., Danis, M., Traore, B. M., and Gentilini, M. (1988). Epidemiology of toxoplasmosis among pregnant women in the Paris area. *Int. J. Epidemiol.*, **17**, 595–602.

17 Williams, K. A. B., Scott, J. M., MacFarlane, D. E., Williamson, J. M. W., Elias-Jones, T. F., and Williams, H. (1981). Congenital toxoplasmosis: a prospective survey in the west of Scotland. *J. Infect.*, **3**, 219–29.

18 Van der Veen, J. and Polak, M. F. (1980). Prevalence of toxoplasma antibodies according to age with comments on the risk of prenatal infection. *J. Hyg.*, **85**, 165–74.

19 Broadbent, E. J., Ross, R., and Hurley, R. (1981). Screening for toxoplasmosis in pregnancy. *J. Clin. Pathol.*, **34**, 659–64.

20 Frenkel, J. K. (1985). Immunity in toxoplasmosis. *Pan Am. Health Organization Bull.*, **19**, 354–67.

21 Krahenbuhl, J. L., Gaines, J. D., and Remington, J. S. (1972). Lymphocyte transformation in human toxoplasmosis. *J. Infect. Dis.*, **125**, 283–8.

22 Sfameni, S. F., Skurrie, I. J., and Gilbert, G. L. (1986). Antenatal screening for congenital infection with rubella, cytomegalovirus and toxoplasma. *Aust. N. Z. J. Obstet. Gynaecol.*, **26**, 257–60.

23 Sever, J. L., Ellenberg, J. H., Ley, A. C., Madden, D. L., Fuccillo, D. A., Tzan, N. R. *et al.* (1988). Toxoplasmosis: maternal and pediatric findings in 23,000 pregnancies. *Pediatrics*, **82**, 181–92.

24 Kimball, A. C., Kean, B. H., and Fuchs, F. (1971). Congenital toxoplasmosis: a prospective study of 4,048 obstetric patients. *Am. J. Obstet. Gynecol.*, **111**, 211–18.

25 Chatterton, J. M. W., Skinner, L. J., Joss, A. W. L., and Ho-Yen, D. O. (1989). Diagnosis of congenital toxoplasmosis in Scotland. *J. Infect.*, **18**, 249–55.

26 Remington, J. S. (1974). Toxoplasmosis in the adult. *Bull. N. Y. Acad. Med.*, **50**, 211–27.

27 Jones, T. C., Kean, B. H., and Kimball, A. C. (1969). Acquired toxoplasmosis. *N. Y. J. Med.*, **15**, 2237–42.

28 Desmonts, G. and Couvreur, J. (1974). Congenital toxoplasmosis. A prospective study of 378 pregnancies. *N. Engl. J. Med.*, **290**, 1110–16.

29 Daffos, F., Forestier, F., Capella-Pavlovsky, M., Thulliez, P., Aufrant, C., Valenti, D. *et al.* (1988). Prenatal management of 746 pregnancies at risk for congenital toxoplasmosis. *N. Engl. J. Med.*, **318**, 271–5.

30 Desmonts, G. and Couvreur, J. (1975). Toxoplasmosis: epidemiologic and serologic aspects of perinatal infection. In *Infections of the fetus and newborn infant. Progress in clinical and biological research.* Vol 3. (ed. S. Krugman and A. A. Gershon). pp. 115–32. Liss, New York.

31 Phillippe, F., Lepetit, D., Grancher, M. F., Morel, A., and Henocq, A. (1988). Pourquoi surveiller les enfants dont les mères ont eu une seroconversion de toxoplasmose pendant la grossesse? *Ann. Pédiatr.*, **35**, 5–10.

32 Couvreur, J., Desmonts, G., and Thulliez, Ph. (1988). Prophylaxis of congenital toxoplasmosis. Effects of spiramycin on placental infection. *J. Antimicrob. Chemother.*, **22**, 193–200.

33 Glasser, L. and Delta, B. G. (1965). Congenital toxoplasmosis with placental infection in monozygotic twins. *Pediatrics*, **35**, 276–83.

34 Garcia, A. G. P., Coutinho, S. G., Amendoeira, M. R. R., Assumpcao, M. R., and Albano, N. (1983). Placental morphology of newborns at risk for congenital toxoplasmosis. *J. Trop. Pediatr.*, **29**, 95–103.

35 Elliot, W. G. (1970). Placental toxoplasmosis: report of a case. *Am. J. Clin. Pathol.*, **53**, 413–7.

36 Cook, G. C. (1990). *Toxoplasma gondii* infection: a potential danger to the unborn fetus and AIDS sufferer. *Q. J. Med.*, **74**, 3–19.

37 Couvreur, J., Desmonts, G., and Girre, J. Y. (1976). Congenital toxoplasmosis in twins. *J. Pediatr.*, **89**, 235–40.

38 Eichenwald, H. F. (1960). A study of congenital toxoplasmosis with particular emphasis on clinical manifestations, sequelae, and therapy. In *Human toxoplasmosis.* Vol 2. (ed. J. C. Siim) pp. 41–9. Munksgaard, Copenhagen.

39 Stray-Pederson, B. (1980). Infants potentially at risk for congenital toxoplasmosis. *Am. J. Dis. Child.*, **134**, 638–42.

40 Koppe, J. G., Kloosterman, G. J., de Roever-Bonnet, H., Eckert-Stroink, J. A., Loewer-Sieger, D. H., and de Bruijne, J. I. (1974). Toxoplasmosis and pregnancy, with a long-term follow-up of the children. *Eur. J. Obstet. Gynaecol. Reprod. Biol.*, **4**, 101–10.

41 Kimball, A. C., Kean, B. H., and Fuchs, F. (1971). The role of toxoplasmosis in abortion. *Am. J. Obstet. Gynecol.*, **111**, 219–26.

42 Couvreur, J. and Desmonts, G. (1962). Congenital and maternal toxoplasmosis. *Dev. Med. Child Neurol.*, **4**, 519–30.

43 Alford, C. A., Stagno, S., and Reynolds, D. W. (1975). Toxoplasmosis: silent congenital infection. In *Infections of the fetus and newborn infant. Progress in clinical and biological Research*. Vol 3. (ed. S. Krugman and A. A. Gershon), pp. 133–57. Liss, New York.

44 Thalhammer, O. (1962). Congenital toxoplasmosis. *Lancet*, **i**, 23–4.

45 Miller, M. J., Aronson, W. J., and Remington, J. S. (1969). Late parasitemia in asymptomatic acquired toxoplasmosis. *Ann. Intern. Med.*, **71**, 139–45.

46 Remington, J. S., Melton, M. L., and Jacobs, L. (1960). Chronic toxoplasma infection in the uterus. *J. Lab. Clin. Med.*, **56**, 879–83.

47 Holliman, R. E. (1990). The diagnosis of toxoplasmosis. *Serodiagn. Immunother. Infect. Dis.*, **4**, 83–93.

48 Langer, H. (1963). Repeated congenital infection with *Toxoplasma gondii*. *Obstet. Gynecol.*, **21**, 318–29.

49 Remington, J. S. (1964). Toxoplasma and chronic abortion. *Obstet. Gynecol.*, **24**, 155–7.

50 Stray-Pedersen, B. and Lorentzen-Styr, A-M. (1977). Uterine toxoplasma infections and repeated abortions. *Am. J. Obstet. Gynecol.*, **128**, 716–21.

51 Remington, J. S., Newell, J. W., and Cavanaugh, E. (1964). Spontaneous abortion and chronic toxoplasmosis. *Obstet. Gynecol.*, **24**, 25–31.

52 Meylan, J. (1971). Toxoplasmosis as a cause of repeated abortion. In *Toxoplasmosis*. (ed. D. Hentsch), pp. 151–7. Huber, Bern.

53 Mahajan, R. C., Gupta, I., Chhabra, M. B., Gupta, A. N., Devi, P. K., and Ganguli, N. K. (1976). Toxoplasmosis — its role in abortion. *Indian J. Med. Res.*, **64**, 797–800.

54 Mahajan, R. C., Gupta, I., Ganguly, N. K., Gupta, A. N., Chhabra, M. B., and Devi, P. K. (1981). Immunological studies of toxoplasmosis in cases of abortion. *Indian J. Pathol. Microbiol.*, **24**, 165–9.

55 Abdel-Hafez, S. K., Shbeeb, I., Ismail, N. S., and Abdel-Rahman, F. (1986). Serodiagnosis of *Toxoplasma gondii* in habitually aborting women and other adults from North Jordan. *Folia Parasitol.*, **3**, 7–13.

56 Southern, P. M. (1972). Habitual abortion and toxoplasmosis. *Obstet. Gynecol.*, **39**, 45–7.

57 Sharf, M., Eibschitz, I., and Eylan, E. (1973). Latent toxoplasmosis and pregnancy. *Obstet. Gynecol.*, **42**, 349–54.

58 Stern, G. A. and Romano, P. E. (1978). Congenital ocular toxoplasmosis. *Arch. Ophthalmol.*, **96**, 615–17.

59 Lou, P., Kazdan, J., and Basu, P. K. (1978). Ocular toxoplasmosis in three consecutive siblings. *Arch. Ophthalmol.*, **96**, 613–14.

60 Wechsler, B., Du, L. T. H., Vignes, B., Piette, J. C., Chomette, G., and Godeau, P. (1986). Toxoplasmose et lupus. *Ann. Méd. Interne*, **137**, 324–30.

61 Cohen-Addad, N. E., Joshi, V. V., Sharer, L. R., Epstein, L. G., Gubitosi, T. A., and Oleske, J. M. (1988). Congenital acquired immunodeficiency syndrome and congenital toxoplasmosis: pathologic support for a chronology of events. *J. Perinatol Med.*, **8**, 328–31.

62 Mitchell, C. D., Erlich, S. S., Mastrucci, M. T., Hutto, S. C., Parks, W. P., and Scott, G. B. (1990). Congenital toxoplasmosis occurring in infants perinatally infected with human immunodeficiency virus I. *Pediatr. Infect. Dis. J.*, **9**, 512–18.

63 Medlock, M. D., Tilleli, J. T., and Pearl, G. S. (1990). Congenital cardiac toxoplasmosis in a newborn with acquired immunodeficiency syndrome. *Pediatr. Infect. Dis. J.*, **9**, 129–32.

64 McCabe, R. E. and Remington, J. S. (1983). The diagnosis and treatment of toxoplasmosis. *Eur. J. Clin. Microbiol.*, **2**, 95–104.

65 Fleck, D. G. (1989). Annotation: diagnosis of toxoplasmosis. *J. Clin. Pathol.*, **42**, 191–3.

66 Wielaard, F., van Gruijthuijsen, H., Duermeyer, W., Joss, A. W. L., Skinner, L., Williams, H. *et al.* (1983). Diagnosis of acute toxoplasmosis by an enzyme immunoassay for specific immunoglobulin M antibodies. *J. Clin. Microbiol.*, **17**, 981–7.

67 Payne, R. A., Joynson, D. H. M., Balfour, A. H., Harford, J. P., Fleck, D. G., Mythen, M. *et al.* (1987). Public Health Laboratory Service enzyme linked immunosorbent assay for detecting toxoplasma specific IgM antibody. *J. Clin. Pathol.*, **40**, 276–81.

68   Joss, A. W. L., Skinner, L. J., Moir, I. L., Chatterton, J. M. W., Williams, H., and Ho-Yen, D. O. (1989). Biotin-labelled antigen screening test for toxoplasma IgM antibody. *J. Clin. Pathol.*, **42**, 206–9.

69   Skinner, L. J., Chatterton, J. M. W., Joss, A. W. L., Moir, I. L., and Ho-Yen, D. O. (1989). The use of an IgM immunosorbent agglutination assay to diagnose congenital toxoplasmosis. *J. Med. Microbiol.*, **28**, 125–8.

70   Sabin, A. B. and Feldman, H. A. (1948). Dyes as microchemical indicators of a new immunity phenomenon affecting a protozoan parasite (toxoplasma). *Science*, **108**, 660–3.

71   Desmonts, G., Daffos, F., Forestier, F., Capella-Pavlovsky, M., Thulliez, Ph., and Chartier, M. (1985). Prenatal diagnosis of congenital toxoplasmosis. *Lancet*, **i**, 500–4.

72   McCabe, R. and Remington, J. S. (1988). Toxoplasmosis: the time has come. *N. Engl. J. Med.*, **318**, 313–15.

73   Piens, M. A. and Garin, J. P. (1989). New perspectives in the chemoprophylaxis of toxoplasmosis. *J. Chemother.*, **1**, 46–51.

74   Hohlfeld, P., Daffos, F., Thulliez, P., Aufrant, C., Couvreur, J., MacAleese, J. *et al.* (1989). Fetal toxoplasmosis: outcome of pregnancy and infant follow up after in utero treatment. *J. Pediatr.*, **115**, 765–9.

75   Grose, C., Itani, O., and Weiner, C. P. (1989). Prenatal diagnosis of fetal infection: advances from amniocentesis to cordocentesis — congenital toxoplasmosis, rubella, cytomegalovirus, varicella virus, parvovirus and human immunodeficiency virus. *Pediatr. Infect. Dis. J.*, **8**, 459–68.

76   Daffos, F., Capella-Pavlovsky, M., and Forestier, F. (1985). Fetal blood sampling during pregnancy with use of a needle guided by ultrasound: a study of 606 consecutive cases. *Am. J. Obstet. Gynecol.*, **153**, 655–60.

77   Grover, C. M., Thulliez, P., Remington, J. S., and Boothroyd, J. C. (1990). Rapid prenatal diagnosis of congenital toxoplasma infection by using polymerase chain reaction and amniotic fluid. *J. Clin. Microbiol.*, **28**, 2297–301.

78   Derouin, F., Thulliez, P., Candolfi, E., Daffos, F., and Forestier, F. (1988). Early prenatal diagnosis of congenital toxoplasmosis using amniotic fluid samples and tissue culture. *Eur. J. Clin. Microbiol. Infect. Dis.*, **7**, 423–5.

79   Naot, Y., Desmonts, G., and Remington, J. S. (1981). IgM enzyme-linked immunosorbent assay test for the diagnosis of congenital toxoplasma infection. *J. Pediatr.*, **98**, 32–6.

80   Holliman, R. E. and Johnson, J. D. (1989). The post-natal serodiagnosis of congenital toxoplasmosis. *Serodiagn. Immunother. Infect. Dis.*, **3**, 323–7.

81   Verhofstede, C., Van Renterghem, L., Plum, J., Vanderschueren, S., and Vanhaesebrouck, P. (1990). Congenital toxoplasmosis and TORCH. *Lancet*, **336**, 622–3.

82   Couvreur, J., Nottin, N., and Desmonts, G. (1980). La toxoplasmose congénitale traitée. *Ann. Pédiatr.*, **27**, 647–52.

83   Couvreur, J., Desmonts, G., and Aron-Rosa, D. (1984). Le pronostic oculaire de la toxoplasmose congénitale: rôle du traitement. *Ann. Pédiatr.*, **31**, 855–8.

84   Wilson, C. B. (1990). Treatment of congenital toxoplasmosis. *Pediatr. Infect. Dis. J.*, **9**, 682–3.

85   Fuith, L. E., Reibnegger, G., Honlinger, M., and Wachter, H. (1988). Screening for toxoplasmosis in pregnancy. *Lancet*, **ii**, 1196.

86   Thorp, J. M., Seeds, J. W., Herbert, W. N. P., Bowes, W. A., Maslow, A. S., Cefalo, R. C. *et al.* (1988). Prenatal management and congenital toxoplasmosis. *N. Engl. J. Med.*, **319**, 372–3.

87   Brand, T. S. (1990). Congenital toxoplasmosis and TORCH. *Lancet*, **336**, 623.

88   Wilson, J. M. G. and Jungner, G. (1968). In *Principles and practice of screening for disease.* WHO Public Health Papers No 34. WHO, Geneva.

89   Burr, M. L. and Elwood, P. C. (1985). Research and development of health promotion services — screening. In *Oxford textbook of public health*. Vol 3. (ed. W. W. Holland, R. Detels, and G. Knox), pp. 373–84. Oxford University Press.

90 Couvreur, J., Desmonts, G., Tournier, G., and Szusterkac, M. (1984). Etude d'une série homogène de 210 cas de toxoplasmose congénitale chez des nourissons âgés de 0 à 11 mois et dépistés de façon prospective. *Ann. Pédiatr.*, **31**, 815–9.

91 Hall, S. M. (1983). Congenital toxoplasmosis in England, Wales and Northern Ireland: some epidemiological problems. *Br. Med. J.*, **287**, 453–5.

92 Hall, S. M. and Glickman, M. (1990). Report from the British paediatric surveillance unit. *Arch. Dis. Child.*, **65**, 807–9.

93 Holliman, R. E., Barker, K. F., and Johnson, J. D. (1990). Selective antenatal screening for toxoplasmosis and the latex agglutination test. *Epidemiol. Infect.*, **105**, 409–14.

94 Joss, A. W. L., Skinner, L. J., Chatterton, J. M. W., Cubie, H. A., Pryde, J. F. D., and Campbell, J. D. (1989). Toxoplasmosis: effectiveness of enzyme immunoassay screening. *Med. Lab. Sci.*, **46**, 107–12.

95 Joynson, D. H. M. and Payne, R. (1988). Screening for toxoplasma in pregnancy. *Lancet*, **ii**, 795–6.

96 Salmon, R. L. (1988). Screening for toxoplasma in pregnancy. *Lancet*, **ii**, 1085–6.

97 Frenkel, J. K. (1990). Diagnosis, incidence and prevention of congenital toxoplasmosis. *Am. J. Dis. Child.*, **144**, 956–7.

98 Joss, A. W. L., Chatterton, J. M. W., and Ho-Yen, D. O. (1990). Congenital toxoplasmosis: to screen or not to screen? *Public Health*, **104**, 9–20.

99 Desmonts, G. (1990). Preventing congenital toxoplasmosis. *Lancet*, **336**, 1017–18.

100 Aspöck, H. (1986). Prevention of congenital toxoplasmosis by serological surveillance during pregnancy: current strategies and future perspectives. In *Parasitic infections immunology, macotic infections, general topics*. Proceedings IXth International Congress of Infectious and Parasitic Diseases, Vol 111. (ed. W. Marget, W. Lang, and E. Gabler-Sandberger), pp. 69–73. MMV Medizin Verlag, München.

101 Jeannel, D., Costagliola, D., Niel, G., Hubert, B., and Danis, M. (1990). What is known about the prevention of congenital toxoplasmosis? *Lancet*, **336**, 359–61.

102 Carter, A. O. and Frank, J. W. (1986). Congenital toxoplasmosis: epidemiologic features and control. *Can. Med. Assoc. J.*, **135**, 618–23.

103 Wilson, C. B. (1990). Treatment of congenital toxoplasmosis during pregnancy. *J. Pediatr.*, **116**, 1003–4.

104 Toxoplasmosis Trust. Leaflets obtainable from: The Toxoplasmosis Trust, 61–71 Collier Street, London N19BE.

105 Teuten, B. (1990). Antenatal screening for toxoplasmosis. *Lancet*, **336**, 818–19.

106 Chevalier, M. (1974). Etude cout-avantage d'un système de prévention de la toxoplasmose congénitale. *Bull. Statistiq. Santé-Sécurité Soc.*, **3**, 71–84.

107 Henderson, J. B., Beattie, C. P., Hale, E. G., and Wright, T. (1984). The evaluation of new services: possibilities for preventing congenital toxoplasmosis. *Int. J. Epidemiol.*, **13**, 65–72.

108 Wilson, C. B. and Remington, J. S. (1980). What can be done to prevent congenital toxoplasmosis? *Am. J. Obstet. Gynecol.*, **138**, 357–63.

109 Roberts, T. and Frenkel, J. K. (1990). Estimating income losses and other preventable costs caused by congenital toxoplasmosis in people in the United States. *J. Am. Vet. Med. Assoc.*, **196**, 249–56.

110 Marteau, T. M. (1989). Psychological costs of screening. *Br. Med. J.*, **299**, 527.

111 Frenkel, J. K. (1973). Toxoplasma in and around us. *Bioscience*, **23**, 343–52.

112 Frenkel, J. K. (1981). Congenital toxoplasmosis: prevention or palliation? *Am. J. Obstet. Gynecol.*, **141**, 359–61.

113 Frenkel, J. K. (1990). Toxoplasmosis in human beings. *J. Am. Vet. Med. Assoc.*, **196**, 240–8.

114 Foulon, W., Naessens, A., Lauwers, S., de Meuter, F., and Amy, J-J. (1988). Impact of primary prevention on the incidence of toxoplasmosis during pregnancy. *Obstet. Gynecol.*, **72**, 363–6.

115   Carter, A. O., Gelmon, S. B., Wells, G. A., and Toepell, A. P. (1989). The effectiveness of a prenatal education programme for the prevention of congenital toxoplasmosis. *Epidemiol. Infect.*, **103**, 539–45.
116   Beattie, C. P. (1980). Congenital toxoplasmosis. *Lancet*, **i**, 873.

7

Immunocompromised patients

DARREL O. HO-YEN

As with all infections, toxoplasmosis is controlled by the body's immune system. Unlike most infections, toxoplasma's ability to encyst and remain dormant allows the organism to be a continual threat to the host (Chapter 1). This threat is minimal in an immunocompetent individual. In those who are immunocompromised there is a threefold threat that involves: control of the initial infection; prevention of the reactivation of dormant infection; and the control of disseminated infection.[1] Predictably, the failure to recognize acquired toxoplasma infection is as common among immunocompromised patients as in normal individuals (Chapter 3); however, the consequences are much worse. In an early review of the literature, the diagnosis of toxoplasma in immunocompromised patients was not made until post-mortem examination in most cases.[2] The position today is probably not greatly improved.

Although toxoplasmosis has been associated with malignant disease since the early 1960s (Chapter 1), the clinical problem was highlighted in 1968.[3] These authors reviewed the literature and concluded that there needed to be an increased clinical awareness of toxoplasma infection in patients with malignancies, especially as it may be treatable. Subsequent work confirmed that among these patients, neurological complications with high mortality were common.[2] In addition there was a high incidence of coincident infection with possibly reactivated viral infections.[2,3] Such infections may contribute to the ease with which toxoplasma is reactivated, or all the reactivated coincident infections may simply reflect the immunocompromised state. In contrast, the role of human immunodeficiency virus (HIV) in reactivating toxoplasma infection in patients with acquired immunodeficiency syndrome (AIDS) is not in doubt. In 1983 an outbreak of ten cases of central nervous system toxoplasmosis was described in Belgium and north America (five were Haitians, two were homosexual, and one was a drug addict).[4] Subsequently, these cases were shown to be AIDS cases and it was recognized that between 25–80 per cent of AIDS patients may develop cerebral toxoplasmosis.[5] It is salutary to recall that it was the identification of the rare *Pneumocystis carinii* pneumonias that led to the Centres for Disease Control proposing a definition of AIDS in 1982; yet, *P. carinii* and toxoplasma infection were both recognized as major infections in the immunocompromised in 1979.[6] In retrospect, the link between toxoplasmosis and the AIDS epidemic could have been made earlier. As the HIV epidemic progresses, it is likely that the epidemiology of cerebral toxoplasmosis will change and that the vast majority of patients will be AIDS patients.

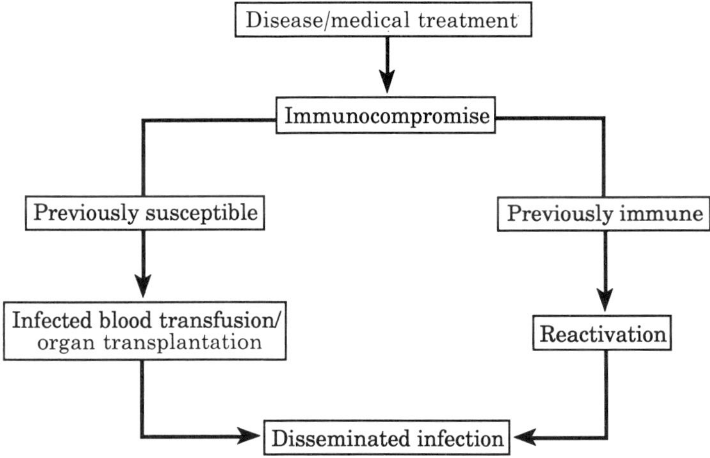

Fig. 7.1   Toxoplasmosis and immunocompromise.

Transplantation can also be complicated with disseminated toxoplasma infection. Either immunosuppressive treatment accompanying organ transplant can cause reactivation of past infection in the recipient; or transplantation of organs from toxoplasma-immune donors to toxoplasma-susceptible individuals may result in disseminated infection (Fig. 7.1). In this chapter, the body's normal immune response to infection will be first considered; then its compromise as a result of malignancies, transplantation, infections, and other conditions. To avoid repetition, common aspects of the clinical features (Chapter 3), diagnosis (Chapter 4), and treatment (Chapter 5) will not be repeated in this chapter. Instead, in these three areas only those observations which are different or important will be highlighted.

IMMUNE RESPONSE

The relationship between *Toxoplasma gondii* and the human host is a delicate one. In an immunocompetent individual, acquired infection is usually followed by an immune response which controls the infection, and results in the parasite encysting and becoming dormant.[7] This state of affairs can remain for decades with the immune response being able to control any subsequent release of bradyzoites from tissue cysts. The balance is destroyed when the immune system is compromised as then any released tachyzoites may proliferate unhindered. Such a situation is to the disadvantage of the parasite as it may result in the death of the individual. Fortunately, modern medicine (which often compromises the immune system) may also redress the balance with early use of anti-toxoplasma therapy (Chapter 5).

Infection by any organism is a result of an interplay of several factors in the host and the parasite (Chapter 1). For toxoplasma, the virulence of the strain, its infecting dose, its stage and route of infection may all be important (Chapter 2). The host defence is equally complex. Intact surface defences, such as skin and mucous membranes, may prevent infection when the infecting dose is low. Nevertheless, it appears that toxoplasma has characteristics that allow its active invasion of cells. There is now good evidence, because of the speed of entry and the energy involved, that toxoplasma's entry into non-phagocytic cells is an active process involving both the parasite and the host cell.[8] Indeed, the tachyzoite may leave the cell by reversing the process involved in invasion.[8] Although this new information needs to be properly interpreted, it is likely that the major factors which determine the outcome of the infection are still the cellular and humoral responses.

Cellular response

Detailed information on the complete, human cellular response to toxoplasma infection is not available. However in some cases, information obtained from other infections or from animal experiments can be applied to human toxoplasmosis.

Macrophages

Macrophages play a crucial role in the host response to infection (Fig. 7.2). They have immediate functions (inflammatory, microbicidal, and tumoricidal activities); as well as presenting and processing antigens for T-lymphocyte activation; and being involved in tissue reorganization and repair.[9] Normally, toxoplasma attaches to the macrophage surface and then becomes engulfed in a phagosome. Peripheral blood monocytes are able to destroy intracellular parasites, and it has been suggested that this high microbicidal activity may explain why most toxoplasma infections are minor.[10] In addition, human splenic macrophages have a similar capacity to kill intracellular toxoplasma.[10] These early responses to infection are probably not specific killing by macrophages. More specific responses are initiated when macrophages act as antigen-presenting cells for T-lymphocytes. Such T-lymphocytes can then produce cytokines which prevents further invasion of macrophages by toxoplasma[9] (Fig. 7.2).

In addition, T-cells can activate macrophages through cytokines[11] (Fig. 7.2). There is also interaction between specific antibody (produced by B-cells) and macrophages. Antibody-coated parasites are unable to prevent fusion of lysosomes with the phagosome and are killed, whereas uncoated parasites are able to inhibit such fusion and remain alive.[9,11] An advantage for the parasite is that it is protected from the body's humoral response within the phagosome. It has also been suggested that hydrocortisone's action in allowing toxoplasma replication is one of stabilization of the lysosomal membrane and thus preventing fusion with the phagosome.[11]

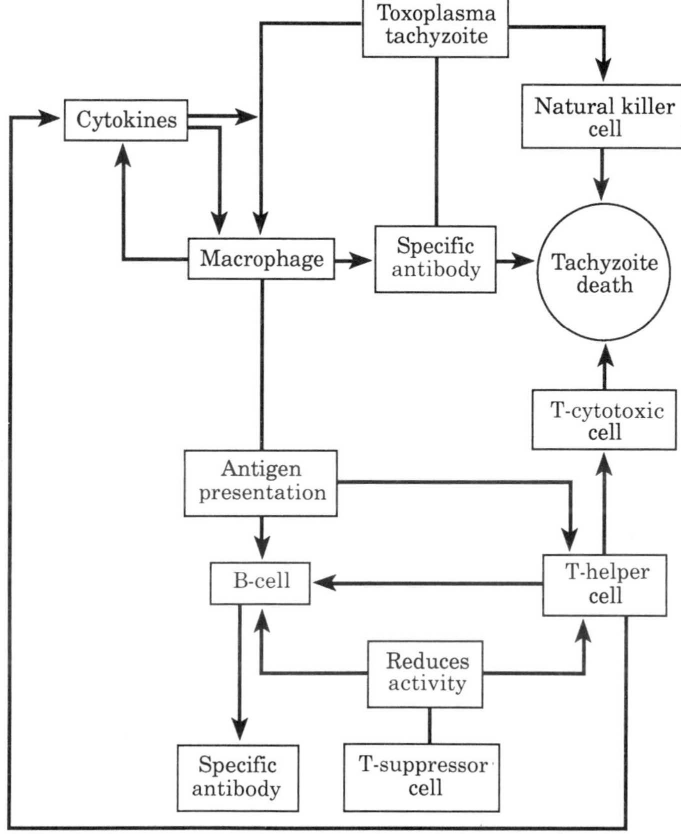

Fig. 7.2   Immune response to toxoplasma infection.

Thus, macrophages can both hasten the destruction of toxoplasma and protect the parasite. Unkilled parasites within macrophages have the added advantage of being distributed by the macrophages throughout the body.

T-cells

Although it was once felt that T-cell responses were unimportant in toxoplasma infection, it is now recognized that several types of T-cell responses are involved (Fig. 7.2). The T-helper cell is critical in the immunological response to infection because of its many roles in stimulating the activities of other cells.[9] In particular, T-helper cells stimulate T-cytotoxic cells which are then able to lyse tachyzoites directly. They also participate in the activation of B-cells which then go on to produce antibody against toxoplasma. The function of T-helper cells is influenced by the T-suppressor cell activity. With asymptomatic lymphadenopathy, there is either an increase in T-suppressor

(CD8) cells, or a decrease in T-helper (CD4) cells, or both.[12] In individuals with prolonged symptoms (>4 weeks), there is a great increase in CD8 cells and a reversal of the CD4:CD8 cell ratio, whereas the ratio is normal in asymptomatic patients with acute infection.[12] These findings are similar to those in patients with a chronic viral infection,[13] and are not a toxoplasma-specific response. Nevertheless, the presence of increased CD8 cells may be related to the severity and time course of the illness.[12] It has also been suggested that low levels of γ-interferon production by toxoplasma-specific T-cells may be related to prolonged symptoms.[14]

In 1948, Frenkel demonstrated that there is a delayed-type hypersensitivity response in the skin of infected individuals to toxoplasma antigen.[15] It was a late development after the acute infection and persisted for many years. The responses are probably due to the TD-lymphocytes.[9] The toxoplasma skin test correlated well with the presence of dye test antibodies and had a low incidence of false positives (0.5%).[16] It could be recommended in countries where laboratory facilities are limited, but the one disadvantage is that it is a late manifestation of infection and may only develop many months later.

An interesting series of experiments in guinea-pigs has shown that resistance to toxoplasma infection was dependent on T-cell rather than B-cell function.[17] T-cell depleted animals could be protected from lethal toxoplasma infection by receiving a transfusion of normal syngeneic T-cells. These experiments and our greater knowledge of the importance of T-cells allow us to be aware that in those situations in which there is a reduction of T-cell function, there is a great likelihood of problems with toxoplasma infection.[18]

Natural killer cells

Like macrophages, natural killer (NK) cells are part of the body's natural immunity and are not specific for toxoplasma.[9] Not surprisingly for cells in the first line of defence, there is greater NK activity in response to membranes of tachyzoites rather than to cytoplasmic components.[19] Although it has been suggested that a reduced NK cell population may be associated with chronic viral infections,[20] this has not been evaluated in patients with persisting toxoplasma symptoms. The immunology is probably complex as the role and activity of NK cells are also influenced by macrophages and T-cells.[9] Therefore NK cells augment the more specific anti-toxoplasma immune response, but the role of NK cells on their own in toxoplasma infection needs to be elucidated.

Humoral response

Humoral responses, especially the production of antibodies are the basis of diagnosis (Chapter 4). Yet, these responses are not as important as cell-mediated immunity in the protection of the host from toxoplasma infection.[17]

Nevertheless, humoral responses cannot be totally separated from those that are cell-mediated,[9] and the interaction of these two types of responses are probably important for the longterm control of toxoplasma infection within the host.

B-cells

After appropriate stimulation, B-cells produce different classes of antibodies,[9] each specific against toxoplasma antigens. Classically, IgM antibodies are produced first, followed by high titres of IgG antibodies which may persist for life (Chapter 4). The IgM antibodies disappear within a few months, but occasionally may persist for a year or more. Nevertheless, because of their early disappearance, they are an important diagnostic marker (Chapter 4). In contrast, IgG antibodies are usually used to denote immunity, but IgG avidity may allow identification of recent infection (Chapter 4). In congenital toxoplasmosis, transient abnormal B-cell proliferation may result in an IgG monoclonal gammopathy.[21] Such immunoglobulin lacks specific antibody activity. Extracellular tachyzoites are quickly killed in the presence of specific antibody and complement, and this is the basis of the Sabin–Feldman dye test (Chapter 4). Antibody-coated tachyzoites are also more easily phagocytosed and are easily destroyed by lysosomes within phagocytes.[9]

Toxoplasma-specific IgA antibodies are particularly concerned with the protection of mucosal surfaces and easily bind to neutrophils, T- and B-lymphocytes.[9] In contrast, IgE antibodies readily bind to mast cells and basophils, and such binding allows these cells to degranulate on contact with specific antigens.[9] IgA antibodies may be useful in the diagnosis of congenital and acute toxoplasmosis.[22] IgE antibodies were present early (at the time of IgM) and before the presence of IgA antibodies.[23] In addition, IgE antibodies do not persist after 4 months. Both IgA and IgE antibodies do not cross the placenta, and their presence in neonates denotes infection.[22,23] An interesting suggestion is that IgE antibodies may be useful in diagnosing reactivated infection.[23] However, the precise roles of IgA and IgE in the immune response to toxoplasma require further study.

Cytokines

Cytokines are a variety of immunological mediators which may be released by activated T-cells, B-cells, macrophages, and NK cells.[9] Their role is to influence and control the immunological response to the infectious agent. Their importance has been underestimated in the past, but their role appears to be principally synergistic.[9] With the ability of modern science to synthesize such mediators, cytokines may have a role in immunotherapy. The demonstration that γ-interferon (produced by T and NK cells) has the ability to enhance antibody synthesis and survival time after toxoplasma infection[24] was the start of interest. These workers then proceeded to show that another cytokine (interleukin-2, usually produced by the T-cells) resulted in significant reduction in mortality of mice infected with toxoplasma.[25] It was concluded that the likely explanation of this action of interleukin-2 was not on B-cells or macrophages, but on NK cells. These initial studies[24,25] were done in mice and there is a case for the utilization of such an approach in humans.

Resistance to infection

A successful parasite does not usually kill its host, but there is a keen battle fought over superiority. Thus different animal species may have different susceptibilities to infection (Chapter 2). The age of the host is also important with infection in the fetus (Chapter 6) being severe compared to the adult (Chapter 3). Similarly adult rats are quite resistant to toxoplasma infection, whereas newborn rats readily succumb.[26] In addition, strain variations also vary the effect of toxoplasma infection (Chapter 2). Predictably, as the parasite is successful it is not without resources. It can resist fusion of its phagosome with lysosomes as has been mentioned, and it may also exhibit several 'escape mechanisms' to avoid the host's immune system.[9] These principally involve the release of antigenic material which can block the receptors of effector cells.[9] Lastly, its ability to encyst and become dormant may be the principal mechanism that has accounted for the continuing success of the parasite.

MALIGNANCIES

The association of toxoplasmosis and malignancy was made by workers at the National Cancer Institute in the USA in 1968.[3] They reported six cases occurring before 1967 and reviewed a further 14 cases already in the literature. Cases were mainly in individuals with haematological malignancies (16/20, 80%, Table 7.1). Similar results were also reported from another cancer hospital in the USA for the period 1972–8.[27] Although these two studies may represent the work in cancer hospitals, a different pattern emerges when toxoplasmosis is considered in other immunocompromised individuals.[2] Excluding the 20 patients reported before 1968,[3] there were a further 61 patients (Table 7.1). These represent cases published and unpublished which were diagnosed in a Toxoplasma Reference Laboratory, mainly for the period 1967–75 and show relatively fewer patients with haematological malignancies. In particular, the number of patients with toxoplasmosis as a result of organ transplant, 11 per cent, is significant; and there are no cases due to infection. Nevertheless, the vast majority of reported cases remain associated with haematological malignancy, especially lymphoid neoplasms (Table 7.1). As the lymphocyte has the major role in controlling toxoplasma infection (Fig. 7.2) this is to be expected. Information from the

Table 7.1   Toxoplasmosis in conditions associated with immunodeficiency (percentage of each series)[2,3,27]

	Literature (before 1967)	Cancer hospital (1972–8)	Literature (1967–75)
Hodgkin's disease	8 (40)	12 (48)	24 (39)
Non-Hodgkin's disease	2 (10)	4 (16)	7 (11)
Acute leukaemia	4 (20)	3 (12)	2 (3)
Chronic leukaemia	2 (10)	4 (16)	7 (11)
Multiple myeloma	–	–	2 (3)
Other malignancies	4 (20)	2 (8)	6 (10)
Collagen vascular disease	–	–	6 (10)
Organ transplant	–	–	7 (11)
Total	20	25	61

Scottish Cancer Registry shows that there has been a steady increase in patients with a diagnosis of non-Hodgkin's lymphoma and leukaemia for the period 1968–82. This is partly a result of better diagnosis, but 5 year survival in these patients has also increased. The position of Hodgkin's lymphoma is different: the incidence has remained virtually the same for 1968–82, but 5 year survival has dramatically increased (18.6% for males and 16.6% for females). Increased survival allows more opportunity for toxoplasma reactivation.

Hodgkin's disease

This disease is by far the most common malignancy associated with toxoplasmosis (Table 7.1). Toxoplasmosis is usually a reactivated infection in patients with Hodgkin's disease,[2,3] and it may occur in the presence or absence of concurrent immunosuppressive chemotherapy for Hodgkin's disease. Nevertheless, it is dubious if the disease process itself is sufficient to cause severe clinical toxoplasmosis. Experiments in hamsters suggest that tumour infiltration is more likely to produce death from other causes rather than from toxoplasmosis.[28] These workers have also made two pertinent observations: there are no cases associating toxoplasmosis with Hodgkin's disease from the prechemotherapy era; and in animals, the drugs used in chemotherapy and irradiation dramatically reduce the immune response to toxoplasma.[28] Certainly, it is plausible that the early and aggressive use of chemotherapy and irradiation may have a cumulative effect in reactivating toxoplasmosis. The occurrence of toxoplasmosis approximately 2–3 years after the onset of Hodgkin's disease[3] is also consistent with such an hypothesis. It must also be emphasized that of the large number of patients with Hodgkin's disease, toxoplasmosis is an infrequent occurrence.

The clinical picture may also be influenced by the underlying disease process and treatment. Among the symptomatic, fever appears universal, however, it is usually attributed to the underlying disease and may remain undiagnosed for many weeks.[27] Other clinical manifestations may be present (Chapter 3), but the more usual presentation is of central nervous system involvement. This may be localized or systemic, and is consistent with a focus of reactivation which may become disseminated.[28] Focal abnormalities include cranial nerve palsies, hemiparesis, aphasia, focal seizures, personality changes, and cerebellar signs.[2,3,27] Equally common are more diffuse findings such as general weakness, myoclonus, lethargy, headache, confusion, and coma.[2,3,27] As with the other clinical manifestations, there may be weeks or months before treatment, death or recovery intervenes. Diagnosis may be particularly difficult as infection is usually reactivated (Chapter 4). The common finding of herpes viruses (cytomegalovirus and herpes simplex) in association with toxoplasmosis in immunocompromised patients appears to be more than fortuitous.[3] It may well be that reactivation of one such infection predisposes to others as these infectious agents are capable of becoming dormant after initial infection. Nevertheless, as all of these infections can produce systemic symptoms such as pyrexia, it can be difficult to decide which infection is predominant. High dye test titres may often be present[27] but toxoplasma-specific IgM is not usually detectable. A cranial computed tomography scan, especially if delayed after a double dose of contrast material, may be particularly helpful. This and other tests are considered in Chapter 4. Brain biopsy is rarely considered but may be definitive.[2]

Although treatment is not usually indicated when acquired toxoplasmosis occurs in the immunocompetent, treatment is critical in the immunocompromised. The main difficulty is that the diagnosis is frequently made at post-mortem. In an important review of the literature, only a quarter of patients with Hodgkin's disease (9/32) received specific anti-toxoplasma therapy (Chapter 5), but 78 per cent (7/9) of those treated had a good or complete clinical response.[2] Such a favourable outcome may not be universal[27] and may be dependent on the stage of the disease, the severity of other treatment, the promptness of anti-toxoplasma therapy, and the degree of marrow suppression. In addition, clinicians should be aware of the importance of toxoplasmosis especially in an immunocompromised patient with fever.[27] The question of routine, prophylactic, anti-toxoplasma therapy has been raised,[29] but when one considers the frequency of toxoplasmosis as a complication, it does not seem justified. Greater vigilance and early anti-toxoplasma therapy (Chapter 5) are the lessons of the past.

Other haematological disorders

Apart from Hodgkin's disease, many other haematological conditions have been associated with toxoplasma infection (Table 7.1). There are two main mechanisms: reactivation of latent toxoplasmosis as a result of immuno-suppressive therapy aimed at the malignancy; and infection in a susceptible individual by transplant or transfusion (Fig. 7.1). As the haematological system is mainly responsible for the body's defences against infection, any dysfunction or disease will increase the likelihood of toxoplasma infection. Non-Hodgkin's lymphomas and chronic leukaemias are probably the most common, but toxoplasmosis associated with acute leukaemias is similar (Table 7.1). One would have expected a greater frequency of chronic leukaemias especially lymphocytic, as these are less aggressive disorders and there is more time for toxoplasma reactivation. However, the acute leukaemias, both lymphoblastic and myeloblastic, are frequently associated with toxoplasmosis (Table 7.1). This may be related to the more aggressive therapy in these conditions. In addition, therapy in these conditions is aimed at the destruction of abnormal cells in the marrow (also destroying normal cells at a reduced rate), so anti-toxoplasma therapy may make matters worse[2] with folinic acid supplements accelerating the haematological disease process.[30] The clinical presentation in these cases is similar to those with Hodgkin's disease, with a preponderance of non-specific findings such as fever which may be irregular[31] or convulsions.[3] Among adults, even after appropriate anti-toxoplasma therapy the outcome may not be favourable.[27] The position in children may be better, although the frequency of toxoplasma infection with haematological disease is high (17/90, 19%).[31] Predictably, in this study because of the age group, 11/17 (65%) had acute lymphoblastic leukaemia. Fortunately, of the 17 patients given anti-toxoplasma treatment, 15/17 (88%) recovered. It is possible that the prognosis in children is better, but it may be more related to the underlying disease and its treatment. Rarely a variety of uncommon haematological conditions have been associated with toxoplasmosis: multiple myeloma,[2] histiocytosis,[31] and hairy cell leukaemia.[32]

Non-haematological malignancies

Toxoplasmosis has been reported in patients with a variety of solid tumours: carcinoma of the breast, ovary, and lung; seminoma, melanoma, chromo-phobe adenoma, neuroblastoma, and thymoma.[2,3] Obviously, single case histories recording such associations are difficult to interpret. One would expect common conditions to be more likely to have a fortuitous toxoplasma reactivation. Therefore, the relative frequency of an association is more reflected in a series of patients rather than in a review of case histories in the literature. The latter may be particularly biased to the unusual. In addition, if one also considered the frequency of the malignancy, the association of toxoplasmosis with ovarian carcinoma may be more significant than with breast carcinoma.[27] To expect a clinician to be aware of the possibility of toxoplasmosis when it may have a frequency of 1:5000 patients[27] is unrealistic. When it does happen, it is not surprising that many of these patients are not given specific anti-toxoplasma therapy.[2] A realistic position is one in which toxoplasmosis is suspected in any patient with unexplained pyrexia or central nervous system involvement, especially if they also have a malignancy.

TRANSPLANTATION

The proposition that one should replace damaged or diseased organs with unaffected ones is attractive. The number and types of organ transplants have increased over the last decade, and there is still a far greater demand than available organs.[33] Although toxoplasmosis is a rare complication of transplantation, it may be fatal. It also appears to complicate some organ transplants more than others, probably related to the degree of immuno-suppression.

Renal transplants

Since the first successful renal transplant in 1954,[33] thousands of these operations have been performed world-wide. In a series of 81 immuno-compromised patients with toxoplasmosis, seven had organ transplants, of which five were renal transplants.[2] The higher proportion of renal trans-plants in this series probably represents the higher numbers of these oper-ations. In an analysis of eight cases of toxoplasmosis in renal transplants, fever was the predominant finding with both generalized and focal neuro-logical signs.[34] Other bacterial infections were present in 2/8 cases, and all cases were fatal with the diagnosis made at post-mortem examination. The first case of generalized toxoplasmosis following renal transplantation sug-gested the possible mechanism of infection.[35] The patient was seronegative before transplantation and developed the disease after the operation prob-ably as a result of transplantation of an infected kidney. Good evidence that the transplant kidney is the risk factor comes from a report in which two toxoplasma seronegative individuals received kidneys from a seropositive donor — both recipients developed toxoplasma infection shortly after trans-plantation.[36] Also, in another situation where two kidneys from the same donor were used, one recipient developed toxoplasma infection whereas another did not.[34] An explanation may be that the healthy recipient was toxoplasma immune. Such a mechanism may explain the small numbers of toxoplasma infections in these patients, as in any one country there will be a majority of either immune or susceptible individuals (Chapter 1). Thus problems within any one country are unlikely to arise. In addition, each transplant may differ in its potential for infection and the individual recipients may differ in their ability to deal with the transplanted infection. Overall, the risk of toxoplasmosis in renal transplant patients is very small. To reduce the risk even further, mismatches between the immunity (either toxoplasma seronegative or seropositive) of the donor and recipient should be avoided.[37]

Cardiac transplants

Since the first cardiac transplant by Dr Christian Barnard in South Africa in 1966,[33] there have been transplants in 22 countries but more than half have been performed at Stanford University.[38] In a series of 400 transplants over 10 years at Stanford, 63 per cent of the infective episodes were due to protozoa.[38] *Pneumocystis carinii* (since reclassified as a fungus) was the cause of 75 per cent (30/40) of these protozoal infections, with a mortality of 33 per cent. In contrast, toxoplasma was responsible for 20 per cent (8/40) of protozoal infections but the mortality was 88 per cent. The most likely source of toxoplasma infection is the transplanted heart which then produces a disseminated primary infection in the recipient host.[39] In this report, two cardiac recipients were toxoplasma seronegative before transplantation, they received hearts from seropositive donors and subsequently developed severe toxoplasma infection. Another study at Stanford provided additional evidence for such a hypothesis.[40] Of 19 patients who were toxoplasma seropositive before transplantation, ten developed serological evidence of reactivated infection with four having toxoplasma-specific IgM antibodies. These ten patients did not have symptoms or signs that could be attributable to toxoplasma infection. Of 31 patients seronegative before transplantation, four received a heart from a seropositive donor and three seroconverted with clinically severe toxoplasmosis.[40] The message is clear: reactivation of toxoplasma infection in the recipient is not clinically significant whereas a primary infection as a result of infection from the donated heart may result in disseminated infection. Such infection may produce a necrotizing encephalitis, myocarditis, pneumonitis, or hepatitis.[39,40] Seroconversion to toxoplasma can be detected 12–35 days (mean 25 days) after transplantation, and precedes clinical symptoms which manifest 26–45 days (mean 36 days) after transplantation.[41] In an excellent study of 250 heart and 35 heart and lung transplants at Papworth Hospital, Cambridge, there was further evidence that primary infection may be fatal whereas reactivated infection is not severe. However the most important conclusion of this study was that prophylaxis with pyrimethamine (25 mg/day for 6 weeks) in toxoplasma seronegative recipients of seropositive hearts was able to prevent severe infection in most cases.[42] Thus, cardiac transplants (the group in which toxoplasma infection is most common) may be preventable by pretransplant testing and prophylactic treatment of those at risk. Although the numbers are small, there seems a greater risk of developing toxoplasmosis after heart and lung transplants (2/35, 5.7%), than after heart transplants (7/250, 2.8%).[42] This may reflect the greater volume of tissue (and therefore potential infection) of the transplant.

Bone marrow transplants

Patients with bone marrow transplants are severely immunocompromised before and after the transplant. In addition, the frequent occurrence of graft versus host disease may itself produce further immunodeficiency.[43] It could be expected that toxoplasmosis in bone marrow recipients would be common as the marrow is such an important part of the body's immune defences. This is not true. Of more than 2000 transplants, the Seattle group identified only 10 cases of toxoplasmosis.[44] In the UK, of 160 transplants, two cases of toxoplasmosis were found.[18] Unlike renal, cardiac, and cardiac and lung transplants, the toxoplasma infection appears to be a result of reactivated infection rather than infection from the donor.[18,43,44] Therefore it is to be expected that in countries with high toxoplasma infection, there should be more problems. Such is the case, and in France in a series of 80 patients, there was serological evidence of toxoplasma reactivation in 20 per cent of patients.[43] In this study reactivated infection resulted in cerebral toxoplasmosis in three patients without serology of current infection. Seropositive recipients from seronegative donors had the most severe infections. There were also two patients in whom the donor and recipient were toxoplasma seronegative, but the patients developed primary toxoplasmosis, presumably infected from another source. The difference between the mechanism in bone marrow transplants compared to others may be due to the type of treatment. Thus if anti-T-cell monoclonal antibody is used, there may be a significant reduction of the body's ability to deal with intracellular pathogens.[18] Infection may be prevented from reactivating by prophylactic chemotherapy in individuals who were toxoplasma seropositive before transplantation. It has been suggested that for prophylaxis in this group, spiramycin or trimethoprim/sulphamethoxazole could be used instead of the more marrow-toxic agents pyrimethamine and sulphadoxine[43] (Chapter 5).

Liver transplants

Infections appear to complicate liver transplants more than any other transplant with about 70 per cent having evidence of bacteraemia or fungaemia.[45] In this series of 102 liver transplants in 93 patients, one case of disseminated toxoplasmosis (involving lung, pleura, kidney, and central nervous system) was identified. Thus toxoplasma infection is rare. The high frequency of other severe infections occurring early after transplant may mean that there is not enough time for toxoplasma to manifest itself. It is believed that the source of most infections is the transplanted liver.[45]

INFECTIONS

Central nervous system toxoplasmosis is unusual, and therefore the demonstration of an outbreak of patients with this manifestation[4] suggested that other factors were involved. Subsequently, human immunodeficiency virus (HIV) producing acquired immune deficiency syndrome (AIDS) was shown to be the explanation of the outbreak. Although HIV is an excellent example of a viral infection producing immunodeficiency, other viruses and infectious agents also demonstrate similar properties. In patients with malignancy, it has been suggested that infection with DNA viruses (herpes simplex and cytomegalovirus) is frequently found in patients with disseminated toxoplasmosis.[3] It may well be that this association represents a symbiotic relationship,[3] with viral and toxoplasma reactivation on a background of immunodeficiency. A similar mechanism may explain *P. carinii* infection and reactivated toxoplasmosis,[46] although this patient was homosexual and may well have had HIV infection. The excellent review of the literature in 1976 which addressed the problem of toxoplasmosis and immunocompromise did not have any cases of toxoplasmosis as a result of other infections.[2] This was because coincident cases of toxoplasmosis and viral infections were believed to reflect underlying malignancy. However, with the subsequent knowledge of HIV and its ability to cause severe immunodeficiency, the possibility that DNA viruses[3] produce immunodeficiency on their own should be re-examined. There is no doubt that in the future, infections as a result of HIV may represent many, if not most, of the patients with toxoplasmosis and immunodeficiency. Although the outlook of patients with AIDS is poor, the prognosis of coincident toxoplasmosis is good if it is recognized.[47] As there is little information on other infections and their relationships with disseminated toxoplasmosis, the remainder of this section will be concerned with toxoplasma and HIV infection.

Human immunodeficiency virus and AIDS

About 30 per cent of AIDS patients who have toxoplasma-specific antibodies suggestive of past infection will develop toxoplasma encephalitis in the course of their illness.[48] It is likely that illness is a result of reactivated latent infection.[47] Thus the frequency of toxoplasma encephalitis is related to the seroprevalence in the country: about 25 per cent of AIDS patients in Europe, and 5–10 per cent in the USA.[48] In terms of a cause of central nervous system complications in AIDS patients, toxoplasma plays a major role. One study showed that toxoplasma accounted for 38 per cent of known central nervous system infections and 36 per cent of focal intracerebral lesions.[49] Infection may also be more common and severe in those AIDS patients of African descent, and in one series toxoplasma encephalitis was the most common cause of death among infections.[50] One possible explanation of this anomaly is the higher seroprevalence of toxoplasmosis in these groups (Chapter 1), however strain variations may also need to be considered.

Toxoplasma encephalitis may be an early or late complication in AIDS patients. The clinical picture is of diffuse or focal neurological signs and symptoms. Initially, non-specific findings, such as headache and disorientations are present.[51] About a third of patients have seizures (focal or generalized); and the majority have focal neurological signs of which hemiparesis is the most common.[51] In comparison with other causes of neurological complications in AIDS patients, infection is more than three times as common as malignant or vascular complications.[49] In addition, toxoplasma encephalitis must be distinguished from dementia associated with HIV infection. It has been suggested that toxoplasma encephalitis can be clinically differentiated from HIV dementia, as the latter evolves slowly and there is preservation of consciousness.[52] Other clinical presentations of acquired toxoplasmosis (Chapter 3) also have to be separated from those caused by HIV infection. Thus, lymphadenopathy is more likely to be due to the HIV infection and a lymph node biopsy is not indicated.[53] Similarly, in patients with HIV infection, cytomegalovirus retinochoroiditis is common and has to be differentiated from toxoplasma retinochoroiditis. In one series of 40 paediatric patients with AIDS, there v as one case of toxoplasma retinochoroiditis and two cases of cytomegalovirus retinochoroiditis.[54] Toxoplasma infection may become disseminated with widespread organ involvement.[55] Rarely there may be testicular toxoplasmosis or panhypopituitarism.[47] Toxoplasma infection may also occur coincidentally with other infections or malignant condition such as Kaposi's sarcoma.[56] Among Africans, brain abscess also appears to be more frequent.[56]

Diagnosis of toxoplasma infection in AIDS patients can be particularly difficult, especially as specific IgM antibody is rarely present (Chapter 4). There is a great need for the development of new tests such as the polymerase chain reaction (Chapter 8). In the meantime, the diagnosis of toxoplasma encephalitis can be made by early clinical suspicion, and the use of the techniques of computed tomography and nuclear magnetic resonance.[55] In some cases, brain biopsy has to be considered,[55] or the trial of specific anti-toxoplasma therapy[47] (Chapter 4). In general, patients with toxoplasma encephalitis who are not treated rapidly die within weeks, whereas with specific treatment the prognosis is good.[47,55] Nevertheless, the result of treatment in one study of 61 AIDS patients with encephalitis was particularly bad.[51] Overall 56/61 (92%) of the patients died with a median time from the initiation of therapy to death of 126 days (range 7 days to 18 months). These researchers concluded that patients who were 'alert' at the time of treatment did significantly better than those who were 'lethargic' or 'stuporous'.[51] It is difficult to determine how representative these results are of the efficacy of treatment. This group called themselves the 'Toxoplasma Encephalitis Study Group', and it appears that 31/157 (20%) of the individuals who were approached took part in the study. It may well be that this study was biased towards those patients with severe disease.

OTHER CONDITIONS

Toxoplasma infection is a rare complication of other conditions. When it does occur, it may be difficult to separate the effects of the underlying disease, its treatment and other infections from that of toxoplasma. In an early analysis of the literature, collagen–vascular disorders accounted for 6/81 (7.4%) of immunocompromised patients with toxoplasma infection,[2] with systemic lupus erythematosus being the most common. Without treatment, central nervous system infection may become disseminated and rapidly fatal.[57] The reactivated infection is possibly a result of treatment with corticosteroids and cyclophosphamide.[57] Other collagen–vascular diseases, such as scleroderma and mixed connective tissue disease, have been rarely associated with toxoplasma infection.[58] The treatment of these disorders with corticosteroids may also be the explanation of the reactivation process. Corticosteroids are able to depress cell-mediated immunity[58] and with these diseases high doses of corticosteroids may also be used. These drugs may also explain the mechanism in the association of autoimmune haemolytic anaemia and reactivated toxoplasma encephalitis.[2]

SUMMARY

Unrecognized toxoplasma infection in the immunocompetent is rarely serious; however, in the immunocompromised, unrecognized disease may be fatal. The normal immune response to toxoplasma is complex with an interplay of all the body's immune cells. Macrophages play a crucial role both in the immediate non-specific immune response and in processing antigens for T-lymphocytes for a toxoplasma-specific response. Toxoplasma also has the ability to remain alive within some macrophages. In this position, toxoplasma is protected from the body's immune response and can be distributed by the macrophages throughout the body. Several types of T-cells respond to toxoplasma infection. Of these, T-helper cells are important in stimulating T-cytotoxic cells and T-suppressor cells moderate the activity of T-helper cells. Natural killer cells have an immediate role in the control of toxoplasmosis and may also be significant in the toxoplasma-specific immune response.

In determining whether or not exposure to toxoplasma results in infection, the B-cell response is probably less important than the T-cell response. Nevertheless, antibody coated cells are more easily phagocytosed and antibody detection (IgG and IgM) is the mainstay of diagnosis. Similarly, the roles of IgA and IgE remain to be elucidated, but both may be helpful in diagnosis. In contrast, synthetic cytokines reduce the mortality of infected animals and may be used to augment the immune response in humans. Thus, the normal immune response is complex and resistance to toxoplasma infection depends on several factors.

The association of toxoplasmosis and malignancy was made 25 years ago. Haematological malignancies are most common, especially Hodgkin's disease. Cases of the latter stem from the post-chemotherapy era and suggest that it is not Hodgkin's disease, but its treatment that predisposed to toxoplasmosis. Infection is reactivated. Similarly, reactivation appears to be the mechanism in other malignancies, and in all malignancies, prophylaxis is probably not indicated as toxoplasmosis as a complication is rare. The position is more varied in transplantation. In most cases (renal, liver, cardiac, and cardiac and lung transplants), serious infection results from a primary infection from the transplanted organ. Reactivated infections are not usually serious. However, in bone marrow transplants, reactivated infections can be fatal. In the transplant group, toxoplasma mismatches should be avoided and prophylaxis is indicated when this occurs.

The importance of other infections as a cause of immunodeficiency has been reassessed with the advent of HIV and AIDS. Toxoplasmosis is usually a result of reactivation and prompt treatment often has a favourable outcome. In the future, most patients with immunocompromise and toxoplasmosis may be HIV/AIDS patients. Such a scenario will increase the medical awareness of toxoplasmosis. The lesson is that toxoplasmosis should be considered in immunocompromised patients with unexplained fever or cerebral symptoms. Vigilance and prophylaxis are critical in dealing with toxoplasmosis in the immunocompromised.

REFERENCES

1 Remington, J. S. (1974). Toxoplasmosis in the adult. *Bull. N. Y. Acad. Sci.*, **50**, 211–27.
2 Ruskin, J. and Remington, J. S. (1976). Toxoplasmosis in the compromised host. *Ann. Intern. Med.*, **84**, 193–9.
3 Vietzke, W. M., Gelderman, A. H., Grimley, P. M., and Valsamis, M. P. (1968). Toxoplasmosis complicating malignancy. *Cancer*, **21**, 816–27.
4 Luft, B. J., Conley, F., and Remington, J. S. (1983). Outbreak of central nervous system toxoplasmosis in Western Europe and North America. *Lancet*, **i**, 781–4.
5 Luft, B. J., Brooks, R. G., Conley, F. K., McCabe, R. E., and Remington, J. S. (1984). Toxoplasmic encephalitis in patients with acquired immune deficiency syndrome. *JAMA*, **252**, 913–17.
6 Ryning, F. W. and Mills, J. (1979). *Pneumocystis carinii, Toxoplasma gondii*, and cytomegalovirus and the compromised host. *West. J. Med.*, **130**, 18–34.
7 Remington, J. S. and Krahenbuhl, J. L. (1976). Immunology of toxoplasma infection. In *Immunology of parasitic infections*. (ed. S. Cohen and E. H. Sadun), pp. 235–67. Blackwell Scientific, Oxford.
8 Werk, R. (1985). How does *Toxoplasma gondii* enter host cells? *Rev. Infect. Dis.*, **7**, 449–57.
9 Roit, I., Brostoff, J., and Male, D. (1989). *Immunology*. (2nd edn). Churchill Livingstone, London.
10 McLeod, R., Bensch, K. G., Smith, S. M. and Remington, J. S. (1980). Effects of human peripheral blood monocytes, monocyte-derived macrophages and spleen mononuclear phagocytes on *Toxoplasma gondii. Cell. Immunol.*, **54**, 330–50.
11 McLeod, R. and Remington, J. S. (1977). Influence of infection with toxoplasma on macrophage function, and role of macrophages in resistance to toxoplasma. *Am. J. Trop. Med. Hyg.*, **26**, 170–86.
12 Luft, B. J., Kansas, G., Engleman, E. G., and Remington, J. S. (1984). Functional and quantitative alterations in T lymphocyte subpopulations in acute toxoplasmosis. *J. Infect. Dis.*, **150**, 761–7.
13 Hamblin, T. J., Hussain, J., Akbar, A. N., Tang, Y. C., Smith, J. L., and Jones, D. B. (1983). Immunological reason for chronic ill health after infectious mononucleosis. *Br. Med. J.*, **287**, 85–8.
14 Sklenar, I., Jones, T. C., Alkan, S., and Erb, P. (1986). Association of symptomatic human infection with *Toxoplasma gondii* with imbalance of monocytes and antigen-specific T cell subsets. *J. Infect. Dis.*, **153**, 315–24.
15 Frenkel, J. K. (1948). Dermal hypersensitivity to toxoplasma antigens (toxoplasmins). *Proc. Soc. Exp. Biol. Med.*, **68**, 634–9.
16 Kaufman, H. E. (1960). Uveitis accompanied by a positive toxoplasma dye test. *Am. Med. Assoc. Arch. Ophthalmol.*, **63**, 767–73.
17 Pavia, C. S. (1986). Protection against experimental toxoplasmosis by adoptive immunotherapy. *J. Immunol.*, **137**, 2985–90.
18 Pendry, K., Tait, R. C., McLay, A., Ho-Yen, D. O., Baird, D., and Burnett, A. K. (1990). Toxoplasmosis after BMT for CML. *Bone Marrow Transplant*, **5**, 65–7.
19 Hauser, W. E., Sharma, S. D., and Remington, J. S. (1983). Augmentation of NK cell activity by soluble and particulate fractions of *Toxoplasma gondii. J. Immunol.*, **131**, 458–63.
20 Whiteside, T. and Herberman, R. B. (1989). The role of natural killer cells in human disease. *Clin. Immunol. Immunopathol*, **53**, 1–23.
21 Van Camp, B., Reynaert, P., and Van Beers, D. (1982). Congenital toxoplasmosis associated with transient monoclonal IgG1-lambda gammopathy. *Rev. Infect. Dis.*, **4**, 173–8.
22 Decoster, A., Darcy, F., Caron, A., and Capron, A. (1988). IgA antibodies against P30 as markers of congenital and acute toxoplasmosis. *Lancet*, **ii**, 1104–7.

23 Pinon, J. M., Toubas, D., Marx, C., Mougeot, G., Bonnin, A., Bonhomme, A. *et al.* (1990). Detection of specific immunoglobulin E in patients with toxoplasmosis. *J. Clin. Microbiol.*, **28**, 1739–43.

24 McCabe, R. E., Luft, B. J., and Remington, J. S. (1984). Effect of murine interferon gamma on murine toxoplasmosis. *J. Infect. Dis.*, **150**, 961–2.

25 Sharma, D. S., Hofflin, J. M., and Remington, J. S. (1985). *In vivo* recombinant interleukin 2 administration enhances survival against a lethal challenge with *Toxoplasma gondii*. *J. Immunol.*, **135**, 4160–3.

26 Lewis, W. P. and Markell, E. K. (1958). Acquisition of immunity to toxoplasmosis by the newborn rat. *Exp. Parasitol.*, **7**, 463–7.

27 Hakes, T. B. and Armstrong, D. (1983). Toxoplasmosis. Problems in diagnosis and treatment. *Cancer*, **52**, 1535–40.

28 Frenkel, J. K., Nelson, B. M., and Arias-Stella, J. (1975). Immunosuppression and toxoplasmic encephalitis. *Hum. Pathol.*, **6**, 97–111.

29 Smith, G. M., Leyland, M. J., Crocker, J., and Geddes, A. M. (1987). Cerebral toxoplasmosis in a patient with Hodgkin's disease. *J. Infect.*, **14**, 243–5.

30 Helmer, R. E. (1975). Hazard of folinic acid with pyrimethamine and sulfadiazine. *Ann. Intern. Med.*, **82**, 124–5.

31 Boguslawska-Jaworska, J., Pisarek, J., and Chybicka, A. (1982). Toxoplasmosis in the course of neoplastic diseases of haematopoietic and lymphatic system in children. *Mater. Med. Pol.*, **48**, 62–7.

32 Mackowiak, P. A., Demian, S. E., Sutker, W. L., Murphy, F. K., Smith, J. W., Thompsett, R. *et al.* (1980). Infections in hairy cell leukemia. *Am. J. Med.*, **68**, 718–24.

33 Diethelm, A. G. (1990). Ethical decisions in the history of organ transplantation. *Ann. Surg.*, **211**, 505–20.

34 Rhodes, R. H., Davis, R. L., Berne, T. V., and Tatter, D. (1977). Disseminated toxoplasmosis with brain involvement in a renal allograft recipient. *Bull. L. A. Neurol. Soc.*, **42**, 16–22.

35 Reynolds, E. S., Walls, K. W., and Pfeiffer, R. I. (1966). Generalized toxoplasmosis following renal transplantation. *Arch. Intern. Med.*, **118**, 401–5.

36 Mejia, G., Leiderman, E., Builes, M., Henao, J., Arbelaez, M., Araugo, J. L. *et al.* (1983). Transmission of toxoplasmosis by renal transplant. *Am. J. Kidney Dis.*, **11**, 615–17.

37 Krick, J. A. and Remington, J. S. (1978). Toxoplasmosis in the adult — an overview. *N. Engl. J. Med.*, **298**, 550–3.

38 Jamieson, S. W., Oyer, P. E., Reitz, B. A., Baumgartner, W. A., Bieber, C. P., Stinson, E. B. *et al.* (1981). Cardiac transplantation at Stanford. *Heart Transplant*, **1**, 86–91.

39 Ryning, F. W., McLeod, R., Maddox, J. C., Hunt., S. and Remington, J. S. (1979). Probable transmission of *Toxoplasma gondii* by organ transplantation. *Ann. Intern. Med.*, **90**, 47–9.

40 Luft, B. J., Naot, Y., Araujo, F. G., Stinson, E. B., and Remington, J. S. (1983). Primary and reactivated toxoplasma infection in patients with cardiac transplants: clinical spectrum and problems in diagnosis in a defined population. *Ann. Intern. Med.*, **99**, 27–31.

41 Wreghitt, T. G., Gray, J. J., and Balfour, A. H. (1986). Problems with serological diagnosis of *Toxoplasma gondii* infections in heart transplant recipients. *J. Clin. Pathol.*, **39**, 1135–9.

42 Wreghitt, T. G., Hakim, M., Gray, J. J., Balfour, A. H., Stovin, P. G. I., Stewart, S. *et al.* (1989). Toxoplasmosis in heart and heart and lung transplant recipients. *J. Clin. Pathol.*, **42**, 194–9.

43 Derouin, F., Gluckman, E., Beauvais, B., Devergie, A., Melo, R., Monny, M. *et al.* (1986). Toxoplasma infection after human allogeneic bone marrow transplantation: clinical and serological study of 80 patients. *Bone Marrow Transplant*, **1**, 67–73.

44 Shepp, D. H., Hackman, R. C., Conley, F. K., Anderson, J. B., and Myers, J. D. (1985). *Toxoplasma gondii* reactivation identified by detection of parasitemia in tissue culture. *Ann. Intern. Med.*, **103**, 218–21.

45 Schroter, G. P. J., Hoelscher, M., Putnam, C. W., Porter, K. A., Hansbrough, J. F., and Starzl, T. E. (1976). Infection complicating orthotopic liver transplantation. *Arch. Surg.*, **111**, 1337–47.

46 Gransden, W. R. and Brown, P. M. (1983). Pneumocystis pneumonia and disseminated toxoplasmosis in a male homosexual. *Br. Med. J.*, **286**, 1614.

47 Holliman, R. E. (1988). Toxoplasmosis and the acquired immune deficiency syndrome. *J. Infect.*, **16**, 121–8.

48 McCabe, R., and Remington, J. S. (1988). Toxoplasmosis: the time has come. *N. Engl. J. Med.*, **318**, 313–15.

49 Snider, W. D., Simpson, D. M., Neilsen, S., Gold, J. W., Metroka, C. E., and Posner, J. B. (1983). Neurological complications of acquired immune deficiency syndrome: analysis of 50 patients. *Ann. Neurol.*, **14**, 403–18.

50 Moskowitz, L. B., Kory, P., Chan, J. C., Haverkos, H. W., Conley, F. K., and Hensley, G. T. (1983). Unusual causes of death in Haitians residing in Miami. *JAMA*, **250**, 1187–91.

51 Haverkos, H. W. (1987). Assessment of therapy for toxoplasma encephalitis. *Am. J. Med.*, **82**, 907–14.

52 Navia, B. A., Petito, C. K., Gold, J. W. M., Cho, E. S., Jordan, B. D., and Price, R. W. (1986). Cerebral toxoplasmosis complicating the acquired immune deficiency syndrome: clinical and neuropathological findings in 27 patients. *Ann. Neurol.*, **19**, 224–38.

53 Farthing, C. F., Henry, K., Shanson, D. C., Taube, M., Lawrence, A. G., Harcourt-Webster, J. N. *et al.* (1986). Clinical investigations of lymphadenopathy, including lymph node biopsies, in 24 homosexual men with antibodies to the human T-cell lymphotropic virus type III (HTLV-III). *Br. J. Surg.*, **73**, 180–2.

54 Dennehy, P. J., Warman, R., Flynn, J. T., Scott, G. B., and Mastrucci, M. T. (1989). Ocular manifestations in pediatric patients with acquired immunodeficiency syndrome. *Arch. Ophthalmol.*, **107**, 978–82.

55 Luft, B. J. and Remington, J. S. (1988). Toxoplasma encephalitis. *J. Infect. Dis.*, **157**, 1–6.

56 Clumeck, N., Sonnet, J., Taelman, H., Mascart-Lemone, F., De Bruyere, M., Vandeperre, P. *et al.* (1984). Acquired immunodeficiency syndrome in African patients. *N. Engl. J. Med.*, **310**, 492–7.

57 Dubin, H. V., Courter, M. H., and Harrell, E. R. (1971). Toxoplasmosis. *Arch. Dermatol.*, **104**, 547–50.

58 Townsends, J. J., Wolinsky, J. S., Baringer, J. R., and Johnson, P. C. (1975). Acquired toxoplasmosis. A neglected cause of treatable nervous system disease. *Arch. Neurol.*, **32**, 335–43.

8

New techniques
MARILYN M. DAVIDSON

Diagnosis of *Toxoplasma gondii* infection in the immunocompetent adult is well established with rising IgG titres and specific IgM confirming an acute infection (Chapter 4). However, during the last decade it has been recognized that there is a need for new techniques to improve the diagnosis in groups of patients such as pregnant women and those with immunocompromise. Specific IgM has been shown to persist in some adults for at least nine months[1,2] therefore its presence during pregnancy may indicate infection prior to conception, which presents little risk to the fetus (Chapter 6). In addition, maternal infection during pregnancy may only be transmitted to about half of fetuses and not all infected fetuses have detectable IgM (Chapter 6). Diagnosis of infection in those who are immunocompromised can be even more difficult as life threatening illness is often due to reactivation of a latent infection (Chapter 7). Dye test titres are rarely high, rising titres are not usually seen, and specific IgM antibody is present in only a few cases.[3]

As an alternative to serology, direct detection by isolation of the parasite in mice or in tissue culture is sometimes attempted.[4,5] However, mouse culture may take 6 weeks and although tissue culture is quicker, it appears to be less sensitive.[5] Also, since latent infection is characterized by the presence of tissue cysts in many tissues and organs, isolation of toxoplasma from tissue does not necessarily indicate active disease. In contrast, the detection of toxoplasma in cerebrospinal fluid (CSF), blood, or amniotic fluid suggests the presence of the tachyzoite and thus active infection. At the moment there are no quick reliable techniques to detect active toxoplasma infection in these samples.

Routine diagnostic assays have used relatively crude antigens and knowledge of their protein composition has been limited.[6] An understanding of the antigenic composition of toxoplasma may lead to better defined and hopefully more specific and sensitive diagnostic tests. Many attempts have been made to characterize these antigens but early results were variable with discrepancies in both the size and number of proteins.[6] There may be several reasons for these discrepancies. Variations in sample preparation and in the antisera used to detect the proteins may have played a part. Also some of the detected proteins were probably of host cell origin, since toxoplasma is an intracellular parasite and it is difficult to obtain preparations which are totally free from host material.[6] The techniques used in these experiments were relatively insensitive and could not accurately determine protein sizes. New techniques based on polyacrylamide gel electrophoresis with sodium dodecyl sulphate (SDS–PAGE) overcame most of these problems.[7] Although first introduced in 1970, it is only in the 1980s that SDS–PAGE has been applied to the study of toxoplasma. When combined with other techniques such as radiolabelling,[8] immunoprecipitation,[8,9] and immunoblotting[10] using monoclonal antibodies,[11] it has allowed more detailed study of the proteins of toxoplasma and the immune response to these proteins.[6,12,13] At the same time, the introduction of techniques to analyse nucleic acids has led to the sequencing of toxoplasma genes[14,15] and to the use of this knowledge in a new test for the detection of toxoplasma DNA, the polymerase chain reaction (PCR).[16] This chapter will describe these new techniques, examine the information about toxoplasma gained from their use and look at the way in which the techniques and new information have been applied to diagnostic problems.

NEW TECHNIQUES

SDS–PAGE

In 1970, SDS–PAGE with a discontinuous buffer system was described[7] and has proven to be an important method for protein analysis. The ability of electrophoresis to separate proteins has been recognized for a long time and was first applied in a clinical setting in 1937.[17] In these early studies, agar or agarose was used as the support medium. PAGE was introduced in 1959 and had several advantages. Polyacrylamide has a high chemical and mechanical stability and can be obtained with very few impurities. It can be used at different concentrations depending on the size of the proteins to be analysed. For toxoplasma, gels in the range of 5–15 per cent are usually used.[12,13] The addition of SDS and a discontinuous buffer system has produced a very sensitive method, capable of resolving 100–200 components.[18] SDS unfolds the proteins to produce rod-shaped particles which pass through the gel at different speeds, where speed is inversely proportional to size. At the end of the run, the smaller proteins are located near the bottom of the gel and the larger ones nearer the top. The discontinuous buffer system using buffers of different pH in the gel and the electrophoresis tank allows sharp separation of protein bands. The protein sample in SDS is applied to a polyacrylamide gel of known pore size and an electrical current passed through the gel.[7] The proteins are detected as discrete bands with protein-specific stains such as Coomassie brilliant blue. By running a set of proteins of known size (mol. wt) on the same gel it is possible to determine the size of the sample proteins. Although SDS–PAGE is a useful technique, it is of limited value on its own. However, when extended by the use of monoclonal antibodies, radiolabelling, radioimmunoprecipitation, and immunoblotting it is a valuable tool for the study of toxoplasma proteins.

Monoclonal antibodies

The normal human immune response to toxoplasma results in the production of antibodies to many toxoplasma antigens,[13] whereas a monoclonal antibody is usually highly specific and recognizes only one antigen.[12] Using a technique first described in 1976, monoclonal antibodies are synthesized in artificially created cells, the result of the fusion of spleen cells from immunized animals with myeloma cells.[11] Cultural conditions are selected which only permit the survival of fused cells, which have the capacity of spleen cells to produce antibody and of myeloma cells to grow indefinitely. By appropriate dilution techniques and culturing from a single cell, colonies of immortal cells are established, each of which produces a pure monoclonal antibody to one antigen. For toxoplasma, mice are immunized with tissue cysts or tachyzoites and cultures of cloned fused cells are tested for the production of specific antibody.[12,19] Since these antibodies only recognize one toxoplasma-specific antigen, they overcome the problem of host cell contamination of antigen preparations. Monoclonal antibodies have been produced against most of the important antigens of toxoplasma[6] and have been used to identify their role in infection.[20]

Radiolabelling and immunoprecipitation

Radiolabelling is the use of the radioactive forms of certain elements e.g. iodine (^{125}I) or sulphur (^{35}S) to label proteins.[8] When the radiolabelled proteins are separated by SDS–PAGE the radioactive bands are detected by exposure to X-ray film. In this way, it is possible to label most of the proteins in a sample or only selected groups. Thus, the surface proteins of tachyzoites have been identified using an enzyme reaction which selectively binds ^{125}I to the amino acid tyrosine exposed on the parasite surface.[12] Radiolabelling can be extended by radioimmunoprecipitation (RIP), where the radioactive proteins are allowed to react with specific antibody, resulting in the formation of a complex of radioactive protein and antibody.[8,9] The complex is separated from unreacted protein and antibody and analyzed by SDS–PAGE. The reaction mixture for electrophoresis disrupts the complex and the radioactive protein bands which have reacted with specific antibody are identified on the gel. Therefore RIP allows the separation and identification of those proteins that act as antigens during toxoplasma infection.[6]

Immunoblotting

The technique of immunoblotting (Western blotting) also detects toxoplasma antigens, but does not use radioactive labels.[6] Proteins are separated by SDS–PAGE, then transferred by an electric current to a sheet of nitrocellulose membrane onto which they become fixed.[10] The membrane can then be subjected to further tests which would not be possible in the gel. The protein bands are either stained directly or are exposed to specific antibody. Monoclonal antibodies have been used to identify toxoplasma antigens[12] and human immune sera have been used to determine which antigens elicit an immune response during toxoplasma infection.[13] Antibodies in the patient's sera which recognize the protein bands form an antibody–antigen complex on the membrane. They are identified by a second antibody linked to an enzyme which causes a coloured reaction on exposure to a chemical substrate. Only those proteins which react with specific antibody are coloured (Fig. 8.1). Immunoblotting may identify antigens which are not recognized by RIP e.g. the 6 kDa surface antigen.[21]

Polymerase chain reaction

The direct detection of very small amounts of specific nucleic acid has been made possible by the introduction in 1985 of the polymerase chain reaction (PCR).[16] This technique results in the amplification (up to a million times) of a specific fragment of DNA from within the parasite genome. Short sequences (usually ~20 base pairs) of DNA (primers) flanking this fragment are used to initiate the reaction. For toxoplasma two genes have been sequenced and fragments within them chosen for use in the PCR, the B1 gene[14] and the P30 surface protein gene.[15] A reaction mixture containing the primers, sample, a heat stable DNA polymerase, and substrates in buffer containing magnesium chloride is subjected to cycles of heating and cooling. During each cycle, heating produces single strands of DNA on which, in the cooling phase, annealing of the primers, then synthesis of complementary DNA occurs. Thus each successive cycle leads to a doubling of the quantity of DNA. At the end of 30 cycles, sufficient quantities of uniform length copies of the specific DNA fragment are usually present to be detected by ethidium bromide staining, following gel electrophoresis.[22] PCR has been shown to be sensitive enough to detect less than 10 tachyzoites in experimental samples.[14] The technique has been useful in detecting toxoplasma nucleic acid in brain biopsy samples from patients with AIDS.[22]

NEW KNOWLEDGE

The use of these new techniques has produced a great volume of new information on many aspects of toxoplasma infection.[6] The different stages of parasite development have been studied and the similarities and differences in antigenic structure evaluated.[23] Equally as important, variations in antigenic proteins have been detected in different strains of toxoplasma.[24,25]

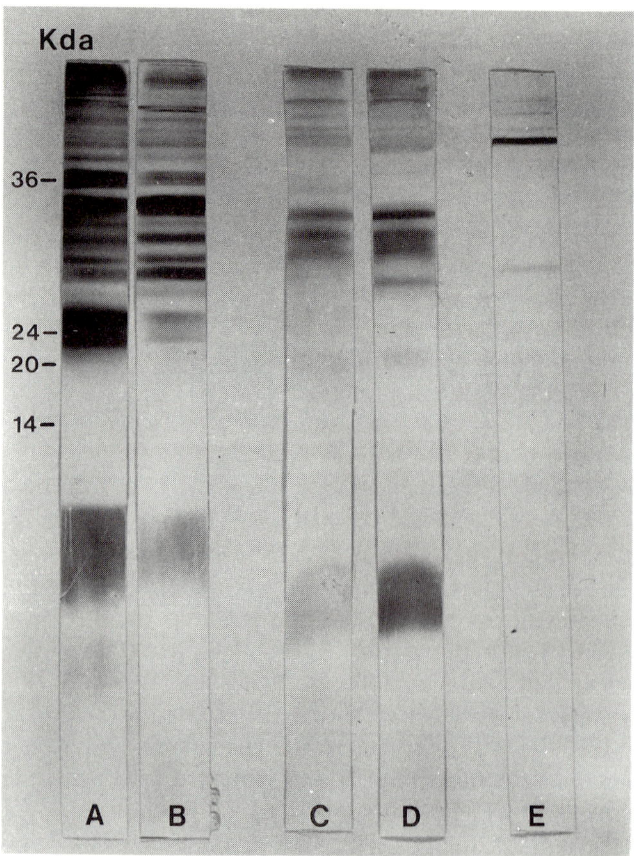

Fig. 8.1 IgG antibody immunoblot analysis of toxoplasma antigens. Antigens were separated by 15 per cent SDS–PAGE, blotted onto nitrocellulose and reacted with human sera. Lane A, B: sera from acute infection (IgM +ve, IgG +ve); lane C, D: sera from latent infection (IgM –ve, IgG +ve); lane E: seronegative (IgM –ve, IgG –ve) sample showing bands due to natural antibodies. Numbers on the left represent mol. wt markers (From Moir, I. L., Davidson, M. M., and Ho-Yen, D. O. (1991) Comparison of IgG antibody profiles by immunoblotting in patients with acute and past *Toxoplasma gondii* infection. *J. Clin. Pathol.*, **44**, 569–72)

Diagnostic assays all use the tachyzoite stage of one strain, the RH strain, and most research studies have also concentrated on the RH strain. The differences between strains and stages of development warrant consideration and may help our understanding of toxoplasma infection. The antigenic composition of the tachyzoite stage has been extensively studied in the last decade.[6] The location of many antigens on the tachyzoite is now known and the immune response to these antigens during infection is better understood.[21] Also the antigens which stimulate antibodies capable of protecting against infections are being extensively studied.[6] A greater understanding of tachyzoite antigens may lead to more sensitive and specific diagnostic assays and the possible production of a vaccine.

Stages of development

Toxoplasma is a complex organism with three separate stages of development (Chapter 2). Two of these, the crescentic tachyzoite and the tissue cyst (containing bradyzoites) (Fig. 8.2) can be found in human infection while the third stage, the oocyst, is the result of sexual reproduction and is only found in the cat family. Tachyzoites actively multiply during acute infection, and recovery is followed by a quiescent, latent phase when tissue cysts are formed which may persist for life (Chapter 2). The tissue cyst can burst to release bradyzoites which may initiate a further cycle of infection. The bradyzoite is smaller and more slender than the tachyzoite, has subterminal nuclei and stains strongly with periodic acid Schiff (PAS) stain for carbohydrates. It is also known that the bradyzoite will survive up to 3 hours in gastric juice, while the tachyzoite is readily killed.[26] In 1983 the first report was published of differences between these two stages at the level of individual proteins which stimulate the production of antibodies in the serum of humans and other animals.[27] This study showed that anti-tachyzoite antibodies reacted in an immunofluorescence assay with tachyzoites, but only partially with bradyzoites. Anti-bradyzoite antibodies, on the other hand, reacted with bradyzoites, but not tachyzoites. This reaction is explained by recent evidence of antigens common to both tachyzoites and bradyzoites and others which were unique to each stage.[23,28] Therefore monoclonal antibody to each stage can show a strong reaction in an enzyme-linked immunosorbent assay (ELISA) with its own stage, but a weaker response to the other stage. Oocysts also possess both shared and unique proteins.[23] This knowledge of antigenic differences between the stages of development has been utilized to produce diagnostic assays which attempt to differentiate the acute and reactivated phases from the latent phase of infection.[29–31]

Strain variation

The ability of the parasite to kill other animals (virulence) has been used to characterize different strains of toxoplasma (Chapter 2). Some strains including the common laboratory strain, the RH strain, are highly virulent for mice, while others (e.g. the Beverley strain) produce an infection which does not kill the mice or cause an obvious illness. The RH strain was isolated from a child and has been used to infect successive generations of laboratory animals.[32] It produces large numbers of tachyzoites in a few days, killing the animals. Examination by radiolabelling of the proteins of toxoplasma strains showed that each strain had unique proteins but the majority of proteins

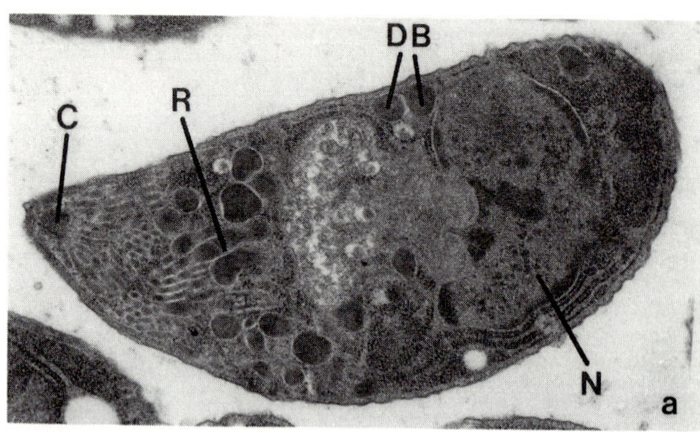

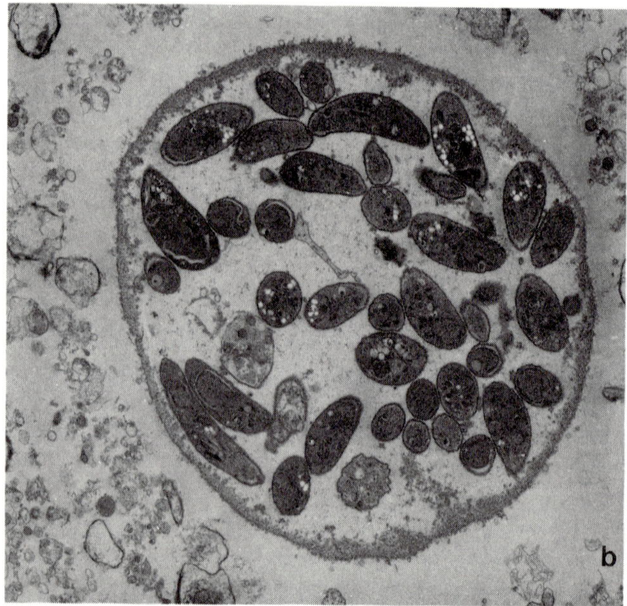

Fig. 8.2 Toxoplasma stages of development. (a) Tachyzoite, C, conoid; R, rhoptry; DB, dense bodies; N, nucleus. Magnification × 20 000. (b) Tissue cyst containing bradyzoites. Magnification × 4000.

were common to all strains.[24,25] Mouse[24] and human[33] antibodies have been shown by immunoblotting to react with similar antigens in different strains, although minor variations, especially in the strength of the immune response, were seen. Consequently, monoclonal antibodies to the RH strain could prevent the growth of the RH strain, but were less effective against other strains.[24] Not surprisingly, the virulent strains, including the RH strain, had a different profile of common enzymes of metabolism from the less virulent ones.[34]

Surface antigens

Surface antigens are thought to be amongst the first to stimulate the host's immune system. Early studies provided conflicting information on both the number and the size of the antigens.[6] More specific techniques such as RIP and PAGE have revealed that there are only four major surface proteins which are susceptible to radiolabelling[6,12,20] (Table 8.1). The most intensely labelled protein has a molecular weight of 30 kDa (P30) and makes up to 3–5 per cent of the total tachyzoite protein.[35] P30 has been purified and tested for use in a diagnostic assay with some success.[36] The largest of the proteins (P43) is a glycoprotein with a molecular weight of 43 kDa.[12,28] However, there is confusion over the size of the other two antigens. Monoclonal antibodies have been described against proteins of 35 kDa (P35), 22 kDa (P22), and 14 kDa (P14),[6,12,20] although each study only describes four antigens. It has been suggested that P14 and P22 may be the same protein and the apparent difference in molecular weight due to discrepancies in measuring size.[12] This latest study also detected a minor protein of 23 kDa (P23).[12] Immunoblotting analysis of the membrane fraction revealed five major surface antigens, the same four proteins previously identified plus an additional antigen of 4–6 kDa.[21] This antigen is largely composed of carbohydrate and stains strongly with PAS. Radioiodination techniques label the amino acid tyrosine which is either absent or not exposed on the surface of the 4–6 kDa antigen and therefore it is not detected by RIP.[8] The 4–6 kDa antigen is a potent antigen and one of the earliest antibody responses of the host is to this antigen[13] (Table 8.1).

Identification of surface antigens has led to interest in their ability to protect against toxoplasma infection. Monoclonal antibodies to surface proteins have been shown to give at least partial protection.[37–39] Some purified surface antigens (P14, P35) have been effective when used to immunize mice,[40] but P30 failed to give protection.[41] Therefore it is likely that surface antigens may have a role in inducing a protective response, but the immune response to toxoplasma infection is not, as yet, fully understood.

Intracellular antigens

Early studies on the soluble antigen fraction of tachyzoites which contains mostly intracellular antigens indicated that there were 9–10 antigens with a molecular weight range from 14 to 133 kDa.[6] Monoclonal antibody, raised against intracellular antigen reacted with a protein of 98 kDa.[42] Since the monoclonal antibody gave positive results in the HA tests but not immunofluorescence assays, it was suggested that the 98 kDa protein was a major component of HA antigen. This resolved a discrepancy between earlier results which had suggested that HA antigen was composed of protein of either 10 kDa or more than 100 kDa.[6] Another antigen of 66 kDa, when

Table 8.1 Major toxoplasma tachyzoite antigens, protein composition unless otherwise stated

Class	Size (kDa)	Location	Chemical characteristics	Use
Surface	6		Carbohydrate	
	22 (14)			Dye test
	30	Parasitophorous vacuole network		and immunofluorescence
	35			assays
	43	?Parasitophorous vacuole network	Glycoprotein	
Intra-cellular	23	Dense body, parasitophorous vacuole network		
	28	Subsurface, parasitophorous vacuole network		
	55–60	Rhoptry	?Penetration enhancing factor	
	58	Subsurface, parasitophorous vacuole network		
	66	Rhoptry, parasitophorous vacuole membrane		
	66	Unknown		Latex agglutination test
	98	Unknown		Haemagglutination test
Excretory/secretory	21	Dense body, subsurface and parasitophorous vacuole network		
	26	Not in whole tachyzoites		
	27	Dense body, parasitophorous vacuole network		?Circulating antigen
	28.5	Dense body, subsurface and parasitophorous vacuole network	Glycoprotein	?Marker of latent infection
	39	Membrane		
	57			?Circulating antigen
	69, 97			?Markers of acute infection
	108	Not in whole tachyzoite		?Marker of latent infection

injected into mice, produced antibodies which reacted exclusively in the latex agglutination tests, suggesting that it may be a major component of latex agglutination antigen[40] (Table 8.1). Neither of these two antigens have been definitely linked to any intracellular structure.

However, other proteins have been characterized and located on specific intracellular organelles (Table 8.1). The rhoptries are sac-like structures located at the anterior end of the tachyzoite (Fig. 8.2) and are thought to play a role in host cell invasion (Chapter 2). They are associated with a family of proteins with major antigens of 55 and 60 kDa and a number of minor antigens.[43,44] Immune electron microscopy using monoclonal antibodies to the 55 and 60 kDa antigens has located these proteins on the bodies of the rhoptries. The monoclonal antibodies are also capable of blocking the activity of a penetration enhancing factor (PEF) which assists in host cell invasion. It is not yet clear if either the 55 or 60 kDa protein is a component of PEF. Another protein of 66 kDa is also associated with the anterior end of the tachyzoite possibly located in the anterior tips of the rhoptries.[45] This 66 kDa protein is secreted on host cell entry and becomes associated with the membrane of the parasitophorous vacuole which is formed on entry to a cell. There is no evidence to indicate whether this 66 kDa protein is related to the 66 kDa protein involved in the latex agglutination reaction (Table 8.1). Two related antigens of 28 and 58 kDa have been detected by a monoclonal antibody.[46] The antibody reacted in immunoperoxidase assays when the tachyzoites had been treated with Triton X100 but not with live parasites. Immune electron microscopy showed no surface labelling, but the antigens were detected in a continuous layer around the tachyzoite just beneath the surface membrane. The antigens are also found in the parasitophorous vacuole network. A 23 kDa protein (P23) has been localized in the dense bodies of the tachyzoite[47] (Fig. 8.2). This protein has been associated with host cell invasion and is secreted into the parasitophorous vacuole.

Excretory secretory antigens

The ability to maintain living tachyzoites in culture free of host cells has allowed the study of excretory/secretory (E/S) antigens without host cell contamination.[48] In such cultures toxoplasma E/S antigens are actively released into the supernate, with perhaps some contamination from dead parasites. Early studies, on the other hand, examined either the antigens released from tissue culture cells infected with toxoplasma (exoantigens),[49] which might be contaminated with proteins from the tissue culture cells, or the antigens found circulating in the blood or urine of infected humans or animals (circulating antigen).[50-53] The first studies of circulating antigen indicated that it contained two of the same antigens (150, 324 kDa) found in haemagglutination antigen; these may be related to the 98 kDa protein.[50]

However, it is likely that circulating antigen contains intracellular antigens from dead tachyzoites as well as E/S antigens. Later more sensitive assays detected more bands with a size range of 45 to 400 kDa in one study[51] and 27 to 57 kDa in another.[52] Different methods of measuring molecular weight may account for the variability in size. Exoantigens from infected tissue cultures have been described with a size range of 63 to 1700 kDa;[49] some of which were probably of host cell origin. Because of the different techniques used in these studies, it is very difficult to make valid comparisons of these results.

Detailed study of E/S antigens from cell free cultures using RIP with sera from latent human infection showed at least nine proteins of 21–108 kDa[48] (Table 8.1). Two of these, the 26 and 108 kDa proteins, were not detected in a preparation of whole tachyzoites, indicating that they were uniquely E/S antigens. For the others, the distinction between intracellular and E/S antigens becomes blurred since some which are detected on internal structures are also actively excreted by the tachyzoite e.g. the 23 kDa dense body antigen.[47] The 21, 27, and 28.5 kDa E/S proteins have been localized in the parasitophorous vacuole[54] which contains a network of proteins composed largely of a 30 kDa protein identified as the P30 surface protein.[55] The surface protein P43 may also be shed into this network.[28] It is likely that the network is important in avoiding the host defence mechanisms, preventing the death of the invading parasite and allowing multiplication. The 66 kDa rhoptry protein is found on the membrane which surrounds the vacuole.[45] The 21, 27, and 28.5 kDa proteins have also been detected on the tachyzoite dense bodies (Fig. 8.2).[54] Two of them (21, 28.5 kDa) have been found in membrane fractions, possibly located just below the surface[54] similar to the intracellular subsurface proteins (28, 58 kDa).[46] The 28.5 kDa E/S protein and the 28 kDa subsurface protein may be identical.[55] Similarly monoclonal antibody to the 27 kDa E/S protein has recognized the 23 kDa dense body protein, suggesting that these proteins may also be the same.[56] The location of dense body proteins just below the surface may reflect a transient step in their secretion. The 39 kDa E/S protein is also found in the membrane fraction but no function is yet known for this protein.[28] Penetration enhancing factor may contain E/S proteins[54] as well as the 55 and 60 kDa rhoptry proteins.[44] It is possible that the two proteins recognized by immunoblotting as components of circulating antigen (27, 57 kDa)[52] may be related to the 27 and 57 kDa proteins, but as yet there is no evidence to substantiate this. In summary, these new investigations are now providing more detailed information on E/S antigens and their sources within the parasite.

The immune response to the E/S antigens shows a very early IgM reaction to the 97 and 69 kDa proteins at a time when the IgM response to surface antigens is still weak.[57] During the course of infection there is a good response to most E/S antigens. Antisera from latent infection detect two proteins of 28.5 and 108 kDa not detected by acute infection sera.[57]

Antibody to these proteins may act as markers of latent infection, whereas IgM antibody to the 69 and 97 kDa proteins may act as indicators of acute infection. Animals immunized with E/S antigens give a good antibody response, indicating that the antigens are highly immunogenic.[48] The response to the 28.5 kDa antigen is particularly strong.[48] Passive transfer of antibodies to E/S antigens to rats was able to delay the time to death when the animals were challenged with a lethal dose of toxoplasma tachyzoites.[48] Further study of E/S antigens is necessary to determine which are important in eliciting a protective response and which may be considered as candidates for vaccine.

Nucleic acid

An increased knowledge of the nucleic acid of toxoplasma is important for a fundamental understanding of the organism.[58] Equally, valuable information which can be used to improve the diagnosis of toxoplasma infection may result. The nucleotide sequences of genes can be used to compare strains of toxoplasma and also identify relationships between toxoplasma and other protozoa. Two genes which code for α and β tubulin have been investigated in this way.[58] α and β tubulin form structures which are assembled into microtubules, organelles which are involved in cell division, motility, and the maintenance of cell shape. When sequenced, these genes show a close relationship between toxoplasma and other protozoa including *Trypanosoma brucei*.[58] Similarly, comparison of the sequence of the gene which codes for ribosomal RNA showed no differences between three strains of toxoplasma or between toxoplasma and *Hammondia hammondi*.[59]

The gene for the 28 kDa protein present just under the surface of tachyzoites and in the parasitophorous vacuole network (Table 8.1) has been isolated and sequenced.[60] It has been shown to be well conserved in three different strains. Attempts are under way to produce synthetic peptides from this gene for use in diagnostic assays. This approach is also being used to produce recombinant proteins from another gene which codes for a component of circulating antigen, nucleoside triphosphate hydrolase.[61] If these synthetic proteins are useful in diagnostic assays, a reproducible source of pure, inexpensive protein is then available. In the future, genes which code for antigens which elicit a protective immune response, could be used to make synthetic proteins for use as vaccines. Sequencing of two genes, the B1 gene[58] and the P30 surface protein gene[15] has allowed synthetic primers for the PCR to be prepared. The B1 gene is repeated 35 times in the genome, but the protein encoded by the gene has, as yet, not been identified.[58] The P30 gene is present as a single copy which codes for a 34.7 kDa polypeptide which is then modified to P30.[15] High levels of mRNA from this gene are found in toxoplasma, probably to produce the quantity of P30 required by the tachyzoite (3–5% of total protein).[35] DNA probes from purified

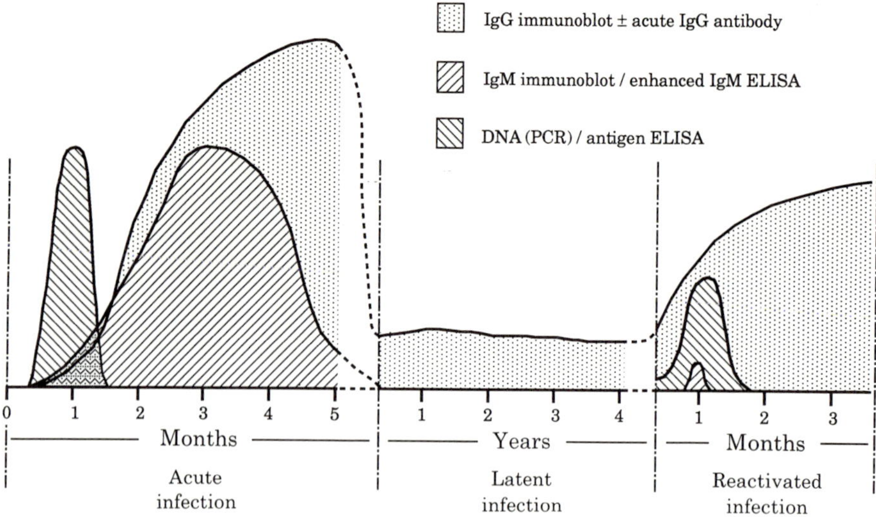

Fig. 8.3   Use of new diagnostic techniques during acute, latent, and reactivated infection.

tachyzoite DNA have been used to identify toxoplasma in mock cultures.[62] Unfortunately the sensitivity of the test was poor ($\approx 10^4$ tachyzoites) and has given way to PCR.

DIAGNOSTIC APPLICATIONS

The new techniques and the knowledge gained from their use have been applied to the diagnosis of toxoplasma infection in an attempt to develop more sensitive and/or specific assays. There is, also, a need for new diagnostic methods which will more accurately detect active toxoplasma infection and differentiate between acute and reactivated infection (Fig. 8.3). Diagnosis can be approached in three ways: the detection of circulating antigen in body fluids, the demonstration of toxoplasma DNA in clinical samples and the evaluation of the antibody response to infection.

Detection of circulating antigen

As the measurement of antibody in some clinical conditions e.g. AIDS, may not be diagnostic, it is probably appropriate to devise and perfect tests which will detect circulating antigen in body fluids.[52,53,63–66] ELISA appears to be the assay method most often used (Chapter 4). An ELISA which used monoclonal antibodies as the capture antibody was not successful.[67] Six monoclonal antibodies were tried, four of which were active against surface antigens and two which reacted with intracellular antigens. The antibodies to intracellular antigens did not trap circulating antigen and the surface antigen monoclonals were less sensitive than a polyclonal antibody fraction.[67] It is likely that the reason for this insensitivity is because the individual monoclonal antibodies could only detect one antigen and even a mixture of four monoclonals did not improve sensitivity. Therefore all subsequent assays have used polyclonal antiserum. Immunoblotting has only been used in two reports.[52,53] In each case it confirmed ELISA results and characterized circulating antigen. It is possibly too cumbersome and time consuming to replace ELISA as the routine assay.

Studies have detected circulating antigen in 20–65 per cent of sera from patients with proven acute toxoplasmosis.[63,64] The antigen is found even in the presence of high levels of specific anti-toxoplasma antibody.[63] Antigen was not detected in sera from patients with latent infection or from seronegative individuals. Detailed follow up of two laboratory acquired infections showed circulating antigen present 18–37 days after infection.[66] In these cases, IgM antibody levels were already high by day 25. However, the IgM titres remained high for a prolonged period whereas circulating antigen was only present for a few days. It appears, therefore, that the presence of circulating antigen is indicative of acute infection, but its absence does not rule out acute infection. Negative results in cases of proven acute disease may be due to either very low levels of antigen or to the short term appearance of antigen in serum. In congenital infection, circulating antigen has been found in the serum of 5 of 12 neonates with congenital toxoplasmosis and in the CSF of 8 of these 12 neonates.[63,64] Amniotic fluid taken at the time of Caesarean section from two cases of congenital disease also contained circulating antigen.[63] Although the numbers examined have been small, antigen detection especially in the CSF may be a valuable addition to the diagnosis of congenital disease.

The detection of circulating antigen in immunocompromised patients could be a valuable aid to diagnosis. Eight of 23 bone marrow or kidney transplant patients with clinical or serological evidence of reactivation showed the presence of circulating antigen either before or at the time of a rise in antibody levels.[65] Antigen levels then fell as antibody levels rose, possibly due to the formation of immune complexes. However, it is difficult to link the presence of circulating antigen with the clinical course of the infection. Of the eight patients with antigenaemia, two had cerebral toxoplasmosis, while the other six remained asymptomatic. In two patients antigen was present before a clinical diagnosis was established. One of these patients died. He developed antigenaemia within 32 days of the transplant and his antibody levels remained stable. In the other patient, antigen was detected transiently 6 months after transplant.

In another study 16 per cent of 125 AIDS patients had antigen detected in their sera for periods of up to one year.[52] Sera from all 125 patients were collected routinely and most patients had no signs of active toxoplasmosis. However, thirteen of the twenty patients who were antigen positive showed clinical symptoms of toxoplasmosis. Therefore the presence of antigenaemia can be helpful in establishing a diagnosis of active toxoplasma infection, especially where there is clinical evidence of disease. Five of those with antigenaemia, two of which had neurological disease, were tested for antigen in the CSF, but were all negative.[52] Another 34 CSFs from those without antigenaemia were all negative for antigen, but there was no clinical history given. In this study, examination of CSF was not helpful. Urine has also been examined for antigen and in experimental infection in mice was shown to be as useful as serum.[53] Antigen was detected in the pooled urine of eight mice five days after infection with RH strain tachyzoites. Serum was positive a day earlier, but by the sixth day, antigen levels were similar in both. Preliminary studies on urine of 20 AIDS patients with signs of toxoplasma encephalitis, showed antigen in the urine of five, but in only three sera.

In conclusion, therefore, detection of circulating antigen in both serum and CSF of congenitally infected infants may be useful.[63,64] In immunocompromised patients serum antigen was sometimes present during reactivated infection (Fig. 8.3), but was absent from the CSF even of patients with toxoplasma encephalitis.[52,65] However, antigen may appear in the urine of these patients.[53] The appearance of circulating antigen suggests active toxoplasma infection, but its relationship to the clinical outcome remains to be evaluated.

Demonstration of nucleic acid

The direct detection of toxoplasma parasites in clinical samples may be helpful especially in situations where other diagnostic techniques are not conclusive, e.g. in toxoplasma encephalitis[3] and congenital infection.[2] The polymerase chain reaction (PCR) is able to detect toxoplasma nucleic acid from parasites quickly.[68] So far, two genes have been used for the amplification; the B1 gene for which no gene product is known[14] and the P30 surface protein gene.[69] Multiple copies of the B1 gene are present in the genome, thereby enhancing the sensitivity of the PCR.[14] Preliminary studies with the B1 gene used pure cultures of toxoplasma and mock clinical samples, and could detect 1–10 organisms.[14] Twenty-one different strains were tested and all proved to be detected by the B1-PCR.[68] It is, therefore, likely that the B1 gene is well conserved in many strains and may be useful for examination of human clinical samples. No reaction was obtained with other organisms tested, e.g. Sarcocystis, Candida and Cryptococcus.[14] The P30 gene has also been used successfully to detect at least seven strains of toxoplasma and gives no false positive results with other related organisms.[69]

Toxoplasma has been demonstrated using PCR in brain biopsies from two AIDS patients in which there were clinical, serological, and CT scan signs of cerebral toxoplasmosis.[22] Biopsies were taken because the patients did not respond well to anti-toxoplasma therapy, but biopsy is not always feasible. PCR is so sensitive it should detect toxoplasma DNA in one cyst, which may indicate latent rather than active infection. It will not differentiate cyst from tachyzoite DNA. Therefore it is necessary to evaluate such results with care, taking into account all the other clinical and laboratory findings. Examination of CSF would be easier and perhaps more relevant to active disease (Fig. 8.3). Herpes simplex virus (HSV) has been shown by PCR in the CSF of 5/6 patients with suspected HSV encephalitis.[70] It is known to be difficult to detect HSV in the CSF of such patients by conventional methods, as is the case with cerebral toxoplasmosis.

Both the B1 and P30 genes have been used in an attempt to diagnose fetal infection during pregnancy.[68,71,72] Abortion products of five ovine cases of suspected toxoplasmosis were examined and toxoplasma was shown by PCR in both fetal tissue and/or placenta in all five.[71] Infection was present in all tissues in three cases, in 2/3 fetal tissues in a fourth, but in only the placenta of the fifth. It may be that infection of the placenta is not always transmitted to the fetus. Infection of amniotic fluid has been demonstrated in cases of suspected human infection.[68,72] In one study 7/40 fluids were positive by PCR whereas mouse inoculation and tissue culture only detected four and five cases respectively.[72] The other study followed prospectively 43 cases of primary toxoplasmosis in pregnancy.[68] PCR detected toxoplasma DNA in five amniotic fluids. Mouse inoculation detected only three of these, and tissue culture only two. An additional five known cases were added and of these ten amniotic fluids, eight were positive by PCR, seven by mouse inoculation, and four by tissue culture. In two instances, mouse inoculation failed to detect toxoplasma when PCR was positive. On the other hand, PCR failed on one occasion to detect infection when mouse inoculation was positive. The tenth case was only diagnosed by inoculation of fetal blood in mice. Fetal blood inoculation detected seven of the ten cases, but PCR was not carried out on these bloods. In only three cases was IgM antibody detected. Therefore PCR shows promise for the rapid detection of fetal infection in amniotic fluid. The negative PCR findings have two possible explanations.[68] Both samples were heavily contaminated with blood which was found to inhibit the PCR. Also only 10 per cent of sample volume for inoculation was used in the PCR; therefore, if very small numbers of parasites were present, the sample for PCR may not have contained any. There has been one report of the detection of toxoplasma DNA in the CSF of a congenitally infected child[73] and CSF may be more widely examined in the future.

PCR appears to be as sensitive as mouse inoculation, but much quicker. It may also be more sensitive than tissue culture. However, its very sensitivity may create problems, since it will detect very small amounts of DNA from latent as well as active infections. In these situations, it may be necessary to choose only those samples which would normally not contain tachyzoites such as blood, CSF, urine, and amniotic fluid. A suitable sample for PCR may be buffy coat because it is possible to detect toxplasma DNA in the buffy coat of experimentally infected mice.[74] PCR shows great promise but is presently an expensive, labour intensive test. It is to be hoped that future developments will allow PCR to become a routine assay.

Detection of toxoplasma antibody

Enhanced ELISAs

Attempts have been made to improve the routine ELISA (Chapter 4) by using either purified P30 surface protein[36] or specific monoclonal antibody to surface proteins.[75–77] P30 was purified from a whole toxoplasma extract with a recovery rate of 5 per cent.[36] It was then used in an ELISA with anti-P30 monoclonal antibody to detect both IgG and IgM antibody to toxoplasma. All 47 acute sera were positive for IgM antibody and all latent sera were positive for IgG antibody. It appears, therefore, that the antibody response to P30 can be used in a diagnostic assay. However, the production of sufficient purified P30 for use in a routine assay is difficult. Purification can be avoided by using crude extract and anti-P30 monoclonal antibody in an IgM capture assay.[75] This assay also discriminates between acute and latent infection, with all 57 acute sera showing positive results and 19/24 latent sera showing negative results. Of the five discrepant results, four were very close to the cut-off value and only one was a high positive. Lack of clinical data meant that acute infection could not be ruled out. Twenty-eight sera with either anti-nuclear factor or rheumatoid factor were also negative.[75] Another IgM capture ELISA using monoclonal antibody to a second surface protein, the 35 kDa protein, has also proved useful in detecting IgM antibody in acute infection.[76,77] As with the P30 assay, no false positives were obtained in sera with anti-nuclear or rheumatoid factor. The use of monoclonal antibody in these assays has led to less non-specific reactions, lower backgrounds, and possibly greater sensitivity.[76,77]

The use of sensitive assays for IgM antibody has meant that IgM can be detected for many months after infection.[1] Therefore, the presence of IgM antibody does not always indicate an early acute infection. IgA antibody may also indicate acute infection. An IgA antibody capture ELISA similar to that for IgM antibody using anti-P30 monoclonal antibody has been set up to test this theory.[78] Samples from all 56 patients with acute infection were positive for IgA antibody while those from 26 patients with latent infection were negative. Levels of IgA and IgM antibody rose in parallel early in the course of infection and, in most cases, disappeared after 3 to 8 months. However, a few samples from those with latent infection, all taken at least one year after infection were still positive for IgM, but not IgA. So, IgA antibody to P30 may be a more reliable marker of acute infection than IgM antibody. IgA to P30 also proved to be useful in the detection of congenital infection.[78] Cord bloods from eight congenitally infected babies contained IgA, whereas only three had IgM. Equally as important, 18 uninfected infants whose mothers were infected in pregnancy showed no IgA, but five had IgM at birth possibly due to contamination by maternal blood. All of the latter were followed up and became seronegative by 6 months. Since the presence of IgM antibody alone should in some cases be evaluated with care, the detection of both IgM and IgA antibodies to P30 may improve the diagnosis of toxoplasmosis especially in congenital infection.

An adaptation of the ELISA, the enzyme-linked immunofiltration assay (ELIFA) has been used to identify differences in specific antibody classes in sera of potentially congenitally infected neonates compared to their mothers.[79] In each test, three sera (the mother at delivery, neonatal cord blood, and a sample from the neonate taken several days or weeks later) are run in parallel in an electrophoresis step with toxoplasma antigen on cellulose acetate membrane. Precipitation lines are formed between specific antibodies in the sera and the toxoplasma antigens. The type and number of antibody classes present are detected by anti-IgG, -IgM, -IgA, or -IgE enzyme labelled conjugates which are filtered through the membrane. Finally, addition of a peroxidase substrate results in a coloured precipitate when a particular antibody class is present. Congenital infection is considered to be present when specific IgM, IgA, or IgE are detected in the precipitation lines with neonatal sera, or when the specific IgG precipitation profiles of mother and neonate are different. Using this assay, 87 per cent of 150 congenital infections were identified.[79] This is better than the 75 per cent which are detected by specific IgM assays (Chapter 6). A proper assessment of ELIFA by other laboratories may show that this technique is of use in the diagnosis of congenital infection.

Acute IgG antibody tests

Formalin and acetone have been used to fix tachyzoites in an attempt to produce a test which will detect anti-toxoplasma IgG antibody which is specific for the active phase of infection[29,30,31] (Fig. 8.3). When sera from more than 20 patients with acute toxoplasmosis were examined in an agglutination assay (Chapter 4) with formalin- and acetone-fixed tachyzoites, high titres were obtained in both.[29] However, in sera from those patients taken in the latent phase, titres against the acetone-fixed parasites were negative whereas titres were still present with formalin-fixed parasites (Table 8.2). These results confirmed those in an earlier report, which suggested that the use of acetone-fixed parasites might differentiate between acute and latent infection.[80] Mouse antibody to acetone-fixed tachyzoites reacted with surface antigens on tachyzoites, which are present during acute or reactivated infection, but not with surface antigens of bradyzoites, which are present in tissue cysts in the latent phase.[29]

The agglutination assay was also applied to the diagnosis of toxoplasma

Table 8.2 Positive results from studies of acute IgG antibody

Study groups		IgM ELISA (%)		Formalin-fixed tachyzoites (%)		Acetone-fixed tachyzoites (%)	
Normal[31]	2 months	7–10/13	(54–77)	10/13	(77)	12/13	(92)
	3–4 months	7–8/8	(88–100)	7/8	(88)	3/8	(38)
	5 months	3/11	(27)	4/11	(36)	1/11	(9)
Normal[29]	Early	20/20	(100)	20/20	(100)	20/20	(100)
	>18/12 later	–		20/20	(100)	0/20	(0)
AIDS[30] patients	Brain lesion	0/31	(0)	20/31	(65)	26/31	(84)
	Brain biopsy proven toxoplasma	0/43	(0)	22/43	(51)	26/43	(61)
	No CNS signs	–		2/58	(3.5)	2/58	(3.5)

encephalitis in AIDS patients[30] (Table 8.2). Two groups with either multiple brain lesions (31 patients) or with biopsy proven toxoplasmosis (43 patients), had significantly higher titres against both acetone- and formalin-fixed tachyzoites than a control group of 58 which had serological evidence of toxoplasma infection but no neurological disease. Only 6.9 per cent of the control group showed a significant result when both test results were combined, but 70–84 per cent of the groups with neurological disease were positive. In contrast, the ratio of results from the formalin fixed agglutination test and the dye test, which had previously been suggested as a useful indicator of active infection in AIDS patients,[81] was of less value on this occasion with only 33–39 per cent of patients with neurological toxoplasmosis showing a significant result. This technique may prove to be a valuable tool in the difficult area of diagnosing toxoplasma encephalitis in AIDS patients.

This approach has been further refined by the development of an ELISA which uses purified acute stage antigens found in acetone-fixed tachyzoites[31] (AC-ELISA). Rabbits were inoculated with acetone-fixed tachyzoites, the IgG antibody extracted and used to purify the acute stage antigens. PAGE analysis of the antigens detected three bands, one of 52 kDa and two of about 6 kDa. Of 13 sera taken within 2 months of the onset of illness, 92 per cent were positive in the AC-ELISA, agreeing with other serological tests such as the dye test and IgM ELISA. However, by 4 months only 38 per cent of sera were positive in the AC-ELISA and after five months only 9 per cent were still positive while 27 per cent still showed IgM antibody (Table 8.2). The AC-ELISA may prove to be a good discriminator of early acute infection (Fig. 8.3).

Immunoblotting and radioimmunoprecipitation
Studying the immune response to toxoplasma infection by immunoblotting techniques may also help identify the antigens which are useful in a

Table 8.3 Common bands detected by immunoblotting of toxoplasma antigens[13,21,82–89]

Bands (kDa)	Acute			Latent			Congenital babies		
	IgA	IgM	IgG	IgA	IgM	IgG	IgA	IgM	IgG
4–6	++	+++	+	+	–	+	–	–	+
20–22	+	+	+	–	–	+	–	–	–
30	+++	+	+	++	–	++	++	+	+
35	++	++	+++	++	–	+++	+	++	+++
115	–	+	++	–	–	++	–	++	++

+++, Strongly stained; ++, moderately stained; +, lightly stained.

diagnostic sense at different stages of infection.[13,21,82–87] The problem with this approach is the disparity of results from different research groups. These may have several explanations. Methods of preparing toxoplasma antigen have varied;[13,85,87] the percentage of acrylamide in the separating gel has differed;[13,84,85,87] and measurement of the size of individual proteins can be difficult. However, there is general agreement on the most commonly detected bands (Table 8.3). In sera from acutely infected adults, a relatively consistent pattern of bands was shown by IgM antibody.[13,21,82,84] The most prominent antibody response was to a 4–6 kDa antigen. This antigen appears to elicit a very early antibody response and in one patient was demonstrated by IgM antibody in an early serum which was still negative in the routine serological tests for IgM.[13] A strong IgM response is also detected to a 35 kDa antigen.[13,83,84] In contrast, more than 20 antigens, ranging from 4–6 to 150 kDa are seen on blots developed with IgG antibody in acute phase sera[13,21,80–85,87] (Fig. 8.1.). The number and size of these bands vary considerably from patient to patient, which may be due to individual patient reactions or to infection with different strains, but some are common. Most of the bands recognized by IgM antibody are also recognized by IgG antibody.[21] However, the relative intensity of staining of some bands differ. The strongest IgG reaction in early acute sera is to the 35 kDa antigen with a weaker response to the 4–6 kDa antigen, which is the opposite of the IgM response.[13,84] With IgA, on the other hand, both the 4–6 kDa and 35 kDa antigens are equally evident, but the most intense reaction is to the 30 kDa antigen, which is also detected by IgM and IgG antibody.[86] In general, more bands are revealed by IgA than by IgM, but fewer than with IgG.

As many bands are seen on IgG blots in the latent phase as the acute phase, but relatively less intense in many cases; more than 20 bands ranging from 4–6 to 150 kDa in size (Fig. 8.1). The earliest studies suggested that the acute and latent phases of infection could be differentiated by the presence or absence of IgG antibody to the 4–6 kDa antigen.[82] IgG to 4–6 kDa was detected in all sera from acute infection, but rarely if at all in latent phase sera. However, further studies have shown that, while IgG antibody to 4–6 kDa is more common and intense in acute phase sera, it is also present in many latent phase sera and cannot be used as a definite acute phase marker.[84,87] One report found no bands less than 27 kDa,[85] but this study was the only one to use tissue culture antigen, while all other work was done on peritoneal exudate antigen. Our own studies suggested that IgG antibody to a doublet of bands 22–25 kDa, which were found in 10/10 acute phase sera but in 0/10 latent phase sera, could be useful as an acute stage marker.[87] Other studies had not identified this[13,85] (Table 8.3). Because IgM is usually detected only in the acute stage of infection, no specific bands were seen in blots developed with IgM antibody from latent phase sera.[84] IgA antibody is also considered as being present only in the acute stage, but one study found IgA specific bands in sera from five cases of latent infection, where routine serology for IgA and IgM was negative.[86] The bands were the same as those detected by IgA in acute phase sera, i.e. 4–6, 30, 35 kDa. In general there are few differences on IgG immunoblots between acute and latent phase sera, except in the intensity of bands which appears to correlate with the level of dye test titres. As dye test titres rise, the intensity of bands, especially the 35 kDa band, increases, then decreases as the dye test titres fall.[13] Therefore, IgG immunoblots of single sera will not help to differentiate the acute and latent phases of infection.

Comparisons of the patterns obtained with sera of congenitally infected babies and sera of their mothers have revealed some distinct differences.[82,88,89] Most striking is the absence of the 4–6 kDa band in IgM blots of the babies. In one group of patients, the 4–6 kDa band was present in 7/7 acutely infected mothers but in only 2/12 congenitally infected babies, where the band was faint.[89] This 4–6 kDa antigen was also not detected by IgA antibody but was found on some IgG blots.[82,86] Babies' immune systems may not respond as well as those of adults to this antigen, perhaps because it is largely carbohydrate. The most consistent finding in IgG blots was a strong band of 35 kDa, and also one at 115 kDa.[89] The 35 kDa band was always stronger than any other band. IgA immunoblots of congenitally infected infants showed bands of 30 and 35 kDa only.[86] In some cases, bands were found in both IgM and IgG blots of sera from the babies which were not present in blots from their mothers' sera.[88,89] These bands ranged from 21–250 kDa in size, but none were consistently present in all the sera which showed differences. In 21 control pairs of sera from mothers acutely infected in pregnancy and their non-infected babies, only one pair showed a difference.[88] These results extend those from the ELIFA tests,[79] giving a further indication that comparison of mother and baby sera can be diagnostic of congenital infection.

It is known that natural antibodies, which give rise to false positive results in some serological tests, are present in the sera of almost all adults and children older than 3 months[90] (Chapter 4). In immunoblotting, bands are frequently detected in sera from healthy seronegative controls[13,90] (Fig. 8.1). Sera from 15 healthy seronegative adults all demonstrated bands of 10–125 kDa which varied in number, molecular weight and staining intensity.[90] No IgG band was detected in all sera, but with IgM antibody the results were more consistent with two bands (80–85, 105 kDa) present in all 15 sera and a 115 kDa band present in 13/15 sera. No reaction was seen with the 4–6 kDa band. Sera from 29 adults with possible acute toxoplasmosis but negative serology revealed similar results.[90] After seroconversion, some of these bands, e.g. 115 kDa, increased greatly in staining intensity and appeared to be a specific response to toxoplasma infection.[13] However, the presence of bands corresponding to natural antibodies does not interfere with the interpretation of blots from acute or latent infection sera.

The immune response to the excretory/secretory antigens may also be of diagnostic value.[57] For the study of this response it was necessary to use radioimmunoprecipitation (RIP) instead of immunoblotting, because preparations of E/S antigens contained large amounts of serum proteins which interfered in immunoblotting. A definite pattern of antibody response to different E/S antigens could be seen as infection progressed. During the first month antibodies to 69 and 97 kDa antigens were dominant in 10/10 sera tested. This was followed during the next two months by the appearance of bands of 34, 39, and 86 kDa. After a year, antibody to more than 20 antigens was detected in 103 sera, with antibody to 28.5, 69, 97, and 108 kDa antigens present in all sera. It is possible, therefore, that the anti-97 kDa antibody may be a marker of acute infection and antibodies to the 28.5 and 108 kDa proteins, markers of latent infection.[57] Congenitally infected babies may have IgM antibody to the 97 kDa antigen, if maternal infection occurred late in pregnancy. It is probably useful to compare the patterns obtained from both mother and baby sera. The detection of antibodies to specific E/S antigens at different stages of infection looks promising as a tool to differentiate acute and latent infection, but these findings still have to be confirmed by further studies.

SUMMARY

New techniques in laboratory methods have led to increased knowledge of toxoplasma and to improved diagnostic assays. Most routine assays used crude antigens and little was known of their composition. Radioimmuno-precipitation and immunoblotting have allowed characterization of many toxoplasma antigens and improved understanding of the immune response to these antigens. Monoclonal antibodies have overcome the problems of host cell contamination and have localized specific antigens in the tachyzoite and the infected cell. Each stage of development has unique antigens and many which are common. Different strains of toxoplasma have minor variations; these do not give rise to significant differences in the immune response.

Three groups of antigens are identified on the tachyzoite. Five major antigens (43, 35, 30, 22, and 4–6 kDa) are exposed on the surface and they are among the first to stimulate the immune system. Some elicit antibodies which are partially protective in animal experiments and thus may be candidates for vaccines. Intracellular antigens have been localized on the rhoptries, the dense bodies and just under the surface. Haemagglutination and latex agglutination antigen are also composed of intracellular antigens which have been partially characterized. Excretory/secretory antigens (at least nine), detected in cell free culture, are important in host cell entry and in the parasitophorous vacuole, which helps the parasite avoid host cell defences. Both surface and intracellular antigens are also shed into the parasitophorous vacuole during infection.

New developments have also been applied to the study of toxoplasma nucleic acid. Comparisons of gene sequences have shown no differences between strains and a close relationship with other protozoa. Sequences from two genes have been used to develop the polymerase chain reaction, which detects very small amounts of toxoplasma DNA. The production of synthetic peptides from specific sequences could provide a cheap, reproducible source of antigens for diagnostic assays.

Knowledge of toxoplasma antigens and nucleic acid has been applied to diagnosis of infection. Circulating antigen is detected in serum, CSF, amniotic fluid, and urine of patients with acute, congenital, or reactivated infection. The presence of antigen indicates an active infection, but does not correlate with severe illness and its absence does not rule out active infection. The use of monoclonal antibodies to specific surface antigens (P30, P35) in routine ELISAs has increased the specificity of these tests; the detection of both IgM and IgA antibody is strong evidence of acute infection. Acetone fixation of tachyzoites exposes antigens which are specific to active infection and, when used in ELISA or agglutination assays, can differentiate between active and latent infection. Study of the immune response to individual antigens by immunoblotting has only a limited value in diagnosis. A strong IgM response to the 4–6 kDa antigen is seen during acute infection, but the IgG response shows few variations in acute and latent infection. Comparison of antibody profiles from mothers and their congenitally infected babies reveal differences which are not present in comparisons with uninfected babies. Antibody to the 97 kDa E/S antigen, detected by RIP, is found early during acute infection. The polymerase chain reaction has identified toxoplasma nucleic acid in brain biopsies of AIDS patients, and from amniotic fluids and CSF in congenital infection. It, like other new techniques described shows considerable promise in solving the diagnostic problems presented by congenital infection and infection in those with immunocompromise.

REFERENCES

1 Joss, A. W. L., Skinner, L. J., Moir, I. L., Chatterton, J. M. W., Williams, H., and Ho-Yen, D. O. (1989). Biotin-labelled antigen screening test for toxoplasma IgM antibody. *J. Clin. Pathol.*, **42**, 206–9.

2 Skinner, L. J., Chatterton, J. M. W., Joss, A. W. L., Moir, I. L., and Ho-Yen, D. O. (1989). The use of an IgM immunosorbent agglutination assay to diagnose congenital toxoplasmosis. *J. Med. Microbiol.*, **28**, 125–8.

3 Holliman, R. E. (1988). Toxoplasmosis and the acquired immune deficiency syndrome. *J. Infect.*, **16**, 121–8.

4 Derouin, F., Mazeron, M. C., and Garin, Y. J. F. (1987). Comparative study of tissue culture and mouse inoculation methods for demonstration of *Toxoplasma gondii*. *J. Clin. Microbiol.*, **25**, 1597–600.

5 Derouin, F., Thulliez, P., Candolfi, E., Daffos, F., and Forestier, F. (1988). Early prenatal diagnosis of congenital toxoplasmosis using amniotic fluid samples and tissue culture. *Eur. J. Clin. Microbiol. Infect. Dis.*, **7**, 423–5.

6 Johnson, A. M. (1985). The antigenic structure of *Toxoplasma gondii*: a review. *Pathology*, **17**, 9–19.

7 Laemmli, U. K. (1970). Cleavage of structural proteins during the assembly of the head of bacteriophage T4. *Nature*, **227**, 680–5.

8 Anders, R. F., Howard, R. J., and Mitchell, G. F. (1982). Parasite antigens and methods of analysis. In *Immunology of parasitic infections*. (2nd edn). (ed. S. Cohen and K. S. Warren), pp. 28–73. Blackwell, Oxford.

9 Kessler, S. W. (1975). Rapid isolation of antigens from cells with a staphylococcal protein A-antibody adsorbent: parameters of the interaction of antibody–antigen complexes with protein A. *J. Immunol.*, **115**, 1617–24.

10 Towbin, H., Staehelin, T., and Gordon, J. (1979). Electrophoretic transfer of proteins from polyacrylamide gels to nitrocellulose sheets: procedure and some applications. *Proc. Natl. Acad. Sci. USA*, **76**, 4350–4.

11 Kohler, G. and Milstein, C. (1975). Continuous cultures of fused cells secreting antibody of predefined specificity. *Nature*, **256**, 495–7.

12 Couvreur, G., Sadak, A., Fortier, B., and Dubremetz, J. F. (1988). Surface antigens of *Toxoplasma gondii*. *Parasitology*, **97**, 1–10.

13 Potasman, I., Araujo, F. G., Desmonts, G., and Remington, J. S. (1986). Analysis of *Toxoplasma gondii* antigens recognized by human sera obtained before and after acute infection. *J. Infect. Dis.*, **154**, 650–7.

14 Burg, J. L., Grover, C. M., Pouletty, P., and Boothroyd, J. C. (1989). Direct and sensitive detection of a pathogenic protozoan, *Toxoplasma gondii*, by polymerase chain reaction. *J. Clin. Microbiol.*, **27**, 1787–92.

15 Burg, J. L., Perelman, D., Kasper, L. H., Ware, P. L., and Boothroyd, J. C. (1988). Molecular analysis of the gene encoding the major surface antigen of *Toxoplasma gondii*. *J. Immunol.*, **141**, 3584–91.

16 Saiki, R. K., Scharf, S., Faloona, F., Mullis, K. B., Horn, G. T., Erlich, H. A. *et al.* (1985). Enzymatic amplification of β-globulin genomic sequences and restriction site analysis for diagnosis of sickle cell anemia. *Science*, **230**, 1350–4.

17 Walker, A. W. (1978). Electrophoresis. In *Scientific foundations of clinical biochemistry*, Vol. 1. (ed. D. L. Williams, R. F. Nunn, and V. Marks) pp. 258–82. Heinemann, London.

18 Johansson, K. -E. (1988). Separation of antigens by analytical gel electrophoresis. In *Handbook of immunoblotting of proteins*, Vol. 1 *Technical descriptions*, (ed. O. J. Bjerrum and N. H. H. Heegaard), pp. 31–50. CRC Press, Boca Raton.

19 Handman, E. and Remington, J. S. (1980). Serological and immunochemical characterization of monoclonal antibodies to *Toxoplasma gondii*. *Immunology*, **40**, 579–88.

20 Handman, E., Goding, J. W., and Remington, J. S. (1980). Detection and characterization of membrane antigens of *Toxoplasma gondii*. *J. Immunol.*, **124**, 2578–83.

21 Sharma, S. D., Mullenax, J., Araujo, F. G., Erlich, H. A., and Remington, J. S. (1983). Western blot analysis of the antigens of *Toxoplasma gondii* recognized by human IgM and IgG antibodies. *J. Immunol.*, **131**, 977–83.

22 Holliman, R. E., Johnson, J. D., and Savva, D. (1990). Diagnosis of cerebral toxoplasmosis in association with AIDS using the polymerase chain reaction. *Scand. J. Infect. Dis.*, **22**, 243–4.

23 Kasper, L. H. (1989). Identification of stage-specific antigens of *Toxoplasma gondii*. *Infect. Immun.*, **57**, 668–72.

24 Ware, P. L., and Kasper, L. H. (1987). Strain-specific antigens of *Toxoplasma gondii*. *Infect. Immun.*, **55**, 778–83.

25 De La Cruz, A. A., Dreesen, D. W., and Evans, D. L. (1989). Western blot analyses and LD_{50} determinations of *Toxoplasma gondii* isolates. *Vet. Immunol. Immunopathol.*, **23**, 355–64.

26 Jacobs, L., Remington, J. S., and Melton, M. L. (1960). The resistance of the encysted form of *Toxoplasma gondii*. *J. Parasitol.*, **46**, 11–21.

27 Lunde, M. N., and Jacobs, L. (1983). Antigenic differences between endozoites and cystozoites of *Toxoplasma gondii*. *J. Parasitol.*, **69**, 806–8.

28 Darcy, F., Charif, H., Caron, D., Deslee, D., Pierce, R. J., Cesbron-Delauw, M. F. *et al.* (1990). Identification and biochemical characterization of antigens of tachyzoites and bradyzoites of *Toxoplasma gondii* with cross-reactive epitopes. *Parasitol. Res.*, **76**, 473–8.

29 Suzuki, Y., Thulliez, P., Desmonts, G., and Remington, J. S. (1988). Antigen(s) responsible for immunoglobulin G responses specific for the acute stage of toxoplasma infection in humans. *J. Clin. Microbiol.*, **26**, 901–5.

30 Suzuki, Y., Israelski, D. M., Danneman, B. R., Stepick-Biek, P., Thulliez, P., and Remington, J. S. (1988). Diagnosis of toxoplasmic encephalitis in patients with acquired immunodeficiency syndrome by using a new serologic method. *J. Clin. Microbiol.*, **26**, 2541–3.

31 Suzuki, Y., Thulliez, P., and Remington, J. S. (1990). Use of acute-stage-specific antigens of *Toxoplasma gondii* for serodiagnosis of acute toxoplasmosis. *J. Clin. Microbiol.*, **28**, 1734–8.

32 Sabin, A. B. (1941). Toxoplasmic encephalitis in children. *JAMA*, **116**, 801–7.

33 Weiss, L. M., Udem, S. A., Tanowitz, H., and Wittner, H. (1988). Western blot analysis of the antibody response of patients with AIDS and toxoplasma encephalitis: antigenic diversity among toxoplasma strains. *J. Infect. Dis.*, **157**, 7–13.

34 Darde, M. L. Bouteille, B., and Pestre-Alexandre, M. (1988). Isoenzymic characterization of seven strains of *Toxoplasma gondii* by isoelectricfocussing in polyacrylamide gels. *Am. J. Trop. Med. Hyg.*, **39**, 551–8.

35 Kasper, L. H., Crabb, J. H., and Pfefferkorn, E. R. (1983). Purification of a major membrane protein of *Toxoplasma gondii* by immunoabsorption with a monoclonal antibody. *J. Immunol.*, **130**, 2407–12.

36 Santoro, F., Afchain, D., Pierce, R., Cesbron, J. Y., Ovlaque, G., and Capron, A. (1985). Serodiagnosis of toxoplasma infection using a purified parasite protein (P30). *Clin. Exp. Immunol.*, **62**, 262–9.

37 Johnson, A. M., McDonald, P. J., and Neoh, S. H. (1983). Monoclonal antibodies to toxoplasma cell membrane surface antigens protect mice from toxoplasmosis. *J. Protozool.*, **30**, 351–6.

38 Sethi, K. K., Endo, T., and Brandis, H. (1981). *Toxoplasma gondii* trophozoites precoated with specific monoclonal antibodies cannot survive within normal murine macrophages. *Immunol. Lett.*, **2**, 343–6.

39 Hauser, W. E., Jr and Remington, J. S. (1981). Effect of monoclonal antibodies on phagocytosis and killing of *Toxoplasma gondii* by normal macrophages. *Infect. Immun.*, **32**, 637–40.

40   Araujo, F. G. and Remington, J. S. (1984). Partially purified antigen preparations of *Toxoplasma gondii* protect against lethal infection in mice. *Infect. Immun.*, **45**, 122–6.

41   Kasper, L. H., Currie, K. M., and Bradley, M. S. (1985). An unexpected response to vaccination with a purified major membrane tachyzoite antigen (P30) of *Toxoplasma gondii*. *J. Immunol.*, **134**, 3426–31.

42   Johnson, A. M., McDonald, P. J., and Neoh, S. H. (1983). Molecular weight analysis of soluble antigens from *Toxoplasma gondii*. *J. Parasitol.*, **69**, 459–64.

43   Sadak, A., Taghy, Z., Fortier, B., and Dubremetz, J. F. (1988). Characterization of a family of rhoptry proteins of *Toxoplasma gondii*. *Mol. Biochem. Parasitol.*, **29**, 203–11.

44   Schwartzman, J. D., and Krug, E. C. (1989). *Toxoplasma gondii*: characterization of monoclonal antibodies that recognize rhoptries. *Exp. Parasitol.*, **68**, 74–82.

45   Kimata, I. and Tanabe, K. (1987). Secretion by *Toxoplasma gondii* of an antigen that appears to become associated with the parasitophorous vacuole membrane upon invasion of the host cell. *J. Cell. Sci.*, **88**, 231–9.

46   Sibley, L. D. and Sharma, S. D. (1987). Ultrastructural localization of an intracellular toxoplasma protein that induces protection in mice. *Infect. Immun.*, **55**, 2137–41.

47   Cesbron-Delauw, M. F., Guy, B., Torpier, G., Pierce, R. J., Lenzen, G., Cesbron, J. Y. *et al.* (1989). Molecular characterization of a 23-kilodalton major antigen secreted by *Toxoplasma gondii*. *Proc. Natl. Acad. Sci. USA*, **86**, 7537–41.

48   Darcy, F., Deslee, D., Santoro, F., Charif, H., Auriault, C., Decoster, A. *et al.* (1988). Induction of a protective antibody-dependent response against toxoplasmosis by *in vitro* excreted/secreted antigens from tachyzoites of *Toxoplasma gondii*. *Parasite Immunol.*, **10**, 553–67.

49   Chumpitazi, B., Ambroise-Thomas, P., Cagnard, M., and Autheman, J. M. (1987). Isolation and characterization of toxoplasma exo-antigens from *in vitro* culture in MRC5 and Vero cells. *Int. J. Parasitol.*, **17**, 829–34.

50   Hughes, H. P. A. (1981). Characterization of the circulating antigen of *Toxoplasma gondii*. *Immunol. Lett.*, **3**, 99–102.

51   Ise, Y., Iida, T., Sato, K., Suzuki, T., Shimada, K., and Nishioka, K. (1985). Detection of circulating antigens in sera of rabbits infected with *Toxoplasma gondii*. *Infect. Immun.*, **48**, 269–72.

52   Hassl, A., Aspöck, H., and Flamm, H. (1988). Circulating antigen of *Toxoplasma gondii* in patients with AIDS: significance of detection and structural properties. *Zentralbl. Bakteriol., Mikrobiol. Hyg.*, **A270**, 302–9.

53   Huskinson, J., Stepick-Biek, P., and Remington, J. S. (1989). Detection of antigens in urine during acute toxoplasmosis. *J. Clin. Microbiol.*, **27**, 1099–101.

54   Charif, H., Darcy, F., Torpier, G., Cesbron-Delauw, M. F., and Capron, A. (1990). *Toxoplasma gondii*: characterization and localization of antigens secreted from tachyzoites. *Exp. Parasitol.*, **71**, 114–24.

55   Sibley, L. D., Krahenbuhl, J. L., Adams, G. M. W., and Weidner, E. (1986). Toxoplasma modifies macrophage phagosomes by secretion of a vesicular network rich in surface proteins. *J. Cell. Biol.*, **103**, 867–74.

56   Darcy, F., Torpier, G., Cesbron-Delauw, M. F., Decoster, A., Ridel, P. R., Duquesne, V. *et al.* (1989). Antigènes de *Toxoplasma gondii* d'intérêt diagnostique et immunoprophylactique potentiel: nouvelles stratégies d'identification. *Ann. Biol. Clin.*, **47**, 451–7.

57   Descoster, A., Darcy, F., and Capron, A. (1988). Recognition of *Toxoplasma gondii* excreted and secreted antigens by human sera from acquired and congenital toxoplasmosis: identification of markers of acute and chronic infection. *Clin. Exp. Immunol.*, **73**, 376–82.

58   Boothroyd, J. C., Burg, J. L., Nagel, S. D., and Perelman, D. (1988). Molecular approaches to the study of *Toxoplasma gondii*. In *Contemporary issues in infectious diseases. Parasitic infections*. (ed. J. H. Leech, M. A. Sande, and R. K. Root), pp. 259–69. Churchill Livingstone, New York.

59 Johnson, A. M., Illana, S., Dubey, J. P., and Dame, J. B. (1987). *Toxoplasma gondii* and *Hammondia hammondi*: DNA comparison using cloned rRNA gene probes. *Exp. Parasitol.*, **63**, 272–8.

60 Prince, J. B., Araujo, F. G., Remington, J. S., Burg, J. L., Boothroyd, J. C. and Sharma, S. D. (1989). Cloning of cDNA's encoding a 28 kilodalton antigen of *Toxoplasma gondii*. *Mol. Biochem. Parasitol.*, **34**, 3–14.

61 Johnson, A. M., Illana, S., McDonald, P. J., and Asai, T. (1989). Cloning, expression and nucleotide sequence of the gene fragment encoding an antigenic portion of the nucleoside triphosphate hydrolase of *Toxoplasma gondii*. *Gene*, **85**, 215–20.

62 Savva, D. (1989). Isolation of a potential DNA probe for *Toxoplasma gondii*. *Microbios*, **58**, 165–72.

63 Araujo, F. G. and Remington, J. S. (1980). Antigenemia in recently acquired acute toxoplasmosis. *J. Infect. Dis.*, **141**, 144–50.

64 Brooks, R. G., Sharma, S. D., and Remington, J. S. (1985). Detection of *Toxoplasma gondii* antigens by a dot-immunobinding technique. *J. Clin. Microbiol.*, **21**, 113–6.

65 Candolfi, E., Derouin, F., and Kien, T. (1987). Detection of circulating antigens in immunocompromised patients during reactivation of chronic toxoplasmosis. *Eur. J. Clin. Microbiol.*, **6**, 44–8.

66 Hermentin, K., Hassl, A., Picher, O., and Aspöck, H. (1989). Comparison of different serotests for specific toxoplasma IgM-antibodies (ISAGA, SPIHA, IFAT) and detection of circulating antigen in two cases of laboratory acquired toxoplasma infection. *Zentralbl. Bakteriol., Mikrobiol. Hyg.*, **A270**, 534–41.

67 Araujo, F. G., Handman, E., and Remington, J. S. (1980). Use of monoclonal antibodies to detect antigens of *Toxoplasma gondii* in serum and other body fluids. *Infect. Immun.*, **30**, 12–16.

68 Grover, C. M., Thulliez, P., Remington, J. S., and Boothroyd, J. C. (1990). Rapid prenatal diagnosis of congenital toxoplasma infection by using polymerase chain reaction and amniotic fluid. *J. Clin. Microbiol.*, **28**, 2297–301.

69 Savva, D., Morris, J. C., Johnson, J. D., and Holliman, R. E. (1990). Polymerase chain reaction for detection of *Toxoplasma gondii*. *J. Med. Microbiol.*, **32**, 25–31.

70 Puchhammer-Stöckl, E., Popow-Kraupp, T., Heinz, F. X., Mandl, C. W., and Kunz, C. (1990). Establishment of PCR for the early diagnosis of Herpes simplex encephalitis. *J. Med. Virol.*, **32**, 77–82.

71 Wheeler, R., Wilmore, H., Savva, D., and Turner, C. B. (1990). Diagnosis of ovine toxoplasmosis using PCR. *Vet. Rec.*, **126**, 249.

72 Dupouy-Camet, J., Laverada de Souza, S., Bougnoux, M. E., Mandelbrot, L., Hennequin, C., Dommergues, M. *et al.* (1990). Preventing congenital toxoplasmosis. *Lancet*, **336**, 1018.

73 Verhofstede, C., Van Renterghem, L., Plum, J., Vanderschueren, S., and Vanhaesbrouck, P. (1990). Congenital toxoplasmosis and TORCH. *Lancet*, **336**, 622–3.

74 Weiss, L. M., Udem, S. A., Salgo, M., Tanowitz, H. B., and Wittner, M. (1991). Sensitive and specific detection of toxoplasma DNA in an experimental murine model: use of *Toxoplasma gondii*-specific cDNA and the polymerase chain reaction. *J. Infect. Dis.*, **163**, 180–6.

75 Cesbron, J. Y., Capron, A., Ovalaque, G., and Santoro, F. (1985). Use of a monoclonal antibody in a double-sandwich ELISA for detection of IgM antibodies to *Toxoplasma gondii* major surface protein (P30). *J. Immunol. Methods*, **83**, 151–8.

76 Balfour, A. H., Harford, J. P., and Goodall, M. (1987). Use of monoclonal antibodies in an ELISA to detect IgM class antibodies specific for *Toxoplasma gondii*. *J. Clin. Pathol.*, **40**, 853–7.

77 Payne, R. A., Joynson, D. H. M., Balfour, A. H., Harford, J. P., Fleck, D. G., Mythen, M. *et al.* (1987). Public Health Laboratory Service enzyme linked immunosorbent assay for detecting toxoplasma specific IgM antibody. *J. Clin. Pathol.*, **40**, 276–81.

78 Decoster, A, Darcy, F., Caron, A., and Capron, A. (1988). IgA antibodies against P30 as markers of congenital and acute toxoplasmosis. *Lancet*, **ii**, 1104–6.

79 Pinon, J. M., Thoannes, H., and Gruson, N. (1985). An enzyme-linked immuno-filtration assay used to compare infant and maternal antibody profiles in toxoplasmosis. *J. Immunol. Methods*, **77**, 15–23.

80 Thulliez, Ph., Remington, J. S., Santoro, F., Ovalaque, G., Sharma, S., and Desmonts, G. (1986). Une nouvelle reaction d'agglutination pour le diagnostic du stade évolutif de la toxoplasmose acquise. *Pathol. Biol. .*, **34**, 173–7.

81 Luft, B. J., Brooks, R. G., Conley, F. K., McCabe, R. E., and Remington, J. S. (1984). Toxoplasmic encephalitis in patients with acquired immune deficiency syndrome. *JAMA*, **252**, 913–17.

82 Erlich, H. A., Rodgers, G., Vaillancourt, P., Araujo, F. G., and Remington, J. S. (1983). Identification of an antigen-specific immunoglobulin M antibody associated with acute toxoplasma infection. *Infect. Immun.*, **41**, 683–90.

83 Partanen, P., Turunen, H. J., Paasivuo, R., Forsblom, E., Suni, J., and Leinikki, P. O. (1983). Identification of antigenic components of *Toxoplasma gondii* by an immunoblotting technique. *FEBS Lett.*, **158**, 252–4.

84 Partanen, P., Turunen, H. J., Paasivuo, R. T. A., and Leinikki, P. O. (1984). Immunoblot analysis of *Toxoplasma gondii* antigens by human immunoglobulins G, M, and A antibodies at different stages of infection. *J. Clin. Microbiol.*, **20**, 133–5.

85 Verhofstede, C., Van Gelder, P., and Rabaey, M. (1988). The infection-stage-related IgG response to *Toxoplasma gondii* studied by immunoblotting. *Parasitol. Res.*, **74**, 516–20.

86 Huskinson, J., Thulliez, P., and Remington, J. S. (1990). Toxoplasma antigens recognized by human immunoglobulin A antibodies. *J. Clin. Microbiol.*, **28**, 2632–6.

87 Moir, I. L., Davidson, M. M., and Ho-Yen, D. O. (1991). Comparison of IgG antibody profiles by immunoblotting in patients with acute and past *Toxoplasma gondii* infection. *J. Clin. Pathol.*, **44**, 569–72.

88 Remington, J. S., Araujo, F. G., and Desmonts, G. (1985). Recognition of different toxoplasma antigens by IgM and IgG antibodies in mothers and their congenitally infected newborns. *J. Infect. Dis.*, **152**, 1020–4.

89 Potasman, I., Araujo, F. G., Thulliez, P., Desmonts, G., and Remington, J. S. (1987). *Toxoplasma gondii* antigens recognized by sequential samples of serum obtained from congenitally infected infants. *J. Clin. Microbiol.*, **25**, 1926–31.

90 Potasman, I., Araujo, F. G., and Remington, J. S. (1986). Toxoplasma antigens recognized by naturally occurring human antibodies. *J. Clin. Microbiol.*, **24**, 1050–4.

9

The future

DARREL O. HO-YEN

Since its discovery nearly 90 years ago, *Toxoplasma gondii* has been a subject for scientific enquiry for many medical practitioners, veterinary surgeons, and scientists. This has resulted in a massive literature, of which we have been able to use only a selection in this book. Interest in toxoplasmosis is not diminishing and the next decades will probably see much of the initial work being brought to fruition. In the last 25 years, the importance of toxoplasmosis in pregnancy (Chapter 6) and in those with immunocompromise (Chapter 7) has been realized. For the next 25 years, control of toxoplasmosis will be achieved by prevention of infection and better management of those who are infected (Fig. 9.1). Preventive measures are critical. In particular, there must be health promotion for pregnant women and medical practitioners must be aware of the risks of toxoplasmosis among the immunocompromised.[1,2] Theoretically, education alone would be able to dramatically control toxoplasmosis. Sadly, this is not realistic: health promotion of pregnant women was suggested in 1981[3] and the need for greater awareness of toxoplasmosis among medical practitioners in 1974.[4] It seems that to be successful, education needs to be combined with other measures. Thus, pregnant women may become more aware of toxoplasmosis when there is health promotion as part of an antenatal screening programme.[5] Similarly, medical practitioners may become more aware of toxoplasmosis if more effective tests for diagnosis, drugs for treatment, and an effective vaccine are developed. Vaccines have historically been a very effective public health measure to control infection, however, because of the need for stringent trials it may be many years before they can be widely implemented.

To look into the future is always fraught with difficulty. The exercise is most likely to be successful if there are limitations on the subject. Many have commented on the need for a greater awareness of toxoplasma infection[1-5] and this chapter will consider the future in terms of the factors that are most likely to increase this awareness. Increased awareness cannot be achieved by effort spent on providing greater information, instead it is likely to be the end result of several factors in the unfolding saga of toxoplasma infection. Thus, this chapter will be in four parts. The future problem of toxoplasma infection will be considered in terms of infection among the healthy, the immunocompromised, and in animals (Chapters 1, 2, 3, 6, and 7). Future developments in technology that will enable infected patients to be better managed will be highlighted (Chapters 4, 5, and 8). The steps that have to be taken for the prevention of infection in individual groups will be stated

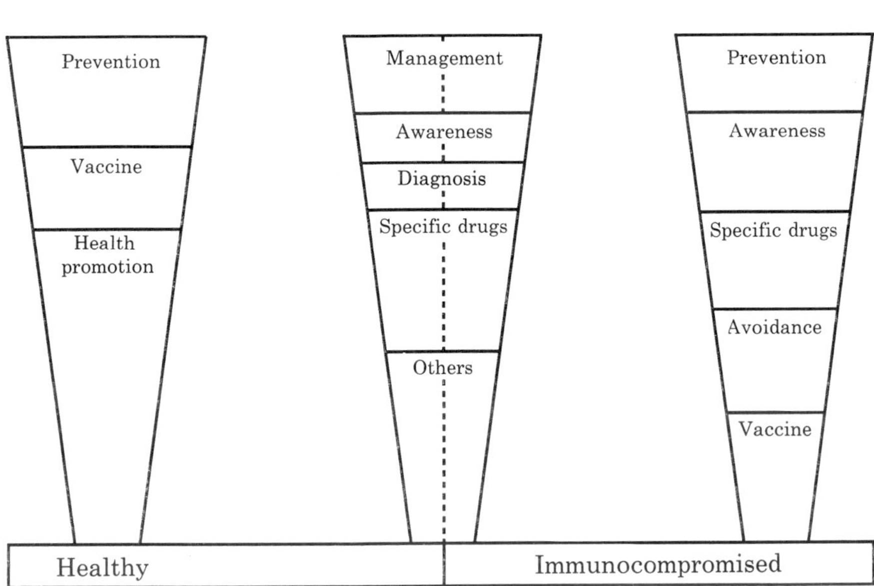

Fig. 9.1 Scheme for the future control of toxoplasmosis. Among the healthy and the immunocompromised, the relative importance of different measures in the management and treatment of infection are illustrated.

(Chapters 6 and 7) and the question of vaccination will be considered in depth. Lastly, an attempt will be made to summarize this book in what the author believes to be a pragmatic sequence of events.

THE PROBLEM

The present problem is that toxoplasma infection is underrecognized.[1–5] In the future, this will not be a difficulty. Instead, a dramatic increase in toxoplasma infection will mean that the problem will be how to manage and prevent this infection.

Human infection

Among those with malignancies, modern medicine is producing longer survival with the use of more aggressive therapy. Such therapy can result in more toxoplasma infection and greater severity of infections. Similarly, transplant surgery is in a time of great growth, and supply cannot satisfy the demand.[6] In a society that feels health is a right, the demand for 'spare-part' surgery will increase (Chapter 7). The knowledge that these attitudes will result in more toxoplasma infection will be accepted by the transplant teams with the same enthusiasm that they have embraced other infections such as *Pneumocystis carinii*, cytomegalovirus, and herpes simplex. Indeed, transplant teams with their efficient protocols and awareness of unusual problems are probably the group best-placed to deal with an upsurge of toxoplasma infections in a dispassionate way. The same cannot be said of obstetricians and pregnant women: the former have too heavy a workload and the latter are emotional. Media attention has resulted in many pregnant women becoming more aware of toxoplasmosis, and are demanding to be tested by their general practitioners and obstetricians (Chapter 6). This will inevitably result in more detected infections. However, the really major factors which will influence future medical practitioners' awareness of toxoplasmosis is human immunodeficiency virus infections (HIV) and the acquired immunodeficiency syndrome (AIDS). The spread of this infection into the heterosexual population not only means that there is a greater potential for acquired infection, but also that there is more congenital infection (Chapter 6). Among those with immunocompromise, this is likely to be the largest group in the future with reactivated toxoplasma infection (Chapter 7). The World Health Organization has estimated that by the year AD 2000, there will be 25–30 million patients with HIV world-wide. This will mean that medical practitioners everywhere will be in increased contact with HIV patients. It is likely that a higher profile of toxoplasmosis among the immunocompromised and pregnant women will also result in a better appreciation of acquired infection in the healthy (Chapter 3). Awareness and appreciation of toxoplasma infection will result in questions in the areas of management and prevention (Fig. 9.1) and will have to be answered.

Animal infection

The prevalence of toxoplasma infection in cats and domestic animals influences the level of human infection (Chapters 1 and 2). Good data on the prevalence of infection in animals are not available, but there is enough information to show that infection is widespread (Chapter 2). Two major results of animal toxoplasmosis are: possible sources of infection for humans, and adverse economic consequences for farmers because of reduced productivity in their animals (Chapter 2). Infection appears greatest in sheep (90–100%)[7] and is implicated in 10–20 per cent of sheep flocks with an abortion problem.[8] There is less infection in cattle (20–40%) and very variable infection in pigs (1–60%) and poultry (0–30%).[7] The maintenance of animals indoors in intensive rearing conditions is associated with reduced levels of infection. Although industrialized feeding tends to reduce toxoplasmosis, careless storage of food can allow greater infection. Thus, 50 grams of infected cat faeces containing 10 million oocysts in 10 tons of feed results in 5–25 sheep-infective doses per kilogram.[8] At the moment there is considerable potential for human infection from ingestion of tissue cysts present in animals and poultry. In the future, a greater demand for 'free-range' animals that have been allowed to graze in pastures could mean that toxoplasma infection among animals will rise. More infection among foodstuffs could have a greater effect on human infection if foreign travel and changed cooking habits produces a penchant for undercooked food (Chapter 1). Changed attitudes to the storage of food may also affect infectivity. Thus freezing of meat reduced its infectivity (Chapter 2). In the future domestic freezers are likely to be more common. Another possibility is the use of irradiation. Irradiation of pork to destroy *Trichinella spiralis* has been approved by the United States Food and Drug Administration. Other bacterial infections would also be killed by irradiation and it is possible that such treatment may become standard.

MANAGEMENT

The diagnosis and treatment of toxoplasmosis are major considerations in the management of the infection among the healthy and those with immunocompromise.

Diagnosis

It is on the subject of diagnosis that one can be most confident that there will be dramatic changes in the future. Today, the Sabin–Feldman dye test, which historically was so useful (Chapter 1) is still the 'gold standard'. This and other traditional antibody tests will be superceded because of the need for early diagnosis. Thus, tests that detect toxoplasma antigens or DNA (Chapter 8) will be more widely used and will allow clinicians the option of early treatment. An enzyme-linked immunosorbent assay (ELISA) for the detection of circulating toxoplasma antigen has proved particularly useful in AIDS patients.[9] Another technique of the future is the polymerase chain reaction (PCR). This technique detects specific DNA and thus the likely presence of the organism (Chapter 8), whereas the detection of antigen may depend on immunolysis.[9] Of interest, is that in AIDS patients the ELISA technique did not detect antigen in the cerebrospinal fluid,[9] but PCR detected DNA.[10] Much needs to be learnt about the specificity and sensitivity of these new techniques. The ability to quantify antigen or DNA may also allow differentiation of the presence of toxoplasma as opposed to active infection. Another area where there are great demands for early diagnosis is in fetal infection (Chapter 6). As with AIDS patients, this is another group in which antibody production may be infrequent or late. Again, the PCR appears sensitive, performing better than isolation studies or detection of specific IgM.[11] In the future, the need for early diagnosis of fetal infection seems unavoidable; and if there is widespread antenatal screening (Chapter 6), early fetal diagnosis will be mandatory.

With each year, laboratory diagnosis becomes more sophisticated. Although it may be possible to relegate the dye test to a lower position, antibody tests are likely to remain influential. This is because antibody production is a major part of the immune response (Chapter 7) and its detection can be an important marker with which to time an infection (Chapters 4 and 8). Thus, the antibody response to excreted/secreted antigens may differentiate acute and chronic infection.[12] Similarly, a more sophisticated analysis of the antibody response as identified by Western blotting may also differentiate acute and past infections.[13] Other developments are in the different classes of antibody. Thus IgA antibody denotes mucosal protection and may be more important than IgG and IgM in certain circumstances (Chapter 4 and 7). The presence of IgE may also be a marker of time of infection which could be useful (Chapters 4 and 7). Best results from laboratory investigations are always achieved when several tests are employed in a laboratory with experience of toxoplasma infection. However, laboratories can do nothing unless the clinician has considered the possibility of toxoplasma infection. Awareness of the manifestations of toxoplasmosis may well be more useful than laboratory expertise (Fig. 9.1).

Treatment

Specific drug treatment is part of the medical approach to both management and prevention of toxoplasmosis (Fig. 9.1). The standard drug combination of pyrimethamine and sulphadiazine has survived for 40 years (Chapter 1). The only widespread, significant modification to this approach was the addition of spiramycin for treatment during pregnancy and for the neonate (Chapter 6). In recent years, several new drugs have appeared to have useful anti-toxoplasma activity (Chapter 5). Although the effectiveness of pyrimethamine, a dihydrofolate reductase inhibitor, is beyond dispute, there are tremendous limitations in its use because of its bone marrow toxicity. Sadly, the new antibiotics with anti-toxoplasma activity (such as clindamycin, roxithromycin, azithromycin) may not be as effective against toxoplasma. There is more hope in newer dihydrofolate reductase inhibitors, trimetrexate, or piritrexim,[14] which because of their greater anti-toxoplasma activity may be used at a lower dose. They may still have to be combined with sulphadiazine to provide synergy. Among the immunocompromised, especially AIDS patients, the lower marrow toxicity could be critical. Another great problem is treatment during pregnancy and the neonate. Here, spiramycin has been used for many years in several countries, yet some have felt that there should be a spiramycin placebo trial.[15] Although we do not believe that such a trial is justified,[16] we are concerned about the cost of spiramycin (Chapter 6). In a cost-benefit analysis, spiramycin treatment can account for 14–27 per cent of the cost depending on the scheme.[17] If newer drugs of equal toxicity are to be introduced, there will need to be a proper drug trial of spiramycin versus the new drug. Such a trial would be far more acceptable than a spiramycin versus placebo trial.

A novel approach to future toxoplasma treatment could be one that goes back into time. The realization that the Chinese herb qinghaosu (artemisinin) had antimalarial properties was significant, although the herb was first used some 1700 years ago. Its effectiveness in toxoplasmosis appears good and it has low toxicity.[18] Derivatives of this drug and others with a trioxane ring[18,19] also show great promise. In particular, their low toxicity make them excellent candidates to replace pyrimethamine in the immuno-compromised. Similarly, if the price is low, they may be worth using in a controlled trial against spiramycin.

Another fascinating new antimalarial which is being developed is hydroxynaphthoquinone (566C80).[20] Apart from its excellent *in vitro* and *in vivo* activity against toxoplasma tachyzoites, this drug also demonstrated remarkably good activity against the tissue cyst stage of toxoplasma.[20] In treated mice, there was a steady decline in the numbers of cysts in their brains compared with untreated controls.[20] Thus, the drug is able to penetrate both the blood–brain barrier and the walls of tissue cysts. Among immunocompromised individuals, reactivation of toxoplasma infection is most important, and it may be that in the future all immunocompromised seropositive patients (especially those with AIDS) will have a course of treatment with a drug active against tissue cysts. In addition, with AIDS patients, hydroxynaphthoquinone has been shown to be effective in prevention and treatment of *Pneumocystis carinii* pneumonia.[21] Therefore, in AIDS patients, the benefits are obvious. However, the same arguments do not apply to pregnant women (Chapter 6). In these individuals, it is the primary infection rather than reactivated toxoplasma infection which produces problems. In these women, tissue cysts may also be useful in stimulating a high level of immunity; if parasites are released from tissue cysts, immunity is stimulated in a similar way to booster vaccinations. Nevertheless, even among women of child-bearing ages, hydroxynaphthoquinone may benefit those individuals with recent infection who are contemplating a pregnancy.

It makes sense to attempt to enhance the immune system of those with immunocompromise. A greater understanding of the normal immune response to toxoplasma (Chapter 7) and the ability of science to synthesize immunological mediators allows enhancement of the immune response. Thus, it has been suggested that γ-interferon may have a role in treatment, and interleukin-2 given prophylactically could prevent the reactivation of toxoplasmosis in AIDS patients.[22] At the moment, such an approach would be both theoretical and expensive. Yet, the concept of immune enhancement is good, even though in the near future it may apply only to patients with special conditions.

PREVENTION

For many infectious diseases, control of the infection has been achieved by prevention. Thus for several common infections (such as rubella and polio) there has been control of the infection although there may not be specific drugs active against the infectious agents.[23] For these mainly childhood infections, control has been achieved by widespread vaccination. Vaccination can also be an opportunity for the lay-person to become aware of an infection; and as they may be questioned, the medical practitioner must have some information on the infectious agent. Thus, vaccination achieves both objectives: health promotion among the lay-public and greater awareness of the infection in medical practitioners (Fig. 9.1). Another approach that also achieves these two objectives is the use of screening programmes. Screening programmes are designed to detect individuals with an illness, or at high risk of developing an illness before patients become concerned enough to present themselves to a medical practitioner. With toxoplasmosis, the risks in a particular situation depend mainly on immunity, but may vary with the situation. Thus, pregnant women who were immune before conception are not a toxoplasma risk to the fetus (Chapter 6), whereas some patients who were immune before immunocompromise are at risk of reactivated infection (Chapter 7). Both vaccination and screening programmes are linked, with the latter identifying individuals who may benefit from the former.

Screening programmes

Two large groups of individuals may benefit considerably from screening programmes: antenatal women and patients with immunocompromise. Now, most of the medical profession would probably accept the arguments for screening in the latter but not in the former. One possible explanation of this is that the latter group is small, whereas antenatal women are a massive group with substantial financial consequences of screening. Although decisions may be easier when costs are not great, the best decisions rely on truth and logic.

Antenatal women

A recent editorial in the *Lancet* concluded that prevention of congenital toxoplasmosis was best achieved by health promotion rather than an antenatal screening programme.[24] The arguments for an antenatal screening programme have been provided in great detail in Chapter 6. The results of several cost-benefit analyses in different countries have shown that screening is of cost-benefit (Chapter 6). These are the facts. Nevertheless, there are some who agree with the *Lancet* editorial. The case for the majority of those who work with toxoplasma is best presented by Dr Georges Desmonts.[25] This wise gentleman with decades of experience in toxoplasmosis depicted the scenario of those who saw education as the sole means of prevention, when asked to explain things to a pregnant woman, they would say:[25]

"There is a parasite, toxoplasma, which is harmless for you but which can cause severe impairment in your child if you become infected during pregnancy. You will notice nothing but we could tell you now if you are immune: if you are not, we can monitor you during pregnancy. If you become infected we can detect the infection in your infant even if he looks normal; we can even detect it *in utero* and treat without delay. Yet we will do nothing, since all this is too expensive. But do not worry. Perhaps you are immune — and if you are not, you will probably escape infection when pregnant. If the worst does happen your fetus might still escape. And if your infant is severely impaired, there is no risk in having another baby because you will now be immune. Nevertheless, try not to get infected. Eat well-cooked meat and wash your hands. Good luck".

It is simplistic to think that a favourable cost-benefit analysis means that cost is no longer a consideration. Like the boomerang it keeps coming back, but in different guises. Even though cost-benefit analyses do not consider the psychological costs of having a mentally-retarded child or that these children make fewer friends and are less likely to marry,[26] many appear unhappy with the results of cost-benefit analyses. Thus in a society that is materialistic and in which central funds are limited, individuals may need to be aware of the risks of toxoplasmosis and take appropriate individual action. As a country, this approach is likely to be less effective and cost more (Chapter 6), however it may reflect the current political approach to health care.

Immunocompromised patients

This is a heterogeneous group of patients whose needs can be very different (Chapter 7). The awareness of the possibility of toxoplasma infection is a major consideration in its control (Fig. 9.1). Fortunately, the financial considerations of screening are not major. Thus, in the future, it is likely that the advantages of avoiding transplants from infected donors to susceptible recipients will be appreciated and incorporated into transplant protocols (Chapter 7). In addition, the problems of reactivated infection as a result of severe immunocompromise, as in bone-marrow transplants or AIDS patients, will be prevented by early treatment or prophylaxis (Chapters 5 and 7). The ability of a drug to be active against tissue cysts[20] offers great hope. In those with immunocompromise, it may be possible in the future to eradicate (or greatly reduce) tissue cysts. Among immunocompromised patients, testing for their immune status to cytomegalovirus and herpes simplex is routine. It would not be a great difficulty to add toxoplasma to this initial serological screen. For those who are suceptible, there is a possibility that an appropriate vaccine (which provided immunity without encysting of the parasite), could be of great value. Such an approach is already in practice with the vaccination against several common infections among those with immunocompromise. Overall, it must be remembered that clinical toxoplasmosis is not common in patients with immunocompromise, however, when it does happen patients may be severely affected. Thus, a heightened awareness of the possibility of this infection is critical.

Vaccines

To the public, vaccines are a universal panacea. In reality, vaccines are only a part of the medical response to a serious infection. They have tremendous limitations as they must be shown to be safe and effective. The objectives for a vaccine can be summarized[27] as follows:

(1)   good humoral, cellular and local immune response;
(2)   protection against clinical disease and reinfection;
(3)   protection over several years, preferably a lifetime;
(4)   minimal immediate and delayed side effects;
(5)   administered simply in a form acceptable to the public;
(6)   satisfies cost-benefit analysis.

Whilst the above objectives may be easily satisfied in many viral and bacterial infections, success against parasitic infections is more difficult to achieve. This is because many parasites, including toxoplasma, have developed ways of living with or avoiding the host's immune system.[28,29] Thus, toxoplasma has the ability to: resist fusion of its phagosome with lysosomes; release antigenic material which blocks receptors of effector cells; and encyst and become dormant (Chapter 7). Unlike many other infections the cat plays an important role in the control of toxoplasmosis (Chapter 2). Vaccination of the cat, especially with the objective of preventing oocyst excretion[28] could dramatically reduce infection of man and other animals (Chapter 2). There may also be sound economic reasons for the vaccination of other animals (Chapter 2). At the very least, important knowledge of the effect of vaccines in animals may subsequently be used to design a human vaccine.[28] The history of possible toxoplasma vaccines spans more than 60 years and almost every possible avenue has been explored[28] (Fig. 9.2). The major problem has been the inability to expand early experimentation. Whereas everyone can agree that a good vaccine would be very useful, it takes considerable financial resources to test the safety and efficacy of a vaccine. With the current vogue for litigation, vaccine manufacturers can easily feel that the rewards of vaccine developments are not sufficient.

Live vaccines

Live vaccines provide specific immunity and are of two groups: those using low-virulent or non-virulent strains of toxoplasma, and those using attenuated toxoplasma organisms (Fig. 9.2). The problem with all live vaccines

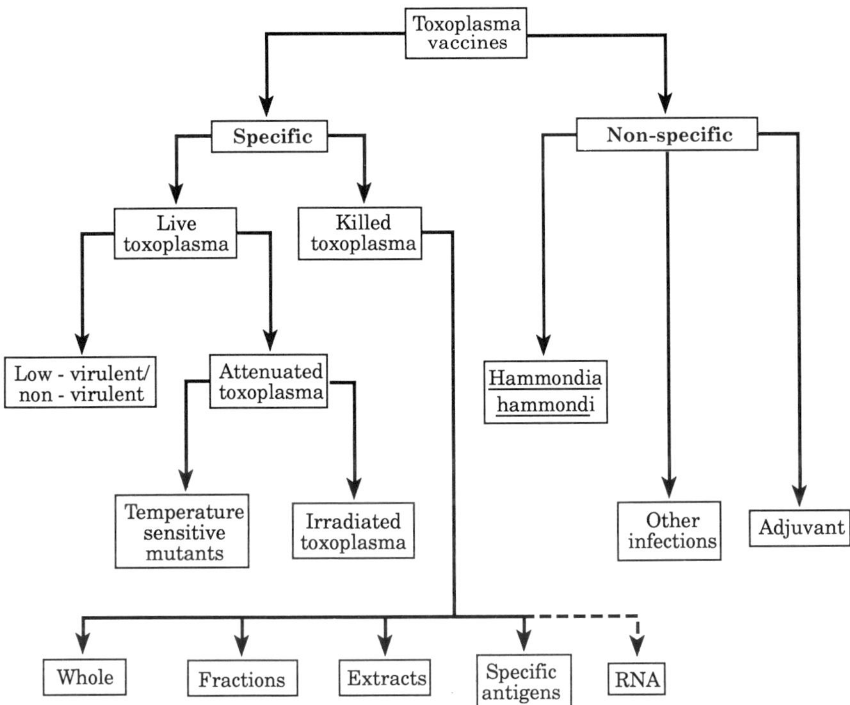

Fig. 9.2   Possible toxoplasma vaccines.

is that they may produce an infection similar to that of the normal toxo-plasma strain. Such an infection could result in vaccine parasites encysting, only to be reactivated later when there is immunocompromise. This may produce similar problems to that of natural infection (Chapter 7). In mice and guinea-pigs, low-virulent vaccines have produced good protection against subsequent challenge by a virulent strain of toxoplasma.[30,31] It is not too surprising that these vaccines provided good protection as they are closest to the natural infection. A refinement of live, low-virulent vaccines is to use the carcinogen, N-methyl-N-nitro-N-nitrosoguanidine, to produce temperature-sensitive mutants of toxoplasma.[32,33] Somewhat surprisingly, one of these mutants, ts-4, did not persist in the host beyond 2 months and no cysts were found.[33] These authors were also unable to explain the elimination of all parasites, and why in some animals the parasites persisted for 1–2 months. If these results were confirmed in a larger study, one of the major arguments against live vaccines (i.e. risk of persistence of the parasite) would be answered. This vaccine has been patented and although vaccinated animals were able to withstand infection with a virulent toxoplasma strain, immunity was not long-lasting.[34] Thus, in vaccinated hamsters, there was 100 per cent protection up to 16 weeks after vaccination, but it had fallen to 60 per cent by week 24.[34]

The objective of vaccination in cats is different and is the prevention of oocyst excretion rather than the prevention of illness. This is because cats do not usually have severe illness (Chapter 2) and oocyst excretion is a major source of infection to humans and other animals (Chapters 1 and 2). The intestinal epithelium of cats may be a separate immune compartment, and thus ts-4 does not prevent the subsequent production of oocysts.[35] Nevertheless, immunity without oocyst shedding could be achieved by giving chemoprophylaxis at the same time as vaccine, in this case a mutant vaccine T-263.[35] It is possible that another mutant vaccine may prevent oocyst shedding without chemoprophylaxis,[35] but larger scale trials are required before this can be totally accepted.

A clever vaccine modification with live parasites is the use of ionizing or ultraviolet irradiation.[36-38] At a certain level of ionizing irradiation, toxoplasma tachyzoites entered cells, but did not proliferate in host cells.[36] Unfortunately, the immunity to challenge with a virulent strain of toxoplasma was not long-lasting, and it may well be that long-lasting immunity can be achieved only by a course of vaccine.[37] Another modification which uses the same principles is ultraviolet irradiation.[38] Again damage to toxoplasma should ensure that there is no multiplication within the host cell. However, it will be necessary to test the efficacy of such vaccines in larger trials.

Killed vaccines

Killed vaccines have always been considered an attractive option as they do not cause infection within the host. Vaccines utilizing killed parasites may do so in a number of ways: whole parasites, fractions of disrupted parasites, extracts, or parasite RNA (Fig. 9.2). Early studies using lysed or formalin-killed organisms did not protect animals from lethal challenge of toxoplasma.[39] Although the animals did not die, toxoplasma was recovered subsequently from the animals' tissues, and thus vaccination did not prevent infection. Disrupted tachyzoites after centrifugation at different speeds have also been used as vaccines.[40] Again, there was protection of mice after a lethal challenge by toxoplasma, and in all fractions except that of 106 000 *g*, the mice developed specific toxoplasma antibody. These fractions provided similar protection whether or not they were administered with Freund's incomplete adjuvant. Nevertheless it appears that in vaccines using killed parasites, fractions of disrupted parasites or soluble extracts are less immunogenic than live vaccines,[28] and may not prevent infection after a severe challenge.

New molecular techniques have greatly changed our approach to vaccination. Thus, it is now possible to synthesize hepatitis B vaccine and many others may follow.[41] There has been a better understanding of toxoplasma antigens as a result of work with these new techniques (Chapter 8). The development of these specific antigens is a promising option for a toxoplasma vaccine (Fig. 9.2). However, some of these antigens because of their restricted composition may not be able to confer immunity.[28] Protection may also vary because antigenic composition is different in the different parasitic stages (Chapter 8), and thus a vaccine may protect against a tachyzoite challenge but not a sporozoite challenge. Success could depend on the judicious choice of antigen(s). Such may be the case with the 14 and 35 kDa antigens. When these were eluted from sodium dodecyl sulphate-polyacrylamide gels (SDS–PAGE) (Chapter 8), and inoculated into mice, the mice were protected from subsequent toxoplasma challenge.[42] Nevertheless, it must be remembered that an SDS–PAGE technique mainly detects protein antigens. Thus such an approach might not identify protective antigens that had a glycolipid composition.

Once a protective antigen has been isolated, it is still necessary to administer it in a safe and efficacious manner. Recently it has been suggested that *Salmonella* could be a carrier of antigens.[43] This has been advocated as *Salmonella sp.* has many similarities to toxoplasma and is an adjuvant in itself.[43] However, it does seem counterproductive to complicate the vaccination process by the requirements of a live organism (such as 'cold chain') when one has isolated protective antigens.

Non-specific immunity

In this category one could consider immunity provided by *Hammondia hammondi*.[44] This organism is closely related to toxoplasma and appears to be non-pathogenic to non-immune animals. Goats vaccinated with *H. hammondi* developed some protection against toxoplasma, possibly because of similar antigens. Thus, there is a cross-protection. Unfortunately this vaccine was not totally protective and toxoplasma was isolated after challenge.[44] Similarly other less related microorganisms (*Corynebacterium parvum*, *Listeria monocytogenes*, *Besnoitia jellisoni* and *Mycobacterium sp.*) may also provide some non-specific protection.[28] Indeed it has been suggested that the adjuvant used with the vaccine may on its own provide some protection.[28] A sophisticated approach is one in which toxoplasma RNA was used.[40] This work stems from knowledge that for some bacterial infections, specific RNA may provide immunity. Sadly, although there was some protection of vaccinated animals, it is possible that this protection was non-specific.[40] It is likely that all of these agents provide some protection against toxoplasma by stimulating the immune system. As such, immunity may only be present as long as the agent is also present.[28] This may not apply to the immunity as a result of *H. hammondi* or toxoplasma RNA vaccination.

Conclusion

Much time and effort has been spent in attempts to develop an effective vaccine. It is likely that success will first be in animals as restrictions and objectives are more clearly defined. A better understanding of the financial consequences of infection may make the case for animal vaccination convincing. The position for a human vaccine is less favourable. A recent review of human vaccines that were needed did not mention toxoplasma.[41] Also mentioned in this review was the fact that the public are more aware of vaccine injuries and therefore vaccines need proper development. Our understanding of toxoplasma immunity needs to be considerably improved. Variations between antigens at different parasite stages, strain differences, factors influencing cyst formation, vaccine delivery and longterm immunity are all important considerations.[28]

The objectives of vaccination are also different in different groups. Thus, among pregnant women, the objective is to prevent infection of the fetus and it does not matter if the parasite encysts in the woman. However, among the immunocompromised, the problem is usually the prevention of reactivated infection. In these situations a vaccine should provide immunity without encysting of the parasite. Similarly, objectives are different among animal vaccines where the economic consequences are usually the major considerations. Yet, one could foresee the situation where an individual pet owner may choose to have a cat vaccinated with the sole object of preventing oocyst excretion. Such a vaccine may be successful at preventing human infection from the cat without preventing maternal–fetal infection in the cat. The possibility that one vaccine could satisfy these very diverse needs is unlikely; however it would be possible to have a cocktail of antigens, each with a different function.

SUMMARY

The future for toxoplasma appears bright. The last decade has produced a vast amount of good quality data from which it is possible to make individual predictions. Sadly, there are many influencing factors, especially economic and political considerations. These considerations are major and would be sufficient to curtail the effects of isolated predictions. Fortunately, this does not appear to be the case with the future of toxoplasma. Instead of isolated events which happen and fade away, toxoplasma's future is one where there is a sequence of related events which can produce an additive effect. Thus, the available data suggests that toxoplasma has a 'domino' future. The momentum generated by individual events could result in greater and faster progress in other areas (Fig. 9.3). Such a momentum would be required to withstand the onslaught of economic and political arguments. A definite and worthy objective for the future is a greater awareness and understanding of toxoplasma infection (Fig. 9.3). Before a diagnosis can be made, it has to be suspected; and before an illness can be prevented, individuals must be aware of its existence. Thus the task is massive and the time-scale for the exercise is more than a decade.

Fig. 9.3 'Domino' future for toxoplasma.

Knowledge of toxoplasma has started to percolate through to the general public. This is quite surprising as the average person is bombarded with information from television, radio, magazines and newspapers. Important issues penetrate this morass mainly because they are given greater exposure. Greater exposure is the explanation of why women are becoming more aware of the problems of toxoplasmosis in pregnancy. The plethora of articles in women's magazines and programmes have been able to target this issue with the result that there is now public pressure on the British National Health Service (Fig. 9.3).

The NHS has responded in several ways: schemes have been set up to try and establish the prevalence of infection, and *ad hoc* testing has been established by many laboratories. These initial responses are poor and likely to produce data which would be difficult to interpret. It would be much better if a large-scale pilot, screening programme could be government-funded to assess the prevalence of infection and the other problems of antenatal screening. Nevertheless, it is likely that much indirect data on the prevalence of infection in pregnancy will be forthcoming. More definite data on prevalence of toxoplasmosis will be obtained in groups with immunocompromise. It is probable that infection will increase among those with malignancies, organ transplants, and human immunodeficiency virus (HIV) infection. When medical practitioners have to deal with large numbers of patients in these groups, their awareness and understanding of toxoplasma will dramatically increase.

A large number of ill patients is a tremendous spur to the laboratory services. If the HIV epidemic progresses as expected, the problems of diagnosis in laboratories will be severe. Fortunately, several techniques, especially the polymerase chain reaction, appear very promising. Thus, it would be safe to predict that, in a decade, the laboratory techniques for diagnosis will be dramatically different from the present. Similarly drug treatment will progress, especially as more diagnosed infection among those with immunocompromise will mean that more individuals required specific drug treatment. Effective, less toxic, and possibly even cheaper drugs may be soon available. Drugs active against tissue cysts are being developed and are a mixed blessing: useful with those with immunocompromise, but perhaps less so in women of childbearing age. The economic climate of the next decade and more realistic policies for the testing, marketing, and dealing with the indemnity of a drug will all help.

For the immunocompromised, more elaborate screening programmes will be developed and implemented. When patchy information on the prevalence of infection among antenatal women is combined with better diagnostic and management techniques (both fetoscopy and newer drugs), an antenatal screening programme would be considered again. At this time, the feelings of the pregnant women will be given greater weight, and perhaps the realization that many are already in screening schemes, many of which may be private. Information from other countries which would have adopted routine antenatal screening may tilt the balance. Will there be a British national screening programme for antenatal women? The answer is probably 'NO'. This is not a heretic statement but one influenced by a major consideration. That consideration is the availability of vaccine, hence the large domino in Fig. 9.3. A major limitation of toxoplasma vaccine is that the vaccine may produce a latent infection by encysting. However this is not a problem in pregnancy where the objective is to prevent a primary infection during pregnancy. Vaccines for pregnant women are available now and such an approach will make large-scale screening during pregnancy unnecessary. This is likely to happen. If a vaccine is developed that does not encyst and is therefore of value to those with immunocompromise, then, vaccination for pregnant women is assured. If these predictions are totally wrong and a vaccine does not become available, then antenatal screening for pregnant women will be adopted.

For decades, the lack of awareness and understanding of toxoplasma infection has been lamented. This book has resulted in a conclusion that this will not continue to be the position. The future is seen as a melting-pot in which the media, the public, the patients, and science will produce a new approach to toxoplasma infection. The arc (sic) will be on high ground.

REFERENCES

1 McCabe, R. and Remington, J. S. (1988). Toxoplasmosis: the time has come. *N. Engl. J. Med.*, **318**, 313–15.

2 Leighty, J. C. (1990). Strategies for control of toxoplasmosis. *J. Am. Vet. Med. Assoc.*, **196**, 281–6.

3 Henderson, J. B., Beattie, C. P., Hale, E. G., and Wright, T. (1984). The evaluation of new services: possibilities for preventing congenital toxoplasmosis. *Int. J. Epidemiol.*, **13**, 65–72.

4 Remington, J. S. (1974). Toxoplasmosis in the adult. *Bull. N. Y. Acad. Med.*, **50**, 211–27.

5 Ho-Yen, D. O. and Chatterton, J. M. W. (1990). Congenital toxoplasmosis — why and how to screen. *Rev. Med. Microbiol.*, **1**, 229–35.

6 Diethelm, A. G. (1990). Ethical decisions in the history of organ transplantation. *Ann. Surg.*, **211**, 505–20.

7 Van Knappen, F. (1989). Toxoplasmosis, old stories and new facts. *Int. Ophthalmol.*, **13**, 371–5.

8 Buxton, D. (1990). Ovine toxoplasmosis: a review. *J. R. Soc. Med.*, **83**, 509–11.

9 Hassl, A., Aspock, H., and Flamim, H. (1988). Circulating antigen of *Toxoplasma gondii* in patients with AIDS: significance of detection and structural properties. *Zentralbl. Bakteriol., Mikrobiol. Hyg., I Abt. A*, **A270**, 302–9.

10 Holliman, R. E., Johnson, J. D., and Savva, D. (1990). Diagnosis of cerebral toxoplasmosis in association with AIDS using the polymerase chain reaction. *Scand. J. Infect. Dis.*, **22**, 243–4.

11 Grover, C. M., Thulliez, P., Remington, J. S., and Boothroyd, J. C. (1990). Rapid prenatal diagnosis of congenital toxoplasma infection by using polymerase chain reaction and amniotic fluid. *J. Clin. Microbiol.*, **28**, 2297–301.

12 Decoster, A., Darcy, F., and Capron, A. (1988). Recognition of *Toxoplasma gondii* excreted and secreted antigens by human sera from acquired and congenital toxoplasmosis: identification of markers of acute and chronic infection. *Clin. Exp. Immunol.*, **73**, 376–82.

13 Moir, I. L., Davidson, M. M., and Ho-Yen, D. O. (1991). Comparison of IgG antibody profiles by immunoblotting in patients with acute and past *Toxoplasma gondii* infection. *J. Clin. Pathol.*, **44**, 569–72.

14 Araujo, F. G., Guptill, D. R., and Remington, J. S. (1987). *In vivo* activity of piritrexim against *Toxoplasma gondii*. *J. Infect. Dis.*, **156**, 828–30.

15 Jeannel, D., Costagliola, D., Niel, G., Hubert, B., and Danis, M. (1990). What is known about the prevention of congenital toxoplasmosis? *Lancet*, **336**, 359–61.

16 Ho-Yen, D. O., Joss, A. W. L., and Chatterton, J. M. W. (1990). Congenital toxoplasmosis and TORCH. *Lancet*, **336**, 624.

17 Joss, A. W. L., Chatterton, J. M. W., and Ho-Yen, D. O. (1990). Congenital toxoplasmosis: to screen or not to screen? *Public Health*, **104**, 9–20.

18 Ou-Yang, K., Krug, E. C., Marr, J. J., and Berens, R. L. (1990). Inhibition of growth of *Toxoplasma gondii* by qinghaosu and derivatives. *Antimicrob. Agents Chemother.*, **34**, 1961–5.

19 Chang, H. R., Jefford, C. W., and Pechere, J-C. (1989). *In vitro* effects of three new 1,2,4-trioxanes (Pentatroxane, Thiahexatroxane, and Hexatroxanone) on *Toxoplasma gondii*. *Antimicrob. Agents Chemother.*, **33**, 1748–52.

20 Araujo, F. G., Huskinson, J., and Remington, J. S. (1991). Remarkable *in vitro* and *in vivo* activities of the hydroxynaphthoquinone 566C80 against tachyzoites and tissue cysts of *Toxoplasma gondii*. *Antimicrob. Agents Chemother.*, **35**, 293–9.

21 Hughes, W. T., Kennedy, W., Shenep, J. L., Flynn, P. M., Hetherington, S. V., Fullen, G. *et al.* (1991). Safety and pharmokinetics of 566C80, a hydroxynaphthoquinone with anti-*Pneumocystis carinii* activity: a phase 1 study in human immunodeficiency virus (HIV) — infected man. *J. Infect. Dis.*, **163**, 843–8.

22 Luft, B. J. and Remington, J. S. (1988). Toxoplasmic encephalitis. *J. Infect. Dis.*, **157**, 1–6.

23 Grist, N. R., Ho-Yen, D. O., Walker, E., and Williams, G. R. (1987). *Diseases of Infection.* Oxford University Press, Oxford.

24 Lancet (1990). Antenatal screening for toxoplasmosis in the UK. **336**, 346–8.

25 Desmonts, G. (1990). Preventing congenital toxoplasmosis. *Lancet*, **336**, 1017–8.

26 Roberts, T. and Frenkel, J. K. (1990). Estimating income losses and other preventable costs caused by congenital toxoplasmosis in people in the United States. *J. Am. Vet. Med. Assoc.*, **196**, 249–56.

27 Evans, A. S. (1982). *Viral infections of humans.* Plenum, New York.

28 Hermentin, K. and Aspock, H. (1988). Efforts towards a vaccine against *Toxoplasma gondii*: a review. *Zentralbl. Bakteriol., Mikrobiol. Hyg., I Abt. A*, **A269**, 423–36.

29 Roit, I., Brostoff, J., and Male, D. (1989). *Immunology.* Churchill Livingstone, London.

30 Krahenbuhl, J. L., Blazkovec, A. A., and Lysenko, M. G. (1971). The use of tissue culture-grown trophozoites of an avirulent strain of *Toxoplasma gondii* for the immunization of mice and guinea pigs. *J. Parasitol.*, **57**, 386–90.

31 Walderand, H. and Frenkel, J. K. (1983). Live and killed vaccines against toxoplasmosis in mice. *J. Parasitol.*, **69**, 60–5.

32 Pfefferkorn, E. R. and Pfefferkorn, L. C. (1976). *Toxoplasma gondii*: isolation and preliminary characterization of temperature-sensitive mutants. *Exp. Parasitol.*, **39**, 365–76.

33 Walderland, H., Pfefferkorn, E. R., and Frenkel, J. K. (1983). Temperature — sensitive mutants of *Toxoplasma gondii*: pathogenicity and persistance in mice. *J. Parasitol.*, **69**, 171–5.

34 Elwell, M. R. and Frenkel, J. K. (1984). Immunity to toxoplasmosis in hamsters. *Am. J. Vet. Res.*, **45**, 2668–74.

35 Frenkel, J. K. (1990). Transmission of toxoplasmosis and the role of immunity in limiting transmission and illness. *J. Am. Vet. Med. Assoc.*, **196**, 233–40.

36 Lunde, E., Lycke, E., and Sourander, P. (1961). Studies of *Toxoplasma gondii* in cell cultures by means of irradiation experiments. *Br. J. Exp. Pathol.*, **42**, 404–7.

37 Chhabra, M. B., Mahajan, R. C., and Ganguli, N. K. (1979). Effects of ^{60}Co irradiation on virulent *Toxoplasma gondii* and its use in experimental immunization. *Int. J. Radiat. Biol.*, **35**, 433–40.

38 Grimwood, B. G. (1980). Infective *Toxoplasma gondii* trophozoites attenuated by ultra-violet irradiation. *Infect. Immun.*, **28**, 532–5.

39 Foster, B. G. and McCulloch, W. F. (1968). Studies of active and passive immunity in animals inoculated with *Toxoplasma gondii. Can. J. Microbiol.*, **14**, 103–10.

40 Araujo, F. G. and Remington, J. S. (1974). Protection against *Toxoplasma gondii* in mice immunized with toxoplasma cell fractions, RNA and synthetic polyribonucleotides. *Immunology*, **27**, 711–21.

41 Robbins, A. (1990). Progress towards vaccines we need and do not have. *Lancet*, **335**, 1436–8.

42 Araujo, F. G. and Remington, J. S. (1984). Partially purified antigen preparations of *Toxoplasma gondii* protect against lethal infection in mice. *Infect. Immun.*, **45**, 122–6.

43 Boothroyd, J. C., Burg, J. L., Nagel, S. D., and Perlman, D. (1988). Molecular approaches to the study of *Toxoplasma gondii*. In *Contemporary issues in infectious disease. Parasitic infections.* (ed. J. H. Leech, M. A. Sande, and R. K. Root), pp. 259–69. Churchill Livingstone, New York.

44 Dubey, J. P. (1981). Prevention of abortion and neonatal death due to toxoplasmosis by vaccination of goats with the non-pathogenic coccidium *Hammondia hammondi. Am. J. Vet. Res.*, **42**, 2155–7.

Appendix 1

Common questions asked in pregnancy
DARREL O. HO-YEN

The Scottish Toxoplasma Reference Laboratory receives a steady stream of telephone calls and letters about toxoplasmosis in pregnancy. After television and radio programmes, or after newspaper and magazine articles, there is usually a dramatic increase in enquiries. The majority of these enquiries are not from pregnant women, but from concerned relatives, friends, and other interested professional people responsible for health care. This appendix covers the main questions that we are asked and the answers that are usually given. It should also be emphasized that answers to some questions are not straightforward; there is disagreement among some specialists and not everyone would give the same advice. It is intended to be easily read by a lay-person. For those individuals who would like a more detailed answer, additional information and detailed references are in the relevant chapters of the book. This appendix specifically deals with toxoplasmosis in pregnancy and is divided into six sections for convenience: general information, sources, results, diagnosis, treatment, and prevention of infection.

GENERAL INFORMATION

Q 1.   What is toxoplasmosis?

Toxoplasmosis is a common infection which infects animals and humans. Although the organism, toxoplasma, has been around for thousands of years, its importance is being increasingly recognized. One area of great concern is toxoplasmosis in pregnancy. The parasite may pass from mother to baby during pregnancy and produce a damaged baby. Recently, it has also been recognized that some infected babies who were apparently normal at birth may develop eye problems, including blindness, many years later.

Q 2.   Who can get it?

Anyone can become infected with toxoplasma. Some individuals have been infected in the past, although they may not know it, and these individuals cannot be infected again. In the UK, about 20 per cent of adults have had a past infection and 80 per cent are susceptible to infection. Individuals who have had close contact with cats or regularly consume undercooked meat are more likely to have had infection. (See questions 6, 33, and 34)

Q 3.   How do you know if you are susceptible?

Your doctor can take a blood sample from you and have it tested. The test is simple and will allow you to know if you have had a previous infection and are therefore immune, or if you have not and are susceptible. Those who are susceptible can take steps to avoid infection. (See questions 6, 7, 33, and 34)

Q 4.   Why is toxoplasmosis serious?

Infection is serious, not because of the effect on the pregnant woman but because of the effect on the fetus. The mother may not even have any complaints or only have a mild illness. The effects on the fetus, especially early in the pregnancy, can be disastrous. A seriously handicapped baby who might require much individual care may result. Such an outcome is fortunately rare, but this and eye problems later in life make toxoplasma infection in pregnancy a serious illness. (See questions 5 and 13)

Q 5.   What is the risk of having an infection in pregnancy?

This is an important question. Overall, the risk is small, about 2/1000 pregnancies. Although 80 per cent of pregnant women are susceptible to infection, the behaviour of these women could greatly influence the likelihood of infection. Yet, risk cannot be removed from pregnancy and is always present: there is risk in driving to the shops for the dubious pleasure of having an ice-cream, but, in the height of summer, the risk of a car accident is believed to be worth it. Risk is also individual. A millionaire would not play Russian roulette for £10 000, but for a poor man whose wife may die unless she has an operation costing £10 000, Russian roulette may be seen as a gift from heaven. Each couple needs to evaluate the risks for themselves. For some, a handicapped child may be a catastrophe; for others, it may not. (See question 13)

SOURCES OF INFECTION

Q 6.   What are the sources of infection?

Infection is principally obtained from the faeces of infected cats, eating undercooked meat, or drinking unpasteurized milk. Hands may be contaminated after cleaning cat litter trays, gardening, playing in the sand-pit, or handling raw meat. Toxoplasma is easily washed off hands and readily destroyed by proper cooking.

Q 7.   Are kittens more of a risk than cats?

Toxoplasma infection is widespread in cats and so many kittens are probably protected by the mother's immunity for the first few months of life. After this time, kittens can become infected from eating other animals such as mice or birds. However, kittens that are well-fed on tinned cat food will not need to hunt and may not become infected. Obviously, kittens may also become infected from the ingestion of toxoplasma in other cats' faeces which may contaminate the environment. Of the well looked after cats that we have tested, about half have been previously infected. After a cat/kitten is infected, millions and millions of toxoplasma eggs are in the faeces for several weeks and can contaminate a large area. Then the faeces become normal and are not a source of infection *unless* the cat/kitten develops some other infection which may stimulate the further excretion of toxoplasma eggs in its faeces. Thus kittens are the greatest risk when several months old, but all cats can remain a risk for life. (See questions 33 and 34)

Q 8.   Are other pets a risk?

No. The faeces of other pets are *not* a risk for toxoplasma infection although they may be a risk for other infectious agents. (See question 33)

Q 9.   My husband (or mother) had toxoplasma in the past, could he/she infect all my future pregnancies?

No. Infection is *not* passed from one adult to another. Infection can be passed from a pregnant woman to an unborn child, but only if the woman has her first toxoplasma infection during pregnancy. Thus infection from a pregnant woman's husband or mother would not affect any future pregnancy.

Q 10.   I am a vegetarian, am I safe?

No. Cats are the main source of infection, and infection may be obtained from consuming unwashed root vegetables which may be contaminated by cat's faeces. Unpasteurized milk, especially goat's milk, is another possible source of infection. (See questions 6 and 7)

RESULTS OF INFECTION

Q 11.   If I am having an infection, will I be sick?

Most patients will have some complaints, but these may be minor such as not feeling well or a fever. Some patients will have a more severe illness especially with enlarged glands in the neck or under the arms. Many patients may not have any symptoms. (See question 4)

Q 12.   If I am not sick, does this mean that the baby will not be infected?

No. The baby may be infected even though the mother has had no evidence of any illness. Therefore, the only way that some infections in the mother can be detected is if her blood is regularly tested. This is not routinely done in the UK.

Q 13.   What are the chances of my baby being infected?

Infection passes to the baby in just over half of women who are infected and not treated. The likelihood of transmission to the baby is influenced by many factors including how many weeks the mother has been pregnant. Of those babies who are infected and whose mothers have not been treated, only about 8 per cent will be severely damaged. A further 17 per cent will have minor illness at birth, but the majority (75 per cent) will be apparently normal at birth. (See questions 5 and 23)

Q 14.   If my baby is normal at birth will it develop any illnesses later in life?

Longterm follow-up studies of apparently normal infected babies suggest that many will develop symptoms, especially eye problems later in life. It may be up to 18 years later. Therefore it is very important that babies are tested at birth to determine if they are infected. If they are infected, babies should be treated even though they appear normal. (See questions 13 and 29)

Q 15.   If I have an infected baby, does this mean that all of my future children will be infected?

No. The only pregnancy that is usually affected is the one in which the mother has had the first toxoplasma infection. Yet, as it can take some time for the infection to settle, it is recommended that a woman who is well waits for six months before she becomes pregnant again. (See question 19)

Q 16.   I had a blood test when I became pregnant, does this mean I am okay?

No. Blood taken when a women becomes pregnant is usually tested for other infections but not for toxoplasmosis. Although some centres are starting to test for toxoplasmosis, you need to ask for a special test in most areas of the country. (See questions 3, 17, and 18)

Q 17.   Can the blood test definitely show if I have a current infection?

In most cases it is possible for the laboratories to say if you have a current infection. However this needs to be confirmed with a second blood test 1–2 weeks later. Very occasionally, it is not possible for the laboratory to be absolutely sure that you were infected during pregnancy. In these cases, subsequent samples or a referral to a Toxoplasma Reference Laboratory will usually clarify the position. (See questions 20 and 21)

DIAGNOSIS

Q 18.   How will doctors be sure that I am having a current infection?

The diagnosis is made in many ways. Initially, doctors will consider your complaints, but the diagnosis is confirmed by a simple blood test. A single blood test is usually sufficient to say that the majority of patients do not have a current infection. If a current infection is suspected a second blood sample needs to be taken to confirm the infection. (See questions 20, 21, and 22)

Q 19.   What happens if I am infected before conception?

There is usually no risk to the baby if a mother has been infected before conception. It does not matter if there was a severe clinical infection, or if the mother was treated for toxoplasmosis. (See question 3)

Q 20.   I have been told that I have a very high toxoplasma IgG antibody level, what does this mean?

This result means that the body's immune system has been exposed to toxoplasma. However it is not possible from the IgG level alone to tell when someone has been infected. It is therefore necessary in these cases to also look for toxoplasma IgM antibody.

Q 21.   I have toxoplasma IgM antibody, what does this mean?

Toxoplasma IgM antibody usually indicates a current infection. Nevertheless, dependent on the types of tests that are used, such an antibody may reflect infection before conception. It is often necessary for a Reference Laboratory, using several tests, to determine when the infection might have occurred. If the infection was before conception, there is no risk. (See questions 15 and 19)

Q 22.   Why is it so difficult to interpret the blood results?

The reason is that there are a vast number of tests, and each test has particular difficulties of interpretation. Most local laboratories use only one test. Thus it is necessary to refer difficult cases to Reference Laboratories where many tests are employed and where there are many years' experience of interpretation of the blood results. (See questions 17, 18, 20, and 21)

Q 23.   How will the doctors know if my baby is infected?

This can be very difficult. During pregnancy, it entails several investigations, such as an examination of the baby by ultrasound and possibly a more detailed examination of the baby by an operation known as fetoscopy. At birth, the baby will also be examined and a sample of blood taken for testing. (See questions 24 and 29)

Q 24.   What is fetoscopy?

This is an operation in which a small tube is passed into the womb and the baby examined. At this time a small blood sample can be taken from the baby. The blood is carefully tested for evidence of toxoplasma infection. The operation is usually done at 17–24 weeks, but it may be necessary to repeat the operation a few weeks later. Unfortunately, there is a small risk of mortality to the fetus associated with the operation.

TREATMENT

Q 25.   Should I have been treated when I had toxoplasmosis diagnosed 2 years ago?

No. In the non-pregnant individual treatment is not necessary in the vast majority of cases. Infection before conception is usually not a risk for that pregnancy. In a few patients, doctors may decide that drug treatment is necessary.

Q 26.   If I have a current toxoplasma infection, what should I do?

If you are diagnosed early enough, even though there is only 60 per cent risk to the fetus, you may be given the option of an abortion. This is an important decision in which you need to carefully consider your particular circumstances. (See questions 5, 13, 23, 24, and 27)

Q 27.   Is there any drug treatment and is it dangerous?

As soon as a current infection is diagnosed in a pregnancy, the woman is offered treatment by a drug called spiramycin. Spiramycin has been given to many thousands of women with very few side effects. Very few women have mild symptoms such as nausea, dry mouth, or skin rashes. If you develop complaints whilst taking spiramycin you should consult your doctor. (See question 28)

Q 28.   Will I be given the same treatment if my baby is infected?

No. If there is evidence that the baby is infected a stronger type of treatment is required. This involves the administration of pyrimethamine, sulphadiazine, and folinic acid. This treatment is alternated with spiramycin treatment. As it is a more toxic treatment, care is required in its administration and patients need to have regular blood tests. (See questions 23 and 24)

Q 29.   What happens when my baby is born?

At birth, the baby is carefully examined for evidence of toxoplasma infection. Blood is also tested. The majority of babies will appear normal. If there is any evidence of infection the baby will be tested again and should be given treatment. This will often be for one year. It is hoped that such treatment will reduce the number of patients that develop eye problems in later life. (See question 14)

Q 30.   Will treatment remove all risk?

No. Treatment during pregnancy is likely to significantly reduce the risk of a baby being infected. However some babies will already have been infected before treatment was started and these babies may be born affected.

Q 31.   I have side effects to spiramycin, is there another drug I can take?

There are several other drugs that are being tested against toxoplasma infection, but there is less experience with these drugs. You and your doctor will have to decide if the side effects are sufficiently severe for you to change to a drug which may not be as useful. (See questions 27 and 28)

Q 32.   I have side effects to pyrimethamine and sulphadiazine, what should I do?

This is a question that should be discussed with your doctor. These drugs are the most effective ones against toxoplasma. A lot depends on the side effect and the stage in pregnancy. In some cases, the doses can be reduced. Your doctor will be able to decide on what should be done. (See question 28)

PREVENTION

Q 33.   I have been told that I am susceptible to toxoplasma infection, what should I do?

Patients who are susceptible should avoid sources of infection. They should not be cleaning cat litter trays. Hands should be washed after gardening or playing in the sand-pit and root vegetables should be washed before consumption. Meat should be well cooked and milk should be pasteurized.

(See questions 6 and 7)

Q 34.   Is it necessary for me to get rid of my cat?

No. It is the faeces from the cat that are infectious. You need not be afraid of touching your cat. If there is no one else to clean the cat litter tray, you should do so daily with gloves and wash your hands afterwards.

Q 35.   Is there a vaccine?

A generally available and tested vaccine has not been produced. However there are several workers who are developing vaccines. In the future, it may be possible to vaccinate your cat. A vaccine may also be available for women, in the next 10 years.

Q 36.   Where can I get more information?

More information is available from the Toxoplasmosis Trust whose address is Room 26, 61–71 Collier Street, London, N1 9BE, Tel: 071 713 0599. Information may be available from the four Toxoplasma Reference Laboratories in the UK. These are in Swansea (Public Health Laboratory, Singleton Hospital, Sgeti, Abertawe, Swansea, SA2 8QA); London (Public Health Laboratory, St George's Hospital, Blackshaw Road, London SW17 OQT); Leeds (Public Health Laboratory, Bridle Path, York Road, Leeds LS15 7TR); and in Inverness (Scottish Toxoplasma Reference Laboratory, Raigmore Hospital, Inverness IV2 3UJ).

Appendix 2

Infection in the laboratory

DARREL O. HO-YEN

This appendix may not be of great interest to many of our readers, but for those of us who are involved in the laboratory, the experiences of our predecessors are both fascinating and instructive. In addition, we still receive telephone enquiries about laboratory acquired infection which may represent a special subgroup of toxoplasma infection. In the 1950s and 1960s, there were more than 21 reported cases of laboratory acquired toxoplasma infection[1] compared to one in the 1980s.[2] Such an improvement in the health of laboratory workers seems quite remarkable, but cannot be attributable to greater expertise. The important difference in the decades is knowledge. We are now aware of the sources of infection and can adopt more realistic working practices (Chapters 1 and 2). Thus, many of the 21 cases may not be due to acquiring infection in the laboratory, and in the 1980s all cases of laboratory acquired toxoplasmosis may not have been reported. As the figure 21 is probably unduly high and the figure one too low, reality is somewhere in between.

A short, but informative publication in 1959 considered 18 cases of laboratory acquired toxoplasmosis.[3] In 10 cases the source of infection was unknown, but in nine of these the explanation was that the individuals had worked with toxoplasma for periods of 18 days to 2 years. It is highly unlikely that just working with toxoplasma is sufficient to explain infection. It is more likely that these nine cases were infected from other sources (Chapter 2). The four cases that were infected as a result of a needle-stick are probably real. Needle-stick injury is serious and not necessarily related to experience. One of us was involved in toxoplasma work for 15 years before experiencing a needle-stick injury. Another has remained seronegative after 16 years working with toxoplasma. In 3/18 cases,[3] the source of infection was believed to be a splash in the face. Such cases are difficult to evaluate and may well be real. Much less likely is the incident of asymptomatic seroconversion of the individual who had a rabbit bite.[3] The suggestion that a laboratory worker may be infected by an aerosol was surprising as was the statement that the RH strain was avirulent.[2] In fact, it is one of the most virulent strains.[4] At present, aerosol transmission is not regarded as common or even likely, but with a virulent strain it is a theoretical possibility. Before the evidence can be accepted, it would need to be in a better documented case and one that did not state 'screening tests for infectious mononucleosis were at first positive'.[2]

Table A2.1   Laboratory infection as a result of a needle-stick injury

Incubation (days)	Clinical	Duration (weeks)	Reference
3	Fever, lymphadenopathy	4	5
9	Fever, lymphadenopathy	3	6
9	Fever, lymphadenopathy and pneumonitis	4	7
8	Lymphadenopathy + rash	Several	8

CLINICAL FEATURES

The best evidence of laboratory acquired toxoplasma infection is in those individuals who suffer a needle-stick injury. In Table A2.1, four such cases are considered.[5–8] The results show that infection after needle-stick injury is severe: the incubation period is short (3–9 days) compared to that in other acquired infections (Chapter 3); the patients are symptomatic; and illness lasts several weeks. This is in sharp contrast to the usual clinical presentation (Chapter 3). Yet, it should not be too surprising. The inoculation of tachyzoites is consistent with a shorter incubation period, and the presence of definite clinical features are also consistent with a high inoculum. Fortunately laboratory personnel are usually of good health and the illness has a favourable outcome.

In an excellent review of the literature, Perkins[1] showed that in 26 cases of laboratory acquired infection there were no cases of focal retinochoroiditis. In one case there was conjunctival hyperaemia and in two there was conjunctivitis. In two cases infection was in pathologists, presumably infected during autopsy, and shown to have an exudative retinitis. The conclusion was that uveitis occurred, in 2/26 (7.7%) of laboratory acquired infection. Unlike congenital infection, ocular involvement was rare and consistent with our understanding of acquired disease (Chapter 3).

TREATMENT

There is good evidence that laboratory acquired infection could be more severe than acquired infection from other routes. When laboratory workers have splashes of infectious material on mucous membranes (especially mouth and eyes), or suffer needle-stick injury they should be treated. The choice of treatment probably depends on the infectious dose. If mucous membranes have been extensively exposed to infectious material, or if there has been a large inoculum, drug treatment should be pyrimethamine, sulphonamide, and folinic acid (Table A2.2). When exposure is more difficult

Table A2.2   Prophylactic treatment for 3 weeks

Severe exposure

Pyrimethamine	25 mg/day
Sulphadiazine	3 gm/day
Folinic acid	6 mg/day

Moderate exposure

Pyrimethamine	25 mg/day
Folinic acid	6 mg/day

Minor exposure

Co-trimoxazole	960 mg b.d.
or Clindamycin	300 mg qid.

to assess, such as aerosol or minor splashes of infected material, the decision should be influenced by the individual concerned. If there is much anxiety, the full treatment of a severe exposure or one of moderate exposure should be given (Table A2.2). Unfortunately, the full treatment does have significant side effects (Chapter 5). Many might feel that the side effects are too severe. In cases with little anxiety, the position should be explained and the individual given the choice of treatment. Alternative treatments (which may be less effective) are co-trimoxazole or clindamycin (Table A2.2). However, as these are less toxic, they have some attraction.

For laboratory personnel, prevention of infection is the ideal. Good technique and care will minimize the risk of infection. Accidents will happen and laboratory workers should be sensitive to when this is likely. They should make their medical practitioners aware of the possibility of laboratory acquired infection, or carry a 'Medical Contact Card'. Work in a medical laboratory has risks and personnel that work in laboratories have different perception of risks. The right decision for the individual will only be made when these different considerations are properly assessed.

REFERENCES

1   Perkins, E. S. (1973). Ocular toxoplasmosis. *Br. J. Ophthalmol.*, **57**, 1–17.
2   Baker, C. C., Farthing, C. P., and Ratnesar, P. (1984). Toxoplasmosis, an innocuous disease? *J. Infect.*, **8**, 67–9.
3   Rawal, B. D. (1959). Laboratory infection with toxoplasma. *J. Clin. Pathol.*, **12**, 59–61.
4   Beattie, C. P. (1985). Toxoplasmosis, an innocuous disease? *Abst. Hyg. Commun. Dis.*, **60**, 206–7.
5   Strom, J. (1951). Toxoplasmosis due to laboratory infection in two adults. *Acta Med. Scand.*, **139**, 244–52.
6   Beverley, J. K. A., Skipper, E., and Marshall, S. C. (1955). Acquired toxoplasmosis with a report of a case of laboratory infection. *Br. Med. J.*, **i**, 577–8.
7   Wright, W. H. (1957). A summary of the newer knowledge of toxoplasmosis. *Am. J. Clin. Pathol.*, **28**, 1–17.
8   Van Soestbergen, A. A. (1957). Het beloop van een laboratorium — infectie met *Toxoplasma gondii. Ned T Geneesk*, **101**, 1649–51.

Index